T0358305

Ring Theory 2019

Proceedings of the Eighth China–Japan–Korea
International Symposium on Ring Theory

Ring Theory 2019

Proceedings of the Eighth China–Japan–Korea
International Symposium on Ring Theory
Nagoya University, Nagoya, Japan 26 – 31 August 2019

editors

Hideto Asashiba
Shizuoka University, Japan

Shigeto Kawata
Nagoya City University, Japan

Nanqing Ding
Nanjing University, China

Nam Kyun Kim
Hanbat National University, South Korea

NEW JERSEY • LONDON • SINGAPORE • BEIJING • SHANGHAI • HONG KONG • TAIPEI • CHENNAI • TOKYO

Published by

World Scientific Publishing Co. Pte. Ltd.
5 Toh Tuck Link, Singapore 596224
USA office: 27 Warren Street, Suite 401-402, Hackensack, NJ 07601
UK office: 57 Shelton Street, Covent Garden, London WC2H 9HE

British Library Cataloguing-in-Publication Data
A catalogue record for this book is available from the British Library.

RING THEORY 2019
Proceedings of the Eighth China–Japan–Korea International Symposium on Ring Theory

Copyright © 2021 by World Scientific Publishing Co. Pte. Ltd.

All rights reserved. This book, or parts thereof, may not be reproduced in any form or by any means, electronic or mechanical, including photocopying, recording or any information storage and retrieval system now known or to be invented, without written permission from the publisher.

For photocopying of material in this volume, please pay a copying fee through the Copyright Clearance Center, Inc., 222 Rosewood Drive, Danvers, MA 01923, USA. In this case permission to photocopy is not required from the publisher.

ISBN 978-981-123-028-8 (hardcover)
ISBN 978-981-123-029-5 (ebook for institutions)
ISBN 978-981-123-030-1 (ebook for individuals)

For any available supplementary material, please visit
https://www.worldscientific.com/worldscibooks/10.1142/12099#t=suppl

Printed in Singapore

Preface

The Eighth China-Japan-Korea International Symposium on Ring Theory (CJK) was held at Nagoya University, Japan, from 26th until 31st of August, 2019. It took the form of a joint symposium with the 52nd Symposium on Ring Theory and Representation Theory in Japan.

The first CJK conference was held in Guilin, China in 1991. Ever since then, the CJK conferences (in Japan it has been referred to as symposiums) have been held once almost every four years: the second CJK in Okayama, Japan (1995), the third in Kyongju, Korea (1999), the fourth in Nanjing, China (2004), and so on with the same hosting rotation by the three countries. This has greatly contributed to the development of ring theory in East Asia, as well as fostering the ties of friendship and mutual understanding among ring theorists of the attending countries, and also welcoming researchers from other parts of the world who are interested in the topics discussed. This time we were delighted to have almost 150 participants from 12 countries and regions, including some globally established ring theorists. It clearly showed a steady growth in the number of attendees over the past three decades, and a growing degree of interest toward ring theory.

We would like to thank all the participants very much for their coming to attend the conference, along with their enthusiasm toward this area of mathematics. We are grateful that the staff members at Nagoya University were so helpful and talented in organizing the venue, and taking care of everything for the smooth progress of the event. The student staffs at the university also worked hard in assisting the participants. Our cordial appreciation goes to the Graduate School of Mathematics at Nagoya University for its financial support (this conference was actually held as its 19th International Conference), and to the persons who kindly permitted us to use their JSPS Grant-in-Aid for Scientific Research (B) and (C). All of these diverse efforts and funding were indispensable to make this a successful meeting. Last but not least, publishing this proceedings volume has been made possible by the authors and referees of the submitted papers, for which we are very thankful.

Anyone is welcome to review the conference, and to discover more about the CJK, by consulting the following webpage:

`https://www.ring-theory-japan.com/cjk2019/`

The next (ninth) CJK is scheduled for 2023 in Korea. We really look forward to further accelerating the scientific achievements through meaningful on-site lectures and face-to-face discussions for the years to come.

The main organizer
Hideto Asashiba
September 2020

Organizing Committee

Editors

Hideto Asashiba (Chief)	– Shizuoka University, Japan
Shigeto Kawata	– Nagoya City University, Japan
Nanqing Ding	– Nanjing University, China
Nam Kyun Kim	– Hanbat National University, Korea

Organizers

Hideto Asashiba (Chief)	– Shizuoka University, Japan
Osamu Iyama	– Nagoya University, Japan
Shigeto Kawata	– Nagoya City University, Japan
Kazutoshi Koike	– National Institute of Technology, Okinawa College, Japan
Yosuke Kuratomi	– Yamaguchi University, Japan
Masahisa Sato	– Aichi University, Japan
Kota Yamaura	– University of Yamanashi, Japan
Nanqing Ding	– Nanjing University, China
Jianlong Chen	– Southeast University, China
Fang Li	– Zhejiang University, China
Quanshui Wu	– Fudan University, China
Nam Kyun Kim	– Hanbat National University, Korea
Tai Keun Kwak	– Daejin University, Korea
Mi Hee Park	– Chung Ang University, Korea

The 8th China-Japan-Korea International Symposium on Ring Theory

Period: August 26–31, 2019
Venue: Sakata–Hirata Hall, Science South Building
Rooms 509, 309, 109, Mathematics Building
Nagoya University, Nagoya, Japan

Program

Day 1: August 26 (Mon), 2019

09:00-10:50 Registration
10:50-11:10 Opening

Invited Lectures: Sakata–Hirata Hall

11:20-12:10 Yuji Yoshino
Auslander-Bridger theory for projective complexes over commutative Noetherian rings

Branch Sessions

13:40-14:00

Room 509 Takuma Aihara
On the weakly Iwanaga–Gorenstein property of gendo algebras

Room 309 Tsiu-Kwen Lee
A note on Skolem–Noether algebras

Room 109 Ziyu Guo
Cohomology rings and application in hypernormal form for a class of 4-dimensional vector fields

14:10-14:30

Room 509 Takahiro Honma
On the gendo-symmetric algebra of a trivial extension algebra

Room 309 Tai Keun Kwak
On CRP rings

Room 109 Jung-Miao Kuo
Partial group actions and partial Galois extensions

14:40-15:00

Room 509 Takahide Adachi
On balanced Auslander–Dlab–Ringel algebras

Day 2: August 27 (Tue), 2019

Invited Lectures: Sakata–Hirata Hall

09:00-09:50 Bernhard Keller
Tate–Hochschild cohomology from the singularity category

10:20-11:10 Tsutomu Nakamura
Pure derived categories and weak balanced big Cohen–Macaulay modules

11:20-12:10 Sei-Qwon Oh
Relationships between quantized algebras and their semiclassical limits

Branch Sessions

13:40-14:00

Room 509 Sota Asai
Wide subcategories and lattices of torsion classes

Room 309 Naoki Endo
Almost Gorenstein Rees algebras

Room 109 Masahisa Sato
Is Ware's problem true or not?

14:10-14:30

Room 509 Toshiya Yurikusa
Density of g-vector cones from triangulated surfaces

Room 309 Ryotaro Isobe
Ulrich ideals in hypersurfaces

Room 109 Yoshiharu Shibata
When is a quasi-discrete module quasi-projective?

14:40-15:00

Room 509 Toshitaka Aoki
g-polytopes of Brauer graph algebras

Room 309 Kazuho Ozeki
The structure of Sally modules and normal Hilbert coefficients

Room 109 Gangyong Lee
Rudimentary rings: Rings have a faithful indecomposable endoregular module

15:10-15:30

Room 509 Yuta Kozakai
Mutations for star-to-tree complexes and pointed Brauer trees

Room 309 Nam Kyun Kim

Characterizations of radicals in skew polynomial and skew Laur polynomial rings

Room 109 Tsunekazu Nishinaka

On Thompson's group F and its group algebra

15:10-15:30

Room 509 Mayu Tsukamoto

Constructions of rejective chains

Room 309 Kui Hu

Some results on Noetherian Warfield domains

Room 109 Diah Junia Eksi Palupi

The construction of a continuous linear representation from a topological group into topological module space over principle ideal domain

16:00-16:20

Room 509 Toshinori Kobayashi

A characterization of local rings of countable representation type

Room 309 Jianlong Chen

Generalized inverses and clean decompositions

Room 109 Nanqing Ding

On Enochs conjecture

16:30-16:50

Room 509 Naoya Hiramatsu

A remark on graded countable Cohen–Macaulay representation type

Room 309 Xiaofeng Chen

The right core inverses of a product and a companion matrix

Room 109 Lixin Mao

Relative coherent modules and semihereditary modules

17:00-17:20

Room 509 Kei-ichiro Iima

On the 2-test modules of projectivity and weakly $\mathfrak{m}$-full ideals

Room 309 Xiankun Du

Triangularization of matrices and polynomial maps

Room 109 Manoj Kumar Patel

A note on *FI*-semi injective modules

Room 309 Naoyuki Matsuoka

Efficient generation of ideals in core subalgebras of the polynomial ring $k[t]$ over a field k

Room 109 Derya Keskin Tütüncü

Baer–Kaplansky classes in categories

16:00-16:20

Room 509 Aaron Chan

Recollement of comodule categories over coalgebra objects

Room 309 Yoshitomo Baba

On two sided Harada rings and QF rings

Room 109 Hwankoo Kim

A new semistar operation on a commutative ring and its applications

16:30-16:50

Room 509 Zhaoyong Huang

The extension dimension of Abelian categories

Room 309 Thomas Dorsey

Questions and counterexamples on strongly clean rings

Room 109 Izuru Mori

Noncommutative matrix factorizations and Knörrer's periodicity theorem

17:00-17:20

Room 509 Na'imah Hijriati

On representation of a ring with unity on a module over a ring with unity

Room 309 Cancelled

Room 109 Kenta Ueyama

Knörrer's periodicity for skew quadric hypersurfaces

Day 3: August 28 (Wed), 2019

Invited Lectures: Sakata–Hirata Hall

09:00-09:50 Pace P. Nielsen

Nilpotent polynomials with non-nilpotent coefficients

10:20-11:10 Haruhisa Enomoto

The Jordan–Hölder property, Grothendieck monoids of exact categories and Bruhat inversions

11:20-12:10 Jiwei He

Noncommutative Auslander theorem and noncommutative quotient singularities

Branch Sessions

13:40-14:00

Room 509 Kunio Yamagata

On a problem of socle-deformations of self-injectve orbit algebras

Room 309 Cancelled

Room 109 Ayako Itaba

Hochschild cohomology of Beilinson algebras of graded down-up algebras

14:10-14:30

Room 509 Hideto Asashiba

2-categorical Cohen–Montgomery duality between categories with I-pseudo-actions and I-graded categories for a small category I

Room 309 Guodong Shi

Q-graded Hopf quasigroups

Room 109 Tomohiro Itagaki

The Hochschild cohomology of a class of exceptional periodic selfinjective algebras of polynomial growth

14:40-15:00

Room 509 Yasuaki Ogawa

General heart construction and the Gabriel-Quillen embedding

Room 309 Linlin Liu

Rota–Baxter H-operators and pre-Lie H-pseudoalgebras over a cocommutative Hopf algebra H

Room 109 Hideyuki Koie

An application of a theorem of Sheila Brenner for Hochschild extension algebras of a truncated quiver algebra

15:10-15:30

Room 509 Xuefeng Mao

DG polynomial algebras and their homological properties

Room 309 Shenglin Zhu

Structures of irreducible Yetter–Drinfeld modules over quasi-triangular Hopf algebras

Room 109 Satoshi Usui

A Batalin–Vilkovisky differential on the complete cohomology ring of a Frobenius algebra

16:00-16:20

Room 509 Haibo Jin

Cohen–Macaulay differential graded modules and negative Calabi–Yau configurations

Room 309 Tongsuo Wu

Boolean graphs — a survey

Room 109 Dong Yeol Oh

The chain conditions on ideals in composite generalized power series rings

16:30-16:50

Room 509 Norihiro Hanihara

Cohen-Macaulay modules over Yoneda algebras

Room 309 Yuehui Zhang

Directed partial orders over non-Archimedean fields

Room 109 Mi Hee Park

Noetherian-like properties in polynomial and power series rings

17:00-17:20

Room 509 Kota Yamaura

Happel's functor and homologically well-graded Iwanaga–Gorenstein algebras

Room 309 Cancelled

Room 109 Andreas Reinhart

On the monoid of ideals of orders in quadratic number fields

18:00-21:00 Conference Banquet

Day 4: August 29 (Thu), 2019

Invited Lectures: Sakata–Hirata Hall

09:00-09:50 Michał Ziembowski

Lie solvability in matrix algebras

10:20-11:10 Hiroyuki Minamoto

On a cubical generalization of preprojective algebras

11:20-12:10 Xianhui Fu

Ideal approximation theory

Day 5: August 30 (Fri), 2019

Invited Lectures: Sakata–Hirata Hall

09:00-09:50 S. Paul Smith
Elliptic algebras

10:20-11:10 Yoshiyuki Kimura
Twist automorphism of quantum unipotent cells and the dual canonical bases

11:20-12:10 Nan Gao
A functorial approach to monomorphism category for species

Branch Sessions

13:40-14:00

Room 509 De Chuan Zhou
On an open problem concerning the small finitistic dimension of a commutative ring

Room 309 Iwan Ernanto
Arithmetic modules over generalized Dedekind domain

Room 109 Ryo Kanda
The characteristic variety of an elliptic algebra

14:10-14:30

Room 509 Kaori Shimada
On the radius of the category of extensions of matrix factorizations

Room 309 Sutopo
Positively graded rings which are maximal orders and generalized Dedekind rings

Room 109 Masaki Matsuno
AS-regularity of geometric algebras of plane cubic curves

14:40-15:00

Room 509 Shinya Kumashiro
The Auslander–Reiten conjecture for non-Gorenstein rings

Room 109 Kazunori Nakamoto
An application of Hochschild cohomology to the moduli of subalgebras of the full matrix ring II

15:10-15:30

Room 509 Maiko Ono

On liftable DG modules over a commutative DG algebra

Room 109 Indah Emilia Wijayanti

On generalized Dedekind modules over generalized Dedekind domain

16:00-16:20

Room 509 Erik Darpö

Functors between higher cluster categories of type A

Room 109 Guisong Zhou

The structure of connected (Hopf) algebras

16:30-16:50

Room 509 Yingying Zhang

Semibricks, wide subcategories and recollements

Room 109 Libin Li

The center subalgebra of the quantized enveloping algebra of a simple Lie algebra revisited

17:00-17:20

Room 509 Xiaojin Zhang

Tilting modules over Auslander–Gorenstein algebras

Room 109 Zhixiang Wu

Classification of Leibniz conformal algebras of rank three

Day 6: August 31 (Sat), 2019

Invited Lectures: Sakata–Hirata Hall

09:00-09:50 Gyu Whan Chang

Unique factorization property of non-UFDs

10:20-11:10 Kenichi Shimizu

Action functor formalism

11:20-12:10 Susumu Ariki

On cyclotomic quiver Hecke algebras of affine type

14:00-14:50 Pu Zhang

Gorenstein-projective and semi-Gorenstein-projective modules

List of Participants

Takahide Adachi	Osaka Prefecture University, Japan
Takuma Aihara	Tokyo Gakugei University, Japan
Toshitaka Aoki	Nagoya University, Japan
Tokuji Araya	Okayama University of Science, Japan
Susumu Ariki	Osaka University, Japan
Sota Asai	Kyoto University, Japan
Hideto Asashiba	Shizuoka University, Japan
Yoshitomo Baba	Osaka Kyoiku University, Japan
Aaron Chan	Nagoya University, Japan
Gyu Whan Chang	Incheon National University, Korea
Jialei Chen	Beijing University of Technology, China
Jianlong Chen	Southeast University, China
Xiaofeng Chen	Southeast University, China
Sangmin Chun	Chung-Ang University, Korea
Yang Chunhua	Weifang University, China
Erik Darpö	Nagoya University, Japan
Nanqing Ding	Nanjing University, China
Thomas Dorsey	CCR-La Jolla, USA
Xiankun Du	Jilin University, China
Naoki Endo	Waseda University, Japan
Haruhisa Enomoto	Nagoya University, Japan
Iwan Ernanto	Universitas Gadjah Mada, Indonesia
Xianhui Fu	Northeast Normal University, China
Nan Gao	Shanghai University, China
Yuichiro Goto	Osaka University, Japan
Ziyu Guo	Beijing University of Technology, China
Norihiro Hanihara	Nagoya University, Japan
Jiwei He	Hangzhou Normal University, China
Akihiko Hida	Saitama University, Japan
Yoshimasa Hieda	Osaka Prefecture University College of Technology, Japan
Na'imah Hijriati	Universitas Gadjah Mada, Indonesia

Naoya Hiramatsu	National Institute of Technology, Kure College, Japan
Chan Yong Hong	Kyung Hee University, Korea
Takahiro Honma	Tokyo University of Science, Japan
Kui Hu	Southwest University of Science and Technology, China, and Kyungpook National University, Korea
Zhaoyong Huang	Nanjing University, China
Kei-ichiro Iima	National Institute of Technology, Nara College, Japan
Ryotaro Isobe	Chiba University, Japan
Ayako Itaba	Tokyo University of Science, Japan
Tomohiro Itagaki	Tokyo University of Science, Japan
Ken-ichi Iwase	Osaka Electro-Communication University, Japan
Osamu Iyama	Nagoya University, Japan
Haibo Jin	Nagoya University, Japan
Ryo Kanda	Osaka University, Japan
Ryoichi Kase	Okayama University of Science, Japan
Kiriko Kato	Osaka Prefecture University, Japan
Shigeto Kawata	Nagoya City University, Japan
Bernhard Keller	University of Paris 7, France
Hwankoo Kim	Hoseo University, Korea
Nam Kyun Kim	Hanbat National University, Korea
Yoshiyuki Kimura	Osaka Prefecture University, Japan
Toshinori Kobayashi	Nagoya University, Japan
Hirotaka Koga	Tokyo Denki University, Japan
Hideyuki Koie	National Institute of Technology, Nagaoka College, Japan
Yuta Kozakai	Shibaura Institute of Technology, Japan
Shinya Kumashiro	Chiba University, Japan
Naoko Kunugi	Tokyo University of Science, Japan
Jung Miao Kuo	National Chung Hsing University, Taiwan
Yosuke Kuratomi	Yamaguchi University, Japan
Tai Keun Kwak	Daejin University, Korea
Gangyong Lee	Chungnam National University, Korea
Tsiu-Kwen Lee	National Taiwan University, Taiwan
Fang Li	Zhejiang University, China
Libin Li	Yangzhou University, China
Guohua Liu	Southeast University, China

Linlin Liu	Southeast University, China
Di-Ming Lu	Zhejiang University, China
Lixin Mao	Nanjing Institute of Technology, China
Xuefeng Mao	Shanghai University, China
Hidetoshi Marubayashi	Naruto University of Education, Japan
Kanzo Masaike	Tokyo Gakugei University, Japan
Hiroki Matsui	University of Tokyo, Japan
Masaki Matsuno	Shizuoka University, Japan
Naoyuki Matsuoka	Meiji University, Japan
Hiroyuki Minamoto	Osaka Prefecture University, Japan
Atsuki Miyauchi	Osaka Prefecture University, Japan
Yuya Mizuno	Osaka Prefecture University, Japan
Izuru Mori	Shizuoka University, Japan
Kota Murakami	Kyoto University, Japan
Yusuke Nakajima	Kavli IPMU, University of Tokyo, Japan
Kazunori Nakamoto	University of Yamanashi, Japan
Tsutomu Nakamura	University of Verona, Italy
Ken Nakashima	Shizuoka University, Japan
Pace P. Nielsen	Brigham Young University, USA
Tsunekazu Nishinaka	University of Hyogo, Japan
Miaomiao Niu	Beijing University of Technology, China
Yasuaki Ogawa	Nagoya University, Japan
Dong Yeol Oh	Chosun University, Korea
Sei-Qwon Oh	Chungnam National University, Korea
Yosuke Ohnuki	National Institute of Technology, Suzuka College, Japan
Maiko Ono	Okayama University, Japan
Kazuho Ozeki	Yamaguchi University, Japan
Diah Junia Eksi Palupi	Universitas Gadjah Mada, Indonesia
Mahdieh Paparizarei	Payame Noor University, Iran
Mi Hee Park	Chung-Ang University, Korea
Manoj Kumar Patel	National Institute of Technology Nagaland, India
Andreas Reinhart	Incheon National University, Korea
Yanli Ren	Nanjing Xiaozhuang University, China
Katsunori Sanada	Tokyo University of Science, Japan
Masahisa Sato	Aichi University, Japan
Yuan Shen	Zhejiang Sci-Tech University, China
Guodong Shi	Southeast University, China

Taiki Shibata	Okayama University of Science, Japan
Yoshiharu Shibata	Yamaguchi University, Japan
Yang Shilin	Beijing University of Technology, China
Kaori Shimada	Meiji University, Japan
Kenichi Shimizu	Shibaura Institute of Technology, Japan
S. Paul Smith	University of Washington, USA
Takao Sumiyama	Aichi Institute of Technology, Japan
Sutopo	Universitas Gadjah Mada, Indonesia
Mayu Tsukamoto	Yamaguchi University, Japan
Derya Keskin Tütüncü	Hacettepe University, Turkey
Akira Ueda	Shimane University, Japan
Taro Ueda	Nagoya University, Japan
Morio Uematsu	Jobu University, Japan
Kenta Ueyama	Hirosaki University, Japan
Katsuhiro Uno	Osaka University, Japan
Satoshi Usui	Tokyo University of Science, Japan
Xian Wang	China University of Mining and Technology, China
Yao Wang	Nanjing University of Information Science and Technology, China
Indah Emilia Wijayanti	Universitas Gadjah Mada, Indonesia
Hon Yin Wong	Nagoya University, Japan
Quanshui Wu	Fudan University, China
Tongsuo Wu	Shanghai Jiao Tong University, China
Zhixiang Wu	Zhejiang University, China
Kunio Yamagata	Tokyo University of Agriculture and Technology, Japan
Kota Yamaura	University of Yamanashi, Japan
Lingling Yao	Southeast University, China
Masayoshi Yoshikawa	Hyogo University of Teacher Education, Japan
Yuji Yoshino	Okayama University, Japan
Michio Yoshiwaki	RIKEN AIP, Japan
Toshiya Yurikusa	Nagoya University, Japan
Xiaochen Zeng	Beijing University of Technology, China
Pu Zhang	Shanghai Jiao Tong University, China
Xiaojin Zhang	Nanjing University of Information Science and Technology, China
Yang Zhang	University of Manitoba, Canada
Yingying Zhang	Hohai University, China

Yuehui Zhang	Shanghai Jiao Tong University, China
Dechuan Zhou	Southwest University of Science and Technology, China, and Hoseo University, Korea
Guisong Zhou	Ningbo University, China
Nan Zhou	Zhejiang Shuren University, China
Shaotao Zhu	Beijing University of Technology, China
Shenglin Zhu	Fudan University, China
Michał Ziembowski	Warsaw University of Technology, Poland

Contents

PART A

Invited Lectures (Survey Articles)

On cyclotomic quiver Hecke algebras of affine type

Susumu Ariki

Department of Pure and Applied Mathematics,
Graduate School of Information Science and Technology,
Osaka University,
1-5 Yamadaoka, Suita, Osaka 565-0871, Japan
E-mail: ariki@ist.osaka-u.ac.jp

We explain the representation theoretic treatment of Fock spaces by the Kyoto school, and their use in the categorification of basic modules of the Kac-Moody Lie algebras of affine Lie type by cyclotomic quiver Hecke algebras. As the (affine) quiver Hecke algebras are also interesting objects for study, I briefly review known results on the representation theory of those algebras of finite Lie type in the final section. This survey is aimed at non-experts.

Keywords: Fock space; categorification; cyclotomic quiver Hecke algebra.

1. Introduction

Let G be a finite group, H a subgroup of G. The vector space $\mathbb{F}X$ over a field $\mathbb{F}$ is the G-module obtained from the G-set $X = G/H$. The opposite algebra of $\mathrm{End}_G(\mathbb{F}X)$ is called the Hecke algebra associated with the pair (G, H). Elements of $\mathrm{End}_G(\mathbb{F}X)$ are described by G-invariant functions $f : G/H \times G/H \to \mathbb{F}$ as

$$xH \mapsto \sum_{yH \in G/H} f(xH, yH)yH$$

where G-action on $G/H \times G/H$ is the diagonal action. Then, the product of the Hecke algebra is given by the convolution product. The characteristic functions of G-orbits on $G/H \times G/H$ form a basis of the Hecke algebra, which is called the Schur basis, and Schur described the structure constant of the product with respect to the Schur basis.

Let $\mathbb{F}_q$ be a finite field, G an algebraic group, H a closed subgroup of G. Then the Hecke algebra associated with the pair $(G(\mathbb{F}_q), H(\mathbb{F}_q))$ may be studied by geometric point of view. As the homogeneous space G/H is a complete variety if and only if H contains a Borel subgroup, the case that G/H is a flag variety has been studied in detail, and the associated

Hecke algebra became famous when it was used in the Kazhdan-Lusztig conjecture. In that case, Iwahori and Matsumoto gave the description of the Hecke algebra in terms of generators and relations. In this description, the algebra is defined by the Weyl group of G only. In particular, the Hecke algebra associated with the general linear group (and its Borel subgroup) is obtained by deforming the group algebra of the symmetric group.

Let S_n be the symmetric group of degree n. The group algebra over a field $\mathbb{F}$ is generated by Coxeter generators $s_1, \dots, s_{n-1}$ and they are subject to the Coxeter relations. In other words, $\mathbb{F}S_n$ is the quotient algebra of the group algebra of the Artin braid group B_n by adding the quadratic relations $s_i^2 = 1$. We may introduce a parameter q to deform $\mathbb{F}S_n$ by changing the quadratic relations to $(s_i - q)(s_i + 1) = 0$. This is the description of the Hecke algebra associated with the general linear group. Those Hecke algebras were generalized to cyclotomic Hecke algebras by us [4] and Broué and Malle [14], independently. More precisely, Broué and Malle defined cyclotomic Hecke algebras for complex reflection groups, and the cyclotomic Hecke algebras we mention here are those associated with the complex reflection groups $G(m, 1, n)$ in the Shephard-Todd notation.

Definition 1.1. The cyclotomic Hecke algebra $H_{G(m,1,n)}(q, Q_1, \dots, Q_m)$ is the $\mathbb{F}$-algebra defined by generators $T_0, T_1, \dots, T_{n-1}$ subject to the relations

$$\begin{gathered}(T_0 - Q_1) \cdots (T_0 - Q_m) = 0, \;\; (T_i - q)(T_i + 1) = 0 \; (1 \le i \le n-1), \\ (T_0T_1)^2 = (T_1T_0)^2, \;\; T_iT_j = T_jT_i \; (j \ne i \pm 1), \\ T_iT_{i+1}T_i = T_{i+1}T_iT_{i+1} \; (1 \le i \le n-2).\end{gathered}$$

We always assume that $\mathbb{F}$ is algebraically closed. Detailed studies of those algebras have been made in the past decades. Those studies were triggered by the author's old work which answered a conjecture by Lascoux, Leclerc and Thibon.

Assume that $q \ne 1$ is a primitive e^{th} root of unity and $Q_i = q^{s_i}$. We denote $\mathcal{H}_n = H_{G(m,1,n)}(q, q^{s_1}, \dots, q^{s_m})$. Let $\mathfrak{g}$ be the affine Kac-Moody Lie algebra of Cartan type $A_{e-1}^{(1)}$, $V(\Lambda)$ its integrable highest weight module with highest weight $\Lambda = \Lambda_{s_1} + \cdots + \Lambda_{s_m}$, where Λ_i, for $i \in \mathbb{Z}/e\mathbb{Z}$, are the fundamental weights. Key observations in the work were as follows.

- Refined induction functors and restriction functors make the direct sum of complexified Grothendieck groups $\oplus_{n \ge 0} K_0(\mathcal{H}_n\text{-mod})_{\mathbb{C}}$ into a $\mathfrak{g}$-module such that we have an isomorphism of $\mathfrak{g}$-modules.

$$\bigoplus_{n \geq 0} K_0(\mathcal{H}_n\text{-mod})_{\mathbb{C}} \simeq V(\Lambda).$$

- We embed $V(\Lambda)$ into the higher level Fock space determined by $(s_1, \ldots, s_m)$, which we will explain in §3. The higher level Fock space has the standard basis indexed by m-tuples of partitions and if we compose the embedding with the above isomorphism, the composition coincides with the direct sum over $n = 0, 1, \ldots$ of the transpose of the decomposition map.
- In particular, if the characteristic of the base field $\mathbb{F}$ is zero, the decomposition numbers are given by the expansion of the canonical basis of $V(\Lambda)$ into the standard basis of the higher level Fock space.
- The collection over $n = 0, 1, \ldots$ of the irreducible $\mathcal{H}_n$-modules are labeled by the highest weight crystal $B(\Lambda)$. Moreover, the labeling is compatible with the labeling given by the Specht module theory for $\mathcal{H}_n$.

If the reader wishes to know more details, papers by myself [1], where the items above are found in Theorem 3.27, and Mathas [24], where cyclotomic q-Schur algebras are also treated, survey the theory.

Hecke algebras of type A and type B are cyclotomic Hecke algebras, and Hecke algebras of type D may be handled by applying the Clifford theory to Hecke algebras of type B with a special choice of parameters. This picture allowed us to study the modular representation theory of Hecke algebras of classical type. We come back to this picture and its modern form in §3 after we introduce other types of Fock spaces.

As one finds in the definition, the cyclotomic Hecke algebras are still defined by Coxeter-like generators. Now we introduce cyclotomic quiver Hecke algebras.[a]

Cyclotomic quiver Hecke algebras first appeared in §3.4 of a paper [21] by Khovanov and Lauda, where they wrote

> The ring $R(\nu; \lambda)$ inherits a grading from $R(\nu)$. These quotient rings should be the analogues of the Ariki-Koike cyclotomic Hecke

[a] We must distinguish between the cyclotomic Hecke algebra and the cyclotomic quiver Hecke algebra, since the definitions are different.

algebras in our framework. Let

$$R(*,\lambda) \stackrel{\text{def}}{=} \bigoplus_{\nu\in\mathbb{N}[I]} R(\nu;\lambda),$$

and switch from $\mathbb{Z}$ to a field $\mathbb{K}$. We expect that, for sufficiently nice Γ and $\mathbb{K}$, the category of graded modules over $R(*;\lambda)$ categorifies the integrable irreducible $U_q(\mathfrak{g})$-representation V_λ with highest weight λ.

The most striking result is the relationship to the classical Hecke algebras. In their paper [10], Brundan and Kleshchev proved the next result. $R^\Lambda(n)$ is the cyclotomic quiver Hecke algebra which we will define in the body of the paper. We will write $R^\Lambda(\beta)$ instead of $R(\beta,\Lambda)$ and $R(*,\Lambda) = \oplus_{n\geq 0} R^\Lambda(n)$.

Theorem 1.1. *Suppose that q is a primitive $(\ell+1)^{th}$ root of unity and $Q_i = q^{s_i}$. We set $A = A_\ell^{(1)}$, $\Lambda = \Lambda_{s_1} + \cdots + \Lambda_{s_m}$ and $t = (-1)^{\ell+1}$. Then*

$$R^\Lambda(n) \simeq H_{G(m,1,n)}(q, Q_1, \ldots, Q_m).$$

The similar result holds for the degenerate cyclotomic Hecke algebra.

A result by Dipper and Mathas [16] tells us that the cyclotomic Hecke algebras for arbitrary nonzero parameters $Q_1, \ldots, Q_m$ are Morita equivalent to direct sum of tensor products of the cyclotomic Hecke algebras whose parameters are powers of q. Hence, we have the following inclusion of classes of algebras up to Morita equivalence.

(Hecke algebras of type A and type B)

(Cyclotomic Hecke algebras)

(Direct sums of tensor products of cyclotomic quiver Hecke algebras)

The key for the proof of Theorem 1.1 was the discovery of the Khovanov-Lauda(-Rouquier) generators. One consequence is the following corollary.

Corollary 1.1. *The group algebra of S_n is a graded algebra.*

I call $R(\nu)$ the affine quiver Hecke algebra and $R(\nu;\lambda)$ the cyclotomic quiver Hecke algebra. Since the introduction of the quiver Hecke algebras, various new results on those algebras have been proven. In the final section, I will briefly review some of the results by Brundan, Kleshchev and their collaborators on $R(\nu)$ and the labeling of irreducible $R(\nu)$-modules. Since my primary goal is to answer basic representation theoretic questions on cyclotomic Hecke algebras, such as the structure of tame block algebras, the decomposition numbers, and so on, my research to the present is to apply this new development to the understanding of block algebras of classical Hecke algebras. As was explained, the Lie type which is relevant for this purpose is the affine type A, and the embedding of the integrable highest weight modules to higher level Fock spaces has played important roles in the proofs. Further, using the τ-tilting theory by Iyama and his school together, we now have good understanding of tame block algebras of Iwahori-Hecke algebras of classical type.[b]

The Fock space for affine type A has been used most successfully for the study of cyclotomic quiver Hecke algebras, but we may also use other Fock spaces to give some useful information on cyclotomic quiver Hecke algebras of affine Lie type. Because of this, we shall begin this survey article by recalling Fock spaces which had been studied by the Kyoto school.

2. Fock spaces

We start with reviewing the soliton theory. Let us begin with the paper

Transformation Groups for Soliton Equations
– Euclidean Lie Algebras and
Reduction of the KP Hierarchy –

by Etsuro Date, Michio Jimbo, Masaki Kashiwara and Tetsuji Miwa, which was published in Publ. RIMS, Kyoto Univ. **18** (1982), 1077–1110. There, the Kadomtsev-Petviashvli equations (the KP equations for short) and their reductions are given in terms of the Hirota bilinear form. For example, the famous KdV equation has the Hirota form $(D_1^4 - 4D_1D_3)\tau \cdot \tau = 0$, where $P(D_1, D_2, \dots)\tau \cdot \tau$ is defined as

$$P\left(\frac{\partial}{\partial y_1}, \frac{\partial}{\partial y_2}, \dots\right) \tau(x_1+y_1, x_2+y_2, \dots)\tau(x_1-y_1, x_2-y_2, \dots)|_{y_1=y_2=\dots=0}.$$

[b]Our next goal will be the understanding of the decomposition numbers of wild block algebras, but it is beyond our hands at this moment. It is also worth investigating some good subcategories of the module category or the derived category of Hecke algebras. Note also that we may consider Hecke algebras for any pair of a finite group and its subgroup. This is another possible research direction to generalize than the cyclotomic quiver Hecke algebras.

Those nonlinear differential equations admit the infinitesimal symmetry $\mathfrak{gl}(\infty)$, i.e. the central extension of

$$\left\{ \sum_{i,j\in\mathbb{Z}} a_{ij}E_{ij} \mid a_{ij} = 0, \text{if } |i-j| \text{ is sufficiently large} \right\},$$

or its reduction (Chevalley generators are infinite sums of E_{ij}'s). Then, the τ-functions $\tau(x) = \langle vac| \exp H(x)|L\rangle$ give solutions of those KP equations, where L runs through the orbit of $|vac\rangle$ in the Fock representation of the infinite dimensional Lie group.

The similar result holds for the BKP equations, where the infinitesimal symmetry is $\mathfrak{go}(\infty)$. The reduction mentioned above is a simple procedure to obtain Chevalley generators of the affine Lie algebras as infinite sums of the generators of $\mathfrak{gl}(\infty)$ or $\mathfrak{go}(\infty)$. In this setup, the affine Lie algebras $A_\ell^{(1)}$ appear as the reduction of the KP hierarchy, $A_{2\ell}^{(2)}$ and $D_{\ell+1}^{(2)}$ appear as the reduction of the BKP hierarchy.

The Fock space for the KP or the BKP equations is based on charged fermions or neutral fermions, respectively. We may rewrite those Fock spaces in terms of partitions or shifted partitions, which form a basis of the Fock space. The nodes of partitions or shifted partitions are given residue $0, 1, \ldots, \ell$ and the action of the Chevalley generator f_i on each of the partitions or shifted partitions adds one node of residue i. Now we can state how the affine Dynkin diagrams arise in the soliton theory. We consider the affine Dynkin diagrams

$$A_{2\ell}^{(2)}(=\widetilde{BC}_\ell), \quad D_{\ell+1}^{(2)}(=\widetilde{B}_\ell), \quad A_\ell^{(1)}(=\widetilde{A}_\ell), \quad C_\ell^{(1)}(=\widetilde{C}_\ell)$$

in this article. (All of them have $\ell+1$ vertices.) In section 3.4, we will introduce cyclotomic quiver Hecke algebras (aka cyclotomic KLR algebras), and use those Fock spaces to obtain dimension formulas for the algebras.

Remark 2.1. $A_{2\ell-1}^{(2)}(=\widetilde{CD}_\ell)$ and $D_\ell^{(1)}$ appear as the reductions from the two component BKP. But we do not use them here.

2.1. *Combinatorial Fock spaces for Lie type* $A_{2\ell}^{(2)}$

The Cartan matrix is given as follows.

$$A_{2\ell}^{(2)} = (a_{ij})_{i,j\in I} = \begin{pmatrix} 2 & -2 & 0 & \dots & 0 & 0 & 0 \\ -1 & 2 & -1 & \dots & 0 & 0 & 0 \\ 0 & -1 & 2 & \dots & 0 & 0 & 0 \\ \vdots & \vdots & \vdots & \ddots & \vdots & \vdots & \vdots \\ 0 & 0 & 0 & \dots & 2 & -1 & 0 \\ 0 & 0 & 0 & \dots & -1 & 2 & -2 \\ 0 & 0 & 0 & \dots & 0 & -1 & 2 \end{pmatrix}.$$

When $\ell = 1$, the Cartan matrix of type $A_2^{(2)}$ is $\left(\begin{smallmatrix} 2 & -4 \\ -1 & 2 \end{smallmatrix}\right)$. Then, the Fock space is the vector space whose basis is the set of shifted partitions. We color the nodes of shifted partitions with the residue pattern which repeats $01\cdots\ell\cdots10$ in each row. For example, if $\ell = 2$ then

0	1	2	1	0	0
	0	1			

2.2. *Combinatorial Fock spaces for Lie type* $D_{\ell+1}^{(2)}$

The Cartan matrix is given as follows.

$$D_{\ell+1}^{(2)} = (a_{ij})_{i,j\in I} = \begin{pmatrix} 2 & -2 & 0 & \dots & 0 & 0 & 0 \\ -1 & 2 & -1 & \dots & 0 & 0 & 0 \\ 0 & -1 & 2 & \dots & 0 & 0 & 0 \\ \vdots & \vdots & \vdots & \ddots & \vdots & \vdots & \vdots \\ 0 & 0 & 0 & \dots & 2 & -1 & 0 \\ 0 & 0 & 0 & \dots & -1 & 2 & -1 \\ 0 & 0 & 0 & \dots & 0 & -2 & 2 \end{pmatrix}.$$

When $\ell = 1$, the Cartan matrix of type $D_2^{(2)}$ is defined to be $A_1^{(1)}$. The Fock space is the same as $A_{2\ell}^{(2)}$ but the residue pattern is different. It repeats $01\cdots\ell\ell\cdots10$ in each row. For example, if $\ell = 2$ then

0	1	2	2	1	0
	0	1			

2.3. *Combinatorial Fock spaces for Lie type* $A_\ell^{(1)}$

The Cartan matrix is given as follows.

$$A_\ell^{(1)} = (a_{ij})_{i,j\in I} = \begin{pmatrix} 2 & -1 & 0 & \dots & 0 & 0 & -1 \\ -1 & 2 & -1 & \dots & 0 & 0 & 0 \\ 0 & -1 & 2 & \dots & 0 & 0 & 0 \\ \vdots & \vdots & \vdots & \ddots & \vdots & \vdots & \vdots \\ 0 & 0 & 0 & \dots & 2 & -1 & 0 \\ 0 & 0 & 0 & \dots & -1 & 2 & -1 \\ -1 & 0 & 0 & \dots & 0 & -1 & 2 \end{pmatrix}.$$

When $\ell = 1$, the Cartan matrix of type $A_1^{(1)}$ is $\left(\begin{smallmatrix} 2 & -2 \\ -2 & 2 \end{smallmatrix}\right)$. The Fock space is the vector space whose basis is the set of partitions. The residue pattern involves $s \in \mathbb{Z}/(\ell+1)\mathbb{Z}$:

> the cell on the rth row and the cth column of a partition has the residue $s - r + c$ modulo $\ell + 1$.

We denote the Fock space with the residue pattern by $\mathcal{F}_s$. For example, if $\ell = 2$ and $s = 1$ then

1	2	0	1	2	0
0	1	2			

2.4. *Combinatorial Fock spaces for Lie type* $C_\ell^{(1)}$

By the folding procedure, we obtain $C_\ell^{(1)}$ from $A_{2\ell-1}^{(1)}$.

Then, the Cartan matrix of the Lie type $C_\ell^{(1)}$ is given as follows.

$$C_\ell^{(1)} = (a_{ij})_{i,j\in I} = \begin{pmatrix} 2 & -1 & 0 & \cdots & 0 & 0 & 0 \\ -2 & 2 & -1 & \cdots & 0 & 0 & 0 \\ 0 & -1 & 2 & \cdots & 0 & 0 & 0 \\ \vdots & \vdots & \vdots & \ddots & \vdots & \vdots & \vdots \\ 0 & 0 & 0 & \cdots & 2 & -1 & 0 \\ 0 & 0 & 0 & \cdots & -1 & 2 & -2 \\ 0 & 0 & 0 & \cdots & 0 & -1 & 2 \end{pmatrix}.$$

Recall that the residue pattern of the Fock space $\mathcal{F}_{s=0}$ for $A_\ell^{(1)}$ is

$$\begin{array}{llllll} 0\,1 \cdots & \ell & 0 & 1 & \cdots & \\ \ell\,0 & 1 & \cdots & \ell & 0 & \cdots \\ \vdots\ \vdots & \ddots & \ddots & \ddots & \ddots & \ddots \end{array}$$

namely, we repeat $01\cdots\ell$ on the rim. We slide the residue sequence on the rim to obtain $\mathcal{F}_s$, where s sits on the corner instead of 0. The Fock space for $C_\ell^{(1)}$ is the same as $A_\ell^{(1)}$, but the residue sequence on the rim repeats $01\cdots\ell\cdots21$. We denote the Fock space by $\mathcal{F}_s$ again.

3. Categorification of integrable highest weight modules

Let $\mathfrak{g}(A)$ be the Kac-Moody Lie algebra associated with the Cartan matrix A and let $U_q(\mathfrak{g}(A))$ be the corresponding quantized enveloping algebra. The $\mathfrak{g}(A)$-module $V(\Lambda_0)$ is the integrable highest weight module whose highest weight is the fundamental weight Λ_0. $V(\Lambda_0)$ is called the basic module. The following features are explained in the references [5][6][8][7].

- In all the cases $A = A_{2\ell}^{(2)}, D_{\ell+1}^{(2)}, A_\ell^{(1)}, C_\ell^{(1)}$ we have considered above, the residue pattern defines the action of $\mathfrak{g}(A)$.
- In the Fock spaces, the vacuum vector generates the highest weight module $V(\Lambda_0)$ of the corresponding affine Lie algebra.
- If $A = A_\ell^{(1)}$ or $A = C_\ell^{(1)}$, we may deform the Fock space to the deformed Fock space which is a $U_q(\mathfrak{g}(A))$-module.
- Moreover, we may define the higher level Fock space associated with a multi-charge $(s_1, \ldots, s_m)$ by

$$\mathcal{F}_{(s_1,\ldots,s_m)} = \mathcal{F}_{s_1} \otimes \cdots \otimes \mathcal{F}_{s_m}.$$

- The vacuum vector of $\mathcal{F}_{(s_1,\ldots,s_m)}$ generates the $U_q(\mathfrak{g}(A))$-module $V_q(\Lambda)$ where the dominant integral weight $\Lambda = \Lambda_{s_1} + \cdots + \Lambda_{s_m}$ is its highest weight.[c]

Remark 3.1. There is another construction of deformed Fock spaces due to Uglov, which is not the tensor product of level one deformed Fock spaces.

The Fock spaces with $m = 1$ and $m = 2$ in type $A_\ell^{(1)}$ are those Fock spaces we have been using for studying Hecke algebras of classical type. In

[c]Brundan and Kleshchev [11] use the dual of the embedding $V_q(\Lambda) \hookrightarrow \mathcal{F}_{(s_1,\ldots,s_m)}$ to give graded decomposition numbers of cyclotomic Hecke algebras. This result is a final form of the picture which we mentioned in the introduction as key observations.

the primitive form, we identify the weight spaces of the integrable module $V(\Lambda)$, where the highest weight Λ is determined by the common parameters $Q_i = q^{s_i}$ of the cyclotomic Hecke algebras $H_{G(m,1,n)}(q, q^{s_1}, \ldots, q^{s_m})$, for $n = 0, 1, \ldots,$ with the complexified Grothendieck groups of the module categories of the cyclotomic Hecke algebras. The current picture adds more nice features by virtue of the cyclotomic categorification theorem by Kang and Kashiwara, and the Chuang-Rouquier derived equivalence which lifts the Weyl group action on $V(\Lambda)$. We will explain the two later in subsections 3.4 and 3.5. As we write in the introduction, the algebras were generalized to cyclotomic Hecke and cyclotomic quiver Hecke algebras. In the next section, we define the cyclotomic quiver Hecke algebra, which categorify the integrable highest weight module $V_q(\Lambda)$ over the quantized enveloping algebra $U_q(\mathfrak{g}(A))$, where $A = (a_{ij})_{i,j \in I}$ is a symmetrizable Cartan matrix.

3.1. *Definition of cyclotomic quiver Hecke algebras*

- Let P be the weight lattice.
- Let $\Pi = \{\alpha_i\}_{i \in I}$ be the set of simple roots.
- We fix a dominant integral weight $\Lambda \in P$.
- Let $\mathbb{F}$ be a field.
- We fix a system of polynomials $Q_{i,j}(u, v) \in \mathbb{F}[u, v]$, for $i, j \in I$:

$$Q_{i,j}(u,v) = \begin{cases} \sum_{p(\alpha_i,\alpha_i)+q(\alpha_j,\alpha_j)+2(\alpha_i,\alpha_j)=0} t_{i,j;p,q} u^p v^q & \text{if } i \neq j, \\ 0 & \text{if } i = j, \end{cases}$$

where $t_{i,j;p,q} \in \mathbb{F}$ are such that $t_{i,j;-a_{ij},0} \in \mathbb{F}^\times$ and

$$Q_{i,j}(u, v) = Q_{j,i}(v, u).$$

The cyclotomic quiver Hecke algebra $R^\Lambda(n)$ associated with polynomials $(Q_{i,j}(u,v))_{i,j \in I}$ and a dominant integral weight $\Lambda \in P$ is the $\mathbb{Z}$-graded unital associative $\mathbb{F}$-algebra generated by

$$\{e(\nu) \mid \nu \in I^n\} \cup \{x_1, \ldots, x_n\} \cup \{\psi_1, \ldots, \psi_{n-1}\}$$

subject to the following relations.

$$e(\nu)e(\nu') = \delta_{\nu,\nu'} e(\nu), \quad \sum_{\nu \in I^n} e(\nu) = 1$$

$$x_r e(\nu) = e(\nu) x_r, \quad x_r x_s = x_s x_r, \quad \psi_r e(\nu) = e(s_r \nu) \psi_r$$

$$\begin{aligned}
\psi_r x_s &= x_s \psi_r \quad (\text{if } s \neq r, r+1) \\
\psi_r \psi_s &= \psi_s \psi_r \quad (\text{if } r \neq s \pm 1) \\
x_r \psi_r e(\nu) &= (\psi_r x_{r+1} - \delta_{\nu_r, \nu_{r+1}}) e(\nu) \\
x_{r+1} \psi_r e(\nu) &= (\psi_r x_r + \delta_{\nu_r, \nu_{r+1}}) e(\nu) \\
\psi_r^2 e(\nu) &= Q_{\nu_r, \nu_{r+1}}(x_r, x_{r+1}) e(\nu)
\end{aligned}$$

$$(\psi_{r+1}\psi_r\psi_{r+1} - \psi_r\psi_{r+1}\psi_r)e(\nu)$$
$$= \begin{cases} \frac{Q_{\nu_r,\nu_{r+1}}(x_r,x_{r+1}) - Q_{\nu_r,\nu_{r+1}}(x_{r+2},x_{r+1})}{x_r - x_{r+2}} e(\nu) & \text{if } \nu_r = \nu_{r+2}, \\ 0 & \text{otherwise,} \end{cases}$$

and the cyclotomic condition $x_1^{\langle \alpha_{\nu_1}^\vee, \Lambda \rangle} e(\nu) = 0$.

Remark 3.2. Define $F_{i,j}(u_1, u_2, v) = \frac{Q_{i,j}(u_1,v) - Q_{i,j}(u_2,v)}{u_1 - u_2} \in \mathbb{F}[u_1, u_2, v]$.

$$\frac{Q_{\nu_r,\nu_{r+1}}(x_r, x_{r+1}) - Q_{\nu_r,\nu_{r+1}}(x_{r+2}, x_{r+1})}{x_r - x_{r+2}}$$

in the definition above is a shorthand notation of $F_{\nu_r,\nu_{r+1}}(x_r, x_{r+2}, x_{r+1})$.

Lemma 3.1. *The algebras $R^\Lambda(n)$ are symmetric algebras. Moreover,*

(1) *The isomorphism classes of cyclotomic quiver Hecke algebras for $A = A_{2\ell}^{(2)}$, $D_{\ell+1}^{(2)}$ and $C_\ell^{(1)}$ do not depend on the choice of $Q_{i,j}(u, v)$.*

(2) *For $A_1^{(1)}$, we may assume that $Q_{i,j}(u, v)$ are of the form*

$$\mathcal{Q}_{0,1}(u, v) = u^2 + tuv + v^2,$$

for a parameter $t \in \mathbb{F}$.

(3) *For $A_\ell^{(1)}$ with $\ell \geq 2$, we may assume that $Q_{i,j}(u, v)$ are of the form*

$$\begin{aligned}
Q_{i,i+1}(u, v) &= u + v \quad (0 \leq i \leq \ell - 1), \\
Q_{\ell,0}(u, v) &= u + tv, \\
Q_{i,j}(u, v) &= 1 \quad (j \not\equiv i \pm 1 \mod (\ell + 1)),
\end{aligned}$$

for a parameter $t \in \mathbb{F}^\times$.

(4) *The isomorphism class of block algebras of $R^\Lambda(n)$ in type $A_\ell^{(1)}$ may depend on the parameter t.*

Proof. The proof that $R^\Lambda(n)$ are symmetric algebras is given in Appendix A of a paper [28] by Shan, Varagnolo and Vasserot. The parts (1), (2), (3) are proved in our papers [5][6][8]. The proofs are simple application of a lemma by Rouquier. In the classification of representation type of block algebras of Hecke algebras in my paper [2], we see the phenomenon that tame block algebras tend to a wild block algebra as the parameter t tends to $(-1)^{\ell+1}$, which implies part (4). □

3.2. *The affine quiver Hecke algebras*

The algebra $R^\Lambda(n)$ is finite dimensional. If we drop the last cyclotomic condition, we denote the algebra by $R(n)$. I shall call it the affine quiver Hecke algebra. It is infinite dimensional.

Remark 3.3. Recently, $R(n)$ are used in the monoidal categorification of cluster algebras $A_q(\mathfrak{n}(w))$ by Kang, Kashiwara, Kim and Oh [19].

3.3. *Grading and central idempotents of $R^\Lambda(n)$ and $R(n)$*

The next two facts are important in applications, where the inner product (α_i, α_j) are given by the symmetrized Cartan matrix.

- The algebra $R^\Lambda(n)$ and $R(n)$ are given $\mathbb{Z}$-grading as follows.

$$\deg(e(\nu)) = 0, \ \ \deg(x_r e(\nu)) = (\alpha_{\nu_r}, \alpha_{\nu_r}),$$

$$\deg(\psi_s e(\nu)) = -(\alpha_{\nu_s}, \alpha_{\nu_{s+1}}).$$

- For $\beta \in Q^+$ with $\mathrm{ht}(\beta) = n$, set

$$I^\beta = \{\nu \in I^n \mid \alpha_{\nu_1} + \cdots + \alpha_{\nu_n} = \beta\}.$$

  Then $e(\beta) = \sum_{\nu \in I^\beta} e(\nu)$ is a central idempotent and

$$R^\Lambda(n) = \bigoplus_{\substack{\beta \in Q^+ \\ \mathrm{ht}(\beta)=n}} R^\Lambda(\beta), \text{ where } R^\Lambda(\beta) = R^\Lambda(n) e(\beta).$$

3.4. *The cyclotomic categorification theorem*

The following theorem is proved by Kang and Kashiwara [18].

Theorem 3.1. *Let A be a symmetrizable Cartan matrix.*

(i) The complexified Grothendieck group of finite dimensional $R^\Lambda(\beta)$-modules may be identified with the weight space $V_q(\Lambda)_{\Lambda-\beta}$ of the highest weight $U_q(\mathfrak{g}(A))$-module $V_q(\Lambda)$.

(ii) There exist exact functors E_i and F_i from the module category of $R^\Lambda(\beta)$ to the module category of $R^\Lambda(\beta \pm \alpha_i)$ which descend to the action of Chevalley generators e_i and f_i on $V_q(\Lambda)$.

The exactness of induction functors, which are left and right adjoint to restriction functors, etc. was established by Kang and Kashiwara [18][20], for Dynkin diagrams Γ which are associated with symmetrizable generalized Cartan matrices. In the categorification picture, namely the picture that the module category $R(*,\Lambda)-$mod categorifies the highest weight module of highest weight Λ, we also know that simple reflections of the Weyl group acting on integrable modules are categorified to derived equivalence by Chuang and Rouquier [15].

3.5. *Chuang-Rouquier derived equivalence*

Since $V(\Lambda)$ is an integrable module, there exists action of the Weyl group. For example, if the Cartan matrix is $A_\ell^{(1)}$ then the Weyl group is the affine symmetric group S_e^{aff}, where $e=\ell+1$. S_e^{aff} is generated by the involutions $\{s_i\}_{i\in\mathbb{Z}/e\mathbb{Z}}$ which are subject to the relations $s_is_js_i=s_js_is_j$ if $j-i\pm1\in e\mathbb{Z}$, $s_is_j=s_js_i$ otherwise.

Remark 3.4. The affine symmetric group S_e^{aff} has another presentation as the bijections $w:\mathbb{Z}\to\mathbb{Z}$ such that $w(i+e)=w(i)+e$, for $i\in\mathbb{Z}$, and $\sum_{i=1}^e w(i)=\frac{e(e+1)}{2}$.

The next theorem is due to Chuang and Rouquier.

Theorem 3.2. *Suppose that two weights $\Lambda-\beta$ and $\Lambda-\beta'$ belong to the same Weyl group orbit. Then, $R^\Lambda(\beta)$ and $R^\Lambda(\beta')$ are derived equivalent.*

In the application to cyclotomic Hecke algebras, e is identified with the quantum characteristic $\min\{k\in\mathbb{N}\mid 1+q+\cdots+q^{k-1}=0 \text{ in } \mathbb{F}\}$. Then, we may use computations on the integrable highest weight modules and their embedding to higher level Fock spaces to obtain results for cyclotomic Hecke algebras.

3.6. *Block algebras and weight spaces of integrable modules*

The theorem below was proven for cyclotomic Hecke algebras by Lyle and Mathas [22]. However, we may state the result in this form because of the Brundan-Kleshchev isomorphism theorem.

Theorem 3.3. *If $A=A_\ell^{(1)}$ and $t=(-1)^{\ell+1}$ if $\ell\geq2$, $t=-2$ if $\ell=1$ then $R^\Lambda(\beta)$ is indecomposable. Namely, it is a block algebra of $R^\Lambda(n)$.*

Thus, the weight spaces of $V_q(\Lambda)$ in this case may be identified with the complexified Grothendieck group of the module category of the block algebras of $R^\Lambda(n)$, where $n = 0, 1, \ldots$.

Conjecture 3.1. *The algebras $R^\Lambda(\beta)$ are indecomposable for any affine Lie type.*

4. Dimension formulas

Let us call the cyclotomic quiver Hecke algebras $R^{\Lambda_0}(n)$ finite quiver Hecke algebras. When the Lie type is $A^{(1)}_{e-1}$ and the parameter is $t = (-1)^e$, then $R^{\Lambda_0}(n)$ is nothing but the finite Hecke algebra in type A, namely the Hecke algebra associated with the symmetric group. Using the embedding of $V(\Lambda_0)$ into the Fock space, we may deduce the dimension formulas [5][6] or graded dimension formulas [11][13][7] of finite quiver Hecke algebras or more general cyclotomic quiver Hecke algebras. The proofs are simple application of the Fock spaces and the categorification we have explained above.

We explained the residue patterns of Fock spaces in section 2. Then, the residue sequence of a standard tableau of a partition or shifted partition of $n \in \mathbb{N}$ is the sequence obtained by reading the residues according to the entries $1, \ldots, n$ of the tableau.

4.1. *Dimension formulas: $A^{(2)}_{2\ell}$*

Theorem 4.1. *We assume that the Cartan matrix is $A^{(2)}_{2\ell}$. Then,*

$$\dim e(\nu')R^{\Lambda_0}(n)e(\nu) = \sum_{\lambda \vdash n} 2^{-\langle d, \mathrm{wt}(\lambda)\rangle - l(\lambda)} K(\lambda, \nu')K(\lambda, \nu),$$

$$\dim R^{\Lambda_0}(\beta) = \sum_{\lambda \vdash n,\ \mathrm{wt}(\lambda) = \Lambda_0 - \beta} 2^{-\langle d, \mathrm{wt}(\lambda)\rangle - l(\lambda)} |\mathrm{ST}(\lambda)|^2,$$

$$\dim R^{\Lambda_0}(n) = \sum_{\lambda \vdash n} 2^{-\langle d, \mathrm{wt}(\lambda)\rangle - l(\lambda)} |\mathrm{ST}(\lambda)|^2,$$

where $\nu, \nu' \in I^n$ and $K(\lambda, \nu)$ is the number of standard tableaux whose residue sequence is ν.

4.2. *Dimension formulas:* $D^{(2)}_{\ell+1}$

Theorem 4.2. *We assume that the Cartan matrix is* $D^{(2)}_{\ell+1}$*. Then,*

$$\dim e(\nu')R^{\Lambda_0}(n)e(\nu) = \sum_{\lambda\vdash n} 2^{-\langle \mathsf{d},\mathrm{wt}(\lambda)\rangle - l(\lambda)} K(\lambda,\nu')K(\lambda,\nu),$$

$$\dim R^{\Lambda_0}(\beta) = \sum_{\lambda\vdash n,\ \mathrm{wt}(\lambda)=\Lambda_0-\beta} 2^{-\langle \mathsf{d},\mathrm{wt}(\lambda)\rangle - l(\lambda)} |\mathrm{ST}(\lambda)|^2,$$

$$\dim R^{\Lambda_0}(n) = \sum_{\lambda\vdash n} 2^{-\langle \mathsf{d},\mathrm{wt}(\lambda)\rangle - l(\lambda)} |\mathrm{ST}(\lambda)|^2.$$

Remark 4.1. Oh and Park [23] have obtained graded dimension formulas for $A^{(2)}_{2\ell}$ and $D^{(2)}_{\ell+1}$ by using Young walls, generalization of Young diagrams.

4.3. *Graded dimension formulas:* $A^{(1)}_{\ell}$ *and* $C^{(1)}_{\ell}$

For $A^{(1)}_{\ell}$ and $C^{(1)}_{\ell}$, we use the embedding of $V_q(\Lambda)$ to the deformed Fock space to deduce graded dimension formulas. We may define the statistics $\deg(T)$ and $K_q(\lambda,\nu)$ is the sum of $q^{\deg(T)}$ over $\mathrm{ST}(\lambda,\nu)$.

Theorem 4.3. *Let the Cartan matrix be* $A^{(1)}_{\ell}$ *or* $C^{(1)}_{\ell}$*. For* $\nu,\nu'\in I^n$*, we have*

$$\dim_q e(\nu)R^{\Lambda}(\beta)e(\nu') = \sum_{\lambda\vdash n,\mathrm{wt}(\lambda)=\Lambda-\beta} K_q(\lambda,\nu)K_q(\lambda,\nu'),$$

$$\dim_q R^{\Lambda}(\beta) = \sum_{\lambda\vdash n,\mathrm{wt}(\lambda)=\Lambda-\beta} K_q(\lambda)^2,$$

$$\dim_q R^{\Lambda}(n) = \sum_{\lambda\vdash n} K_q(\lambda)^2,$$

where $\dim_q M := \sum_{k\in\mathbb{Z}} \dim(M_k)q^k$ *for* $M=\oplus_{k\in\mathbb{Z}}M_k$*.*

5. Applications

In this section, we explain two applications of the higher level Fock space and the deformed Fock space to Hecke algebras and Specht module theory of the cyclotomic quiver Hecke algebras in affine type C.

5.1. *Tame block algebras of Hecke algebras*

We may use the graded dimension formula in type $A^{(1)}_{e-1}$ to analyze the idempotent truncation of block algebras $R^{\Lambda}(\beta)$. The advantage of the KLR generators compared with the classical Coxeter generators is that it is easy to construct idempotents and compute the Gabriel quiver of the idempotent truncation.

Suppose that the characteristic of the base field is odd. Computation based on the above tools plus various results such as Ohmatsu's theorem, Rickard's star theorem, criterion of tilting discreteness by Adachi, Aihara and Chan for Brauer graph algebras, we may show that the basic algebras of tame block algebras are very restricted. Here, the cellularity in the sense of Graham and Lehrer plays an important role.

Note that block algebras are cellular by old results by Dipper, James and Murphy for type A and B, by Geck for type D. The next theorem [3], which is based on another paper [2], determines the Morita classes of tame block algebras of Hecke algebras of classical type.

Theorem 5.1. *We consider block algebras of Hecke algebras of classical type over an algebraically closed field of odd characteristic and* $q \neq 1$, $e \geq 2$. *If it is finite, then it is Morita equivalent to a Brauer line algebra. If it is infinite-tame, then it is Morita equivalent to one of the algebras below.*

(1) In type A *or* D, *we must have* $e = 2$ *and it is a Brauer graph algebra whose Brauer graph is* (2)—○—(2) *or* (2)—(2)—○.
(2) In type B *with* $e = 2$, *it is either one of the Brauer graph algebras in (1), or the symmetric Kronecker algebra, which is the Brauer graph algebra with one non-exceptional vertex and one loop. Otherwise, we must have* $e \geq 4$ *is even and* $Q = -1$ *and it is the Brauer graph algebra whose Brauer graph is* (2)—(2)—(2).

5.2. *Specht modules in affine type C*

The following result [9] is another application of the categorification. The definition of Specht module in affine type C requires new interpretation of Garnir elements.

Theorem 5.2. *We may define Specht modules* S^λ, *for multi-partitions* $\lambda \vdash n$. *If* n *is small enough such that the height* n *part of* $V(\Lambda)$ *for* $C_\ell^{(1)}$ *and* C_∞ *are the same, then the set* $\mathrm{ST}(\lambda)$ *of standard tableaux of shape* λ *form a basis of* S^λ *and we have the graded character formula:*

$$\mathrm{ch}_q S^\lambda = \sum_{T \in \mathrm{ST}(\lambda)} q^{\deg(T)} \mathrm{res}(T).$$

Conjecture 5.1. *Suppose that the Cartan matrix is of affine type C. Then, the Specht modules give a cellular algebra structure on $R^\Lambda(n)$.*

6. Results on affine quiver Hecke algebras of finite Lie type

In this final section, we briefly review some results on the representation theory of affine quiver Hecke algebras. As Khovanov and Lauda [21] proved, the categories of finite dimensional graded modules over $R(n)$, $n = 0, 1, \ldots,$ categorify the negative half $U_q^-(\mathfrak{g})$ of the quantized enveloping algebra. The canonical basis and the dual canonical basis in $U_q^-(\mathfrak{g})$ have representation theoretic meaning. Namely, in good cases [29], the former is identified with basis elements of the split Grothendieck group of finitely generated graded projective modules given by indecomposables, and the latter is identified with basis elements of the Grothendieck group of finite dimensional graded modules given by irreducibles. On the other hand, the works by Kato [17], Brundan, Kleshchev and McNamara [12][25] concern the Poincaré-Birkhoff-Witt bases. In finite Lie type, the PBW bases are in bijection with reduced expressions of the longest element w_0. Let N be the number of positive roots and let $\{s_i\}_{i\in I}$ be simple roots. For a reduced expression $w_0 = s_{i_1} \cdots s_{i_N}$, we associate a total order on the set of positive roots by

$$\alpha_{i_1} < s_{i_1}(\alpha_{i_2}) < \cdots < s_{i_1} \cdots s_{i_{N-1}}(\alpha_{i_N}).$$

The order is a convex order in the sense that if β, γ, $\beta+\gamma$ are positive roots and $\beta < \gamma$ then $\beta < \beta+\gamma < \gamma$. Every convex order is obtained this way. For each convex order, they constructed standard modules which categorify the corresponding PBW basis, and proper standard modules which categorify the dual PBW basis. Moreover, irreducible $R(\alpha)$-modules are labeled by Kostant partitions $(\beta_1 \ge \cdots \ge \beta_l)$ of α. That is, β_i are positive roots such that $\sum_{i=1}^{l} \beta_i = \alpha$.

Definition 6.1. Let α be a positive root. We consider the embedding

$$R(\beta) \otimes R(\gamma) \to R(\alpha),$$

for $\beta, \gamma \in Q^+$ with $\beta+\gamma = \alpha$, and the restriction $\mathrm{Res}_{\beta,\gamma}$ of an $R(\alpha)$-module to an $R(\beta) \otimes R(\gamma)$-module. An $R(\alpha)$-module L is cuspidal if $\mathrm{Res}_{\beta,\gamma}(L) \neq 0$ implies

(i) β is a sum of roots less than or equal to α,
(ii) γ is a sum of roots greater than or equal to α.

There is a unique cuspidal module $L(\alpha)$, for a positive root α. Then the proper standard module $\overline{\Delta}(\lambda)$, for a Kostant partition $\lambda = (\beta_1 \geq \cdots \geq \beta_l)$ of $\alpha \in Q^+$ is defined to be

$$\mathrm{Ind}_{R(\beta_1)\otimes\cdots\otimes R(\beta_s)}^{R(\alpha)} L(\beta_1) \otimes \cdots \otimes L(\beta_s)$$

up to grading shift. Each proper standard modules has unique head $L(\lambda)$. The irreducible modules $L(\lambda)$ give a complete set of self-dual irreducible $R(\beta)$-modules. To define standard modules, we introduce root modules $\Delta(\alpha)$, for positive roots α, as the projective limit of $\Delta_n(\alpha)$, which satisfy

$$0 \to q^{(n-1)(\alpha,\alpha)} L(\alpha) \to \Delta_n(\alpha) \to \Delta_{n-1}(\alpha) \to 0,$$

$$0 \to q^{(\alpha,\alpha)} \Delta_{n-1}(\alpha) \to \Delta_n(\alpha) \to L(\alpha) \to 0,$$

where a power of q means shift of the grading. Then, we introduce divided power modules $\Delta(\alpha^m)$ and define the standard module $\Delta(\lambda)$, for a Kostant partition $\lambda = (\gamma_1^{m_1}, \ldots, \gamma_s^{m_s})$ with $\gamma_1 > \cdots > \gamma_s$ of α, by

$$\Delta(\lambda) = \mathrm{Ind}_{R(\gamma_1^{m_1})\otimes\cdots\otimes R(\gamma_s^{m_s})}^{R(\alpha)} \Delta(\gamma_1^{m_1}) \otimes \cdots \otimes \Delta(\gamma_s^{m_s}).$$

For the details, see the paper by Brundan, Kleshchev and McNamara [12].

Remark 6.1. For affine Lie type, we also have the notion of convex order. It is a pre-order on the set of positive roots which is a total order on the set of positive real roots. For the precise definition, see Definition 2.1 and Remark 2.4 of a paper by McNamara and Tingley [27] or Definition 3.1 by McNamara [26]. If α is a real positive root, then there is a unique self-dual cuspidal irreducible $R(\alpha)$-module. For imaginary positive roots, there is no cuspidal $R(n\delta)$-module if $n \geq 2$, and the set of cuspidal irreducible $R(\delta)$-modules is given in Theorem 17.3 of McNamara's paper [26].

References

[1] S. Ariki, Finite dimensional Hecke algebras, Trends in representation theory of algebras and related topics, *EMS Ser. Congr. Rep.*, Eur. Math. Soc., 1-48 (2008).

[2] S. Ariki, Representation type for block algebras of Hecke algebras of classical type, *Adv. Math.* **317**, 823-845 (2017).

[3] S. Ariki, Tame block algebras of Hecke algebras of classical type, *J. Aust. Math. Soc.*, arXiv:1805.0834.

[4] S. Ariki and K. Koike, A Hecke algebra of $(\mathbb{Z}/r\mathbb{Z}) \wr S_n$ and construction of its irreducible representations, *Adv. Math.* **106**, 216-243 (1994).

[5] S. Ariki and E. Park, Representation type of finite quiver Hecke algebras of type $A_{2\ell}^{(2)}$, *J. Algebra* **397**, 457-488 (2014).

[6] S. Ariki and E. Park, Representation type of finite quiver Hecke algebras of type $D_{\ell+1}^{(2)}$, *Trans. Amer. Math. Soc.* **368**, 3211-3242 (2016).

[7] S. Ariki and E. Park, Representation type of finite quiver Hecke algebras of type $C_{\ell}^{(1)}$, *Osaka J. Math.* **53**, 463-488 (2016).

[8] S. Ariki, K. Iijima and E. Park, Representation type of finite quiver Hecke algebras of type $A_{\ell}^{(1)}$ for arbitrary parameters, *Int. Math. Res. Not.* **IMRN 2015**, 6070-6135 (2015).

[9] S. Ariki, E. Park and L. Speyer, Specht modules for quiver Hecke algebras of type C, *Publ. Res. Inst. Math. Sci.* **55**, 565-626 (2019).

[10] J. Brundan and A. Kleshchev, Blocks of cyclotomic Hecke algebras and Khovanov-Lauda algebras, *Invent. Math.* **178**, 451-484 (2009).

[11] J. Brundan and A. Kleshchev, Graded decomposition numbers for cyclotomic Hecke algebras, *Adv. Math.* **222**, 1883-1942 (2009).

[12] J. Brundan, A. Kleshchev, and P. McNamara, Homological properties of finite-type Khovanov-Lauda-Rouquier algebras, *Duke Math. J.* **163**, 1353-1404 (2014).

[13] J. Brundan, A. Kleshchev, and W. Wang, Graded Specht modules, *J. reine angew. Math.* **655**, 61-87 (2011).

[14] M. Broué and G. Malle, Zyklotomische Heckealgebren (German), *Astérisque* **212**, 119-189 (1993).

[15] J. Chuang and R. Rouquier, Derived equivalences for symmetric groups and sl_2-categorification, *Ann. of Math. (2)* **167**, 245-298 (2008).

[16] R. Dipper and A. Mathas, Morita equivalences of Ariki-Koike algebras, *Math. Z.* **240**, 579-610 (2002).

[17] S. Kato, Poincaré-Birkhoff-Witt bases and Khovanov-Lauda-Rouquier algebras, *Duke Math. J.* **163**, 619-663 (2014).

[18] S.-J. Kang and M. Kashiwara, Categorification of highest weight modules via Khovanov-Lauda-Rouquier Algebras, *Invent. Math.* **190**, 699-742 (2012).

[19] S.-J. Kang, M. Kashiwara, Myungho Kim and Se-jin Oh, Monoidal categorification of cluster algebras, *J. Amer. Math. Soc.* **31**, 349-426 (2018).

[20] M. Kashiwara, Biadjointness in cyclotomic Khovanov-Lauda-Rouquier algebras, *Publ. Res. Inst. Math. Sci.* **48**, 501-524 (2012).

[21] M. Khovanov and A. Lauda, A diagrammatic approach to categorification of quantum groups I, *Represent. Theory* **13**, 309-347 (2009).

[22] S. Lyle and A. Mathas, Blocks of cyclotomic Hecke algebras, *Adv. Math.* **216**, 854-878 (2007).

[23] Se-jin Oh and E. Park, Young walls and graded dimension formulas for finite quiver Hecke algebras of type $A_{2\ell}^{(2)}$ and $D_{\ell+1}^{(2)}$, *J. Algebraic Combin.* **40**, 1077-1102 (2014).

[24] A. Mathas, The representation theory of the Ariki-Koike and cyclotomic q-Schur algebras, Representation theory of algebraic groups and quantum groups, *Adv. Stud. Pure Math.* **40**, 261-320 (2004).

[25] P. McNamara, Finite dimensional representations of Khovanov-Lauda-Rouquier algebras I: Finite type, *J. reine angew. Math.* **707**, 103-124 (2015).

[26] P. McNamara, Representations of Khovanov-Lauda-Rouquier algebras III: symmetric affine type, *Math. Z.* **287**, 243-286 (2017).

[27] P. McNamara and P. Tingley, Face functors for KLR algebras, *Represent. Theory* **21**, 106-131 (2017).

[28] P. Shan, M. Varagnolo and E. Vasserot, On the center of quiver Hecke algebras, *Duke Math. J.* **166**, 1005-1101 (2017).

[29] M. Varagnolo and E. Vasserot, Canonical bases and KLR algebras, *J. reine angew. Math.* **659**, 67-100 (2011).

Divisibility properties of the quotient ring of the polynomial ring $D[X, Y, U, V]$ modulo $(XV - YU)$

Gyu Whan Chang

Department of Mathematics Education, Incheon National University, Incheon 22012, Republic of Korea
E-mail: whan@inu.ac.kr

Let D be an integral domain, X, Y, U, V be indeterminates over D, $D[X, Y, U, V]$ be the polynomial ring over D, $(XV - YU)$ be the ideal of $D[X, Y, U, V]$ generated by $XV - YU$, and $R = D[X, Y, U, V]/(XV - YU)$. In this paper, we study some properties of the t-operation on R which turn out to be similar to those of the polynomial ring over D. Among other things, we show that (i) R is a PvMD (resp., ring of Krull type, t-SFT PvMD) if and only if D is, and (ii) if D is a PvMD, then $Cl(R) = Cl(D) \oplus \mathbb{Z}$, where $\mathbb{Z}$ is the additive group of integers.

Keywords: $D[x, y, u, v]$; t-operation; PvMD; class group; polynomial ring.

0. Introduction

Let D be an integral domain with quotient field K. Let Λ be an index set, $\mathbb{Z}^{(\Lambda)}$ be the direct sum of Λ-copies of the additive group of integers, $\{X_i, Y_i, U_i, V_i \mid i \in \Lambda\}$ be a set of indeterminates over D, and

$$R = D[\{X_i, Y_i, U_i, V_i\}]/(\{X_iV_i - Y_iU_i\}).$$

Hence, if we let x_i, y_i, u_i, v_i be the images of X_i, Y_i, U_i, V_i in R, respectively, then $R = D[\{x_i, y_i, u_i, v_i\}]$ and $R_{D-\{0\}} = K[\{x_i, y_i, u_i, v_i\}]$ with $x_iv_i = y_iu_i$. Claborn shows that $R_{D-\{0\}}$ is a Krull domain with $Cl(R_{D-\{0\}}) = \mathbb{Z}^{(\Lambda)}$, and he then uses this ring $R_{D-\{0\}}$ with some additional results to prove that if G is an abelian group, then there is a Krull domain A such that $Cl(A) = G$ [8, Proposition 6]. Finally, he shows that if $\{X_i\}_{i=1}^{\infty}$ is a set of indeterminates over A and if Q is a prime ideal of $A[\{X_i\}]$ with ht$Q \geq 2$, then Q contains a nonconstant prime polynomial f_Q; so if S is the multiplicative set of $A[\{X_i\}]$ generated by such prime polynomials, then $A[\{X_i\}]_S$ is a Dedekind domain with $Cl(A[\{X_i\}]_S) = G$ [8, Theorem 7].

Note that if D is a Krull domain, then $t = v$ on D. (The definitions related to the t-operation will be reviewed in Section 1.) It is also known that the following six conditions are equivalent (see, for example, [19]): (i) D is a Krull domain, (ii) every nonzero ideal of D is t-invertible, (iii) every nonzero prime ideal of D is t-invertible, (iv) every t-ideal I of D is a t-product of prime ideals, i.e., $I = (P_1 \cdots P_n)_t$ for some prime ideals P_i of D, (v) every nonzero principal ideal of D is a t-product of prime ideals, and (vi) every nonzero prime ideal of D contains a t-invertible prime ideal. Hence, the t-operation is a very important notion when we study the properties of Krull domains. It is known that if D is a Krull domain, then $R = D[\{x_i, y_i, u_i, v_i\}]$ is a Krull domain with $Cl(R) = Cl(D) \oplus \mathbb{Z}^{(\Lambda)}$ [11, Proposition 14.5 and Corollary 14.9]. More generally, if D is a Prüfer v-multiplications domain (PvMD), then R is a PvMD with $Cl(R) = Cl(D) \oplus \mathbb{Z}^{(\Lambda)}$ [7, Theorem 2.6].

Let $R = D[X, Y, U, V]/(XV - YU)$, and let x, y, u, v be the images of X, Y, U, V in R, respectively; so $R = D[x, y, u, v]$ and $R_{D-\{0\}} = K[x, y, u, v]$ with $xv = yu$. It is easy to see that x, y, u are algebraically independent over D, $R[\frac{1}{x}] = D[x, y, u][\frac{1}{x}]$, but x, y, u are not prime elements of R. In this paper, we study some ring-theoretic properties of R focusing on the t-operation (including PvMDs and their class groups). In Section 2, we study some properties of the t-operation on R which turn out to be similar to those of the t-operation on the polynomial ring over D. For example, we show that (i) $t\text{-Max}(R) = \{PR \mid P \in t\text{-Max}(D)\} \cup \{Q \in t\text{-Max}(R) \mid Q \cap D = (0)\}$, (ii) if $Q \in t\text{-Max}(R)$ with $Q \cap D = (0)$, then $\text{ht}Q = 1$ and Q is t-invertible, (iii) R is t-linked over D, and (iv) the map $\varphi : Cl(D) \to Cl(R)$ given by $\varphi(cl(I)) = cl((IR)_t)$ is a well-defined group monomorphism. In Section 3, we study several ring-theoretic properties of R. Precisely, we prove that (v) R is a PvMD (resp., a ring of Krull type, an independent ring of Krull type, a TV-PvMD, a t-SFT PvMD, a t-almost Dedekind domain, an SM domain, a UMT-domain, a weakly Krull domain, a generalized Krull domain) if and only if D is, and (vi) if D is a PvMD, then $Cl(R) = Cl(D) \oplus \mathbb{Z}$. Some of the results are already proved in [7], but we give proofs for the completeness of this paper.

1. Definitions related to the t-operation

In this section, we review definitions related to the t-operation on an integral domain for the reading of this paper including introduction.

Let D be an integral domain with quotient field K. A nonzero D-submodule I of K is called a *fractional ideal* if $dI \subseteq D$ for some $0 \neq d \in D$. Let $\mathbf{F}(D)$ be the set of nonzero fractional ideals of D. For $I \in \mathbf{F}(D)$, let

$I^{-1} = \{x \in K \mid xI \subseteq D\}$; then $I^{-1} \in \mathbf{F}(D)$, whence we can define the v-, t-, and w-operations as follows:

- $I_v = (I^{-1})^{-1}$,
- $I_t = \bigcup\{J_v \mid J \subseteq I \text{ and } J \in \mathbf{F}(D) \text{ is finitely generated}\}$, and
- $I_w = \{x \in K \mid xJ \subseteq I \text{ for some finitely generated } J \in \mathbf{F}(D) \text{ with } J_v = D\}$.

Clearly, $I \subseteq I_w \subseteq I_t \subseteq I_v$, and if I is finitely generated, then $I_t = I_v$. Let $* = v, t$, or w. An $I \in \mathbf{F}(D)$ is called a *$*$-ideal* if $I_* = I$, and a $*$-ideal is called a *maximal $*$-ideal* if it is maximal among proper integral $*$-ideals of D. Let $*$-Max(D) denote the set of maximal $*$-ideals of D. While v-Max$(D) = \emptyset$ if D is a rank-one nondiscrete valuation domain, it is well known that t-Max$(D) \neq \emptyset$ when D is not a field; a maximal t-ideal is a prime ideal; each prime ideal minimal over a t-ideal is a t-ideal; t-Max$(D) = w$-Max(D); $I_w = \bigcap_{P \in t\text{-Max}(D)} ID_P$ for all $I \in \mathbf{F}(D)$; and $P_w = P$ for all nonzero prime ideals P of D with $P_w \neq D$.

An $I \in \mathbf{F}(D)$ is said to be *invertible* (resp., *t-invertible*) if $II^{-1} = D$ (resp., $(II^{-1})_t = D$). It is easy to see that an invertible ideal is a t-invertible t-ideal. A t-ideal I of D is said to be *of finite type* if $I = J_v$ for some nonzero finitely generated ideal J of D. It is known that I is t-invertible if and only if I_t is of finite type and ID_P is principal for all $P \in t$-Max(D) [18, Proposition 2.6]. Let $T(D)$ be the group of t-invertible fractional t-ideals of D under the t-multiplication $I * J = (IJ)_t$, and $Inv(D)$ (resp., $Prin(D)$) be its subgroup of invertible (resp., nonzero principal) fractional ideals of D. Then $Cl(D) = T(D)/Prin(D)$, called the *class group of D*, is an abelian group and $Pic(D) = Inv(D)/Prin(D)$, the *Picard group of D*, is a subgroup of $Cl(D)$. For $I \in T(D)$, let $cl(I) \in Cl(D)$ denote the equivalence class of $T(D)$ containing I. Hence, if $I, J \in T(D)$, then $cl(I) = cl(J)$ if and only if $J = xJ$ for some $0 \neq x \in K$. Clearly, if D is a Krull domain, then $Cl(D)$ is the usual divisor class group of D, and if D is a Dedekind domain or a Prüfer domain, then $Cl(D) = Pic(D)$.

An integral domain D is said to be of *finite t-character* if each nonzero nonunit of D is contained in only finitely many maximal t-ideals of D. We say that D is a *Prüfer v-multiplication domain* (PvMD) if every nonzero finitely generated ideal of D is t-invertible. It is known that D is a PvMD if and only if D_P is a valuation domain for all $P \in t$-Max(D) [13, Theorem 5], if and only if the polynomial ring $D[X]$ over D is a PvMD [18, Theorem 3.7]. We say that D is a *ring of Krull type* if D is a locally finite intersection of essential valuation overrings of D; equivalently, D is a PvMD of finite

t-character [13, Theorem 7]. A ring of Krull type is called an *independent ring of Krull type* if no two distinct maximal t-ideals contain a nonzero prime ideal. A ring of Krull type is a *generalized Krull domain* if each prime t-ideal is a maximal t-ideal. As in [15], we say that D is a TV-domain if $v = t$ on D, i.e., $I_v = I_t$ for all $I \in \mathbf{F}(D)$. Following [20], we say that a nonzero ideal A of D is a *t-SFT-ideal* if there exist a nonzero finitely generated ideal $B \subseteq A$ and a positive integer k such that $a^k \in B_v$ for all $a \in A_t$ and that D is a *t-SFT-ring* if each nonzero ideal of D is a t-SFT-ideal. It is known that D is a t-SFT-ring if and only if each prime t-ideal of D is a t-SFT-ideal [20, Proposition 2.1]. We say that D is a *strong Mori domain* (SM-domain) if D satisfies the ascending chain condition on integral w-ideals of D. An integral domain D is *t-locally Noetherian* if D_P is Noetherian for all $P \in t\text{-Max}(D)$. It is known that D is an SM-domain if and only if D is a t-locally Noetherian domain of finite t-character [22, Theorem 1.9] and that D is a Krull domain if and only if D is an integrally closed SM-domain [22, Theorem 2.8]. We mean by $t\text{-dim}(D) = 1$ that every prime t-ideal of D is a maximal t-ideal. A *weakly Krull domain* D is an integral domain such that $t\text{-dim}(D) = 1$ and D is of finite t-character. Hence, Krull domains are weakly Krull domains.

2. The t-operation on $D[x, y, u, v]$

Let D be an integral domain with quotient field K, X, Y, U, V be indeterminates over D, $D[X, Y, U, V]$ be the polynomial ring over D, and

$$R = D[X, Y, U, V]/(XV - YU).$$

Note that $R_{D-\{0\}} = K[X, Y, U, V]/(XV - YU)$. Also, $D[X, Y, U, V] = D[Y, U, V][X]$ and $VD[Y, U, V] \cap (YU)D[Y, U, V] = (YUV)D[Y, U, V]$; so $R \cong D[Y, U, V][\frac{YU}{V}]$ [21, Proposition 7.6], and thus R is an integral domain. Hence, if we let x, y, u, v be the images of X, Y, U, V in R, respectively, then $R = D[x, y, u, v]$, $R_{D-\{0\}} = K[x, y, u, v]$, $xv = yu$, and $\{x, y, u\}$ (resp., $\{x, y, \frac{u}{x}\}$, $\{x, y, v\}$, $\{x, y, \frac{v}{y}\}$, e.t.c.) are algebraically independent over D.

Lemma 2.1. *Let $R = D[x, y, u, v]$ and I a nonzero fractional ideal of D.*

(1) $R[\frac{1}{x}] = D[x, y, u][\frac{1}{x}] = D[x, y, \frac{u}{x}][\frac{1}{x}]$.
(2) $R = D[x, y, \frac{u}{x}] \cap R_{(x,y)}$ and $IR[\frac{1}{x}] \cap K = IR \cap K = I$.
(3) D is integrally closed if and only if R is integrally closed.
(4) $ID[x, y, \frac{u}{x}] \cap K[x, y, u, v] = IR$.
(5) If $I \subseteq D$, then $ID[x, y, \frac{u}{x}][\frac{1}{x}] \cap K[x, y, u, v] = ID[x, y, \frac{u}{x}][\frac{1}{x}] \cap R = IR$.

(6) I is a prime ideal if and only if IR is a prime ideal of R.

Proof. (1) The equalities follow because $v = \frac{yu}{x}$.

(2) For the equality of $R = D[x, y, \frac{u}{x}] \cap R_{(x,y)}$, see [11, Lemma 14.6]. Also, $I \subseteq IR \cap K \subseteq IR[\frac{1}{x}] \cap K = I$. Thus, $IR[\frac{1}{x}] \cap K = IR \cap K = I$.

(3) Note that $R_{(x,y)}$ is a rank-one DVR [11, Lemma 14.6]. Hence, if D is integrally closed, then $D[x, y, \frac{u}{x}]$ is integrally closed, and thus R is integrally closed by (2). Conversely, assume that R is integrally closed. Then $R[\frac{1}{x}]$ is integrally closed, and so $D[x, y, u][\frac{1}{x}]$ is integrally closed by (1). Clearly, $D[x, y, u][\frac{1}{x}] \cap K = D$. Thus, D is integrally closed.

(4) Let $f \in ID[x, y, \frac{u}{x}] \cap K[x, y, u, v]$. Then f can be written uniquely as a polynomial in $D[x, y, \frac{u}{x}]$, and hence the coefficients of f are in I. Thus, $f \in ID[x, y, u, v]$. The reverse containment is clear.

(5) The equalities follow because

$$\begin{aligned} ID[x, y, u, v] &\subseteq ID[x, y, u, v][\frac{1}{x}] \cap D[x, y, u, v] \\ &\subseteq ID[x, y, u, v][\frac{1}{x}] \cap K[x, y, u, v] \\ &= (ID[x, y, u, v][\frac{1}{x}] \cap K[x, y, \frac{u}{x}]) \cap K[x, y, u, v] \\ &= ID[x, y, \frac{u}{x}] \cap K[x, y, u, v] \\ &= ID[x, y, u, v], \end{aligned}$$

where the last equality follows from (4).

(6) Let I be a prime ideal of D. Since $x, y, \frac{u}{x}$ are algebraically independent over D, $ID[x, y, \frac{u}{x}]$ is a prime ideal of $D[x, y, \frac{u}{x}]$. Thus, IR is a prime ideal of R by (5). Conversely, assume that IR is a prime ideal of R. Then

$$IR \cap D = (IR \cap K) \cap D = I \cap D = I$$

by (2), and thus I is a prime ideal of D. □

Let S be a multiplicative set of D and I be a nonzero fractional ideal of D. Then $(ID_S)_t = (I_tD_S)_t$, and if I is finitely generated or if I_t is of finite type, then $(ID_S)^{-1} = I^{-1}D_S$ [18, Lemma 3.4]. Hence, if I is t-invertible, then $(ID_S)_v = I_vD_S = (ID_S)_t = I_tD_S$. In particular, if D is a PvMD, then $(ID_S)_t = I_tD_S$ for all nonzero ideals I of D. From now on, we use this result without a comment.

Lemma 2.2. *Let $f_1, \dots, f_n$ be nonzero elements of D such that $(f_1, \dots, f_n)_v = D$. If Q is a proper t-ideal of D, then $(QD[\frac{1}{f_i}])_t \subsetneq D[\frac{1}{f_i}]$ for at least one of $i = 1, \dots, n$.*

Proof. Assume that $(QD[\frac{1}{f_i}])_t = D[\frac{1}{f_i}]$ for $i = 1, \ldots, n$. Then there is a nonzero finitely generated ideal I of D such that $I \subseteq Q$ and $D[\frac{1}{f_i}] = (ID[\frac{1}{f_i}])^{-1} = I^{-1}D[\frac{1}{f_i}]$ for $i = 1, \ldots, n$. Hence, $I^{-1} \subseteq \bigcap_{i=1}^{n} D[\frac{1}{f_i}] = D$ (cf. [1, Lemma 2.1] for the equality), and thus $D = I_v \subseteq Q_t \subseteq D$ or $Q_t = D$, a contradiction. □

Let S be a multiplicative set of D, and let I be a nonzero ideal of D. If ID_S is a t-ideal of D_S, then $ID_S \cap D$ is a t-ideal of D [18, Lemma 3.17]. Hence, if P is a maximal t-ideal of D such that PD_S is a t-ideal of D_S, then PD_S is a maximal t-ideal of D_S.

Lemma 2.3. *Let $R = D[x, y, u, v]$.*

(1) $xR \cap vR = xvR$ and $yR \cap uR = yuR$. Hence, $(x, v)_t = (y, u)_t = R$.
(2) If Q is a maximal t-ideal of R, then $QR[\frac{1}{x}]$ or $QR[\frac{1}{v}]$ is a maximal t-ideal.
(3) (x, y), (x, u), (v, y), and (v, u) are height-one t-invertible prime t-ideals of R, and hence maximal t-ideals.
(4) $cl(x, y) = -cl(x, u) = -cl(v, y) = cl(v, u)$ in $Cl(R)$.

Proof. (1) $xR = (X, YU)/(XV - YU)$ and $vR = (V, YU)/(XV - YU)$, and note that $(X, YU) \cap (V, YU) = (XV, YU)$ in $D[X, Y, U, V]$ because X, V are algebraically independent over $D[Y, U]$. Thus, $xR \cap vR = xvR$. The same argument also shows that $yR \cap uR = yuR$.

(2) This follows directly from (1), Lemma 2.2 and the remark before this lemma.

(3) It suffices to show the case of (x, y). Note that

$$(x, y) = (X, Y, XV - YU)/(XV - YU) = (X, Y)/(XV - YU)$$

and $\text{ht}(X, Y) = 2$ in $D[X, Y, U, V]$. Thus, $\text{ht}(x, y) = 1$. Next, note that $\frac{v}{y} = \frac{u}{x} \in (x, y)^{-1}$; so $v = y \cdot \frac{u}{x} \subseteq (x, y)(x, y)^{-1}$. Thus, $(x, v)_t \subseteq ((x, y)(x, y)^{-1})_t$, and hence $((x, y)(x, y)^{-1})_t = D[x, y, u, v]$ by (1).

(4) This follows because $(x, u) = \frac{u}{v}(y, v)$, $(x, y) = \frac{y}{v}(v, u)$, and $((x, y)(x, u))_t = (x^2, xu, xy, yu)_t = (x^2, xu, xy, xv)_t = x(x, u, y, v)_t = xR$. □

Let $\{X_\alpha\}$ be a nonempty set of indeterminates over D and I be a nonzero fractional ideal of D. It is well-known that $(ID[\{X_\alpha\}])^{-1} = I^{-1}D[\{X_\alpha\}]$, $(ID[\{X_\alpha\}])_v = I_vD[\{X_\alpha\}]$, $(ID[\{X_\alpha\}])_t = I_tD[\{X_\alpha\}]$, and $(ID[\{X_\alpha\}])_w = I_wD[\{X_\alpha\}]$ ([14, Lemma 4.1 and Proposition 4.3] and [10, Lemma 2.1]). We next show that this is true of $D[x, y, u, v]$.

Proposition 2.1. *Let I be a nonzero fractional ideal of D.*

(1) $(ID[x,y,u,v])^{-1} = I^{-1}D[x,y,u,v]$.
(2) $(ID[x,y,u,v])_v = I_vD[x,y,u,v]$.
(3) $(ID[x,y,u,v])_t = I_tD[x,y,u,v]$.
(4) $(ID[x,y,u,v])_w = I_wD[x,y,u,v]$.

Proof. (1) Clearly, $I^{-1}D[x,y,u,v] \subseteq (ID[x,y,u,v])^{-1}$. For the reverse containment, let $h \in (ID[x,y,u,v])^{-1}$. Then $hI \subseteq D[x,y,u,v] \subseteq D[x,y,\frac{u}{x}]$, and since $x, y, \frac{u}{x}$ are algebraically independent over D, $h \in (ID[x,y,\frac{u}{x}])^{-1} = I^{-1}D[x,y,\frac{u}{x}]$. Also, $h \in K[x,y,u,v]$, and thus $h \in I^{-1}D[x,y,\frac{u}{x}] \cap K[x,y,u,v] = I^{-1}D[x,y,u,v]$ by Lemma 2.1(4).

(2) This is an immediate consequence of (1).

(3) If A is a nonzero finitely generated subideal of $ID[x,y,u,v]$, there is a nonzero finitely generated subideal J of I such that $A \subseteq JD[x,y,u,v]$. Hence, by (2),

$$A_v \subseteq (JD[x,y,u,v])_v = J_vD[x,y,u,v] \subseteq I_tD[x,y,u,v],$$

and thus $(ID[x,y,u,v])_t \subseteq I_tD[x,y,u,v]$. For the reverse containment, let $0 \neq a \in I_t$. Then $a \in H_v$ for some nonzero finitely generated subideal H of I, and hence $a \in H_vD[x,y,u,v] = (HD[x,y,u,v])_v \subseteq (ID[x,y,u,v])_t$. Thus, $I_t \subseteq (ID[x,y,u,v])_t$, and so $I_tD[x,y,u,v] \subseteq (ID[x,y,u,v])_t$.

(4) If $0 \neq a \in I_w$, there is a nonzero finitely generated ideal J of D such that $J_v = D$ and $aJ \subseteq I$. Note that $JD[x,y,u,v]$ is finite generated, $(JD[x,y,u,v])_v = J_vD[x,y,u,v] = D[x,y,u,v]$ by (2), and $aJD[x,y,u,v] \subseteq ID[x,y,u,v]$; hence $a \in (ID[x,y,u,v])_w$. Thus, $I_wD[x,y,u,v] \subseteq (ID[x,y,u,v])_w$. For the reverse containment, let $f \in (ID[x,y,u,v])_w$, and let A be a nonzero finitely generated ideal of $D[x,y,u,v]$ such that $A^{-1} = D[x,y,u,v]$ and $fA \subseteq ID[x,y,u,v]$. Then $AD[x,y,u,v][\frac{1}{x}]$ is a nonzero finitely generated ideal of $D[x,y,u,v][\frac{1}{x}]$ such that $(AD[x,y,u,v][\frac{1}{x}])^{-1} = A^{-1}D[x,y,u,v][\frac{1}{x}] = D[x,y,u,v][\frac{1}{x}]$ and $fAD[x,y,u,v][\frac{1}{x}] \subseteq ID[x,y,u,v][\frac{1}{x}]$. Hence,

$$f \in (ID[x,y,u,v][\frac{1}{x}])_w \cap K[x,y,u,v].$$

Note that

$$I_wD[x,y,\frac{u}{x}] = (ID[x,y,\frac{u}{x}])_w = (ID[x,y,\frac{u}{x}][\frac{1}{x}])_w \cap K[x,y,\frac{u}{x}]$$
$$= (ID[x,y,u,v][\frac{1}{x}])_w \cap K[x,y,\frac{u}{x}],$$

where the first and second equalities follow because $x, y, \frac{u}{x}$ are algebraically independent over D. Thus, $f \in I_w D[x, y, \frac{u}{x}] \cap K[x, y, u, v] = I_w D[x, y, u, v]$ by Lemma 2.1(4). □

The next result is the $D[x, y, u, v]$-analogue of the fact that a nonzero ideal I of D is a prime t-ideal if and only if $ID[\{X_\alpha\}]$ is a prime t-ideal.

Corollary 2.1. *Let I be a nonzero ideal of D and $R = D[x, y, u, v]$.*

(1) I is a prime t-ideal of D if and only if IR is a prime t-ideal of R.
(2) I is invertible if and only if IR is invertible.
(3) I is t-invertible if and only if IR is t-invertible.

Proof. (1) This is an immediate consequence of Lemma 2.1(6) and Proposition 2.1.

(2) and (3). Note that $(IR)(IR)^{-1} = (IR)(I^{-1}R) = (II^{-1})R$ by Proposition 2.1(1), $((IR)(IR)^{-1})_t = (II^{-1})_t R$ by Proposition 2.1(3), and $JR \cap K = J$ for any fractional ideal J of D by Lemma 2.1(2). Thus, I is invertible (resp., t-invertible) if and only if IR is invertible (resp., t-invertible). □

We next study the maximal t-ideals of $D[x, y, u, v]$ which is the analogue of [10, Proposition 2.2] that if Q is a maximal t-ideal of $D[\{X_\alpha\}]$, then either $Q \cap D = (0)$ or $Q \cap D \neq (0)$ and $Q = (Q \cap D)D[\{X_\alpha\}]$.

Proposition 2.2. *Let $R = D[x, y, u, v]$ and Q be a maximal t-ideal of R.*

(1) If $Q \cap D \neq (0)$, then $Q \cap D$ is a maximal t-ideal of D and $Q = (Q \cap D)R$.
(2) If $Q \cap D = (0)$, then $htQ = 1$ and Q is t-invertible.

Proof. (1) By Lemma 2.3(2), we may assume that $QR[\frac{1}{x}]$ is a maximal t-ideal, and since x, y, u are algebraically independent over D, $QR[\frac{1}{x}] \cap D[x, y, u]$ is a maximal t-ideal. Note that $(QR[\frac{1}{x}] \cap D[x, y, u]) \cap D = Q \cap D \neq (0)$; so $Q \cap D$ is a maximal t-ideal of D and $QR[\frac{1}{x}] \cap D[x, y, u] = (Q \cap D)D[x, y, u]$. Thus, $QR[\frac{1}{x}] = (Q \cap D)R[\frac{1}{x}]$, and hence $Q = QR[\frac{1}{x}] \cap R = (Q \cap D)R[\frac{1}{x}] \cap R = (Q \cap D)R$.

(2) Let Q be a maximal t-ideal of R such that $Q \cap D = (0)$. Again, by Lemma 2.3(2), we assume that $QR[\frac{1}{x}]$ is a maximal t-ideal of $R[\frac{1}{x}]$. Hence, if we let

$$Q' = QR[\frac{1}{x}] \cap D[x, y, \frac{u}{x}],$$

then Q' is a maximal t-ideal of $D[x, y, \frac{u}{x}]$, $Q' \cap D = (0)$, and $Q'D[x, y, \frac{u}{x}][\frac{1}{x}] = QR[\frac{1}{x}]$; so ht$Q$ = ht(Q') = 1 and Q' is t-invertible [7, Lemma 1.3].

Note that $QR[\frac{1}{x}]$ is t-invertible, so there is a nonzero finitely generated subideal A of Q such that $QR[\frac{1}{x}] = (AR[\frac{1}{x}])_v$. Assume that M is a maximal t-ideal of R. If $M \cap D \neq (0)$, then $M = PR$ for some $P \in t$-Max(D) by (1). Hence, $MR[\frac{1}{x}]$ is a prime t-ideal, and hence $AA^{-1} \not\subseteq MR[\frac{1}{x}]$ or $AA^{-1} \not\subseteq M$. Next, assume that $M \cap D = (0)$. If $x \notin M$, then it is clear that $AA^{-1} \not\subseteq M$. Assume that $x \in M$. Since $M \cap D = (0)$ and ht$M = 1$, $M_{D-\{0\}}$ is a prime t-ideal of $K[x, y, u, v]$, and since $K[x, y, u, v]$ is a Krull domain [11, Proposition 14.9], there are only finitely many maximal t-ideals of R, say $M_1, \ldots, M_k$ such that $x \in M_i$ and $M_i \cap D = (0)$. Choose $g \in Q - \bigcup_{i=1}^{k} M_i$, and let $B = A + gR$. Then $QR[\frac{1}{x}] = (BR[\frac{1}{x}])_v$. Replacing A with B if necessary, we can assume that $A \not\subseteq M$ for $M \in t$-Max(R) with $M \cap D = (0)$ and $x \in M$. Thus, $(AA^{-1})_t = R$, and so $(AR[\frac{1}{x}])_v = A_v R[\frac{1}{x}]$ and $A_v R_M = QR_M$ for all maximal t-ideals M of R. Thus, $Q = A_v$ [18, Proposition 2.8(3)] and Q is t-invertible. □

Corollary 2.2. *If $R = D[x, y, u, v]$, then*

$$t\text{-}Max(R) = \{PR \mid P \in t\text{-}Max(D)\} \cup \{Q \in t\text{-}Max(R) \mid Q \cap D = (0)\}.$$

Proof. If Q is a maximal t-ideal of R such that $Q \cap D \neq (0)$, then $Q \cap D$ is a maximal t-ideal of D and $Q = (Q \cap D)R$ by Proposition 2.2. Thus, t-Max$(R) \subseteq \{PR \mid P \in t$-Max$(D)\} \cup \{Q \in t$-Max$(R) \mid Q \cap D = (0)\}$. For the reverse containment, note that if P is a maximal t-ideal of D, then PR is a prime t-ideal of R by Corollary 2.1. Let Q be a maximal t-ideal of R with $PR \subseteq Q$. Then $P \subseteq Q \cap D \neq (0)$, and hence $Q = (Q \cap D)R$ and $Q \cap D$ is a maximal t-ideal of D by Proposition 2.2. Thus, $P = Q \cap D$, whence PR is a maximal t-ideal. □

The next result is the analogue of [17, Proposition 4.2] that D is of finite t-character if and only if $D[X]$ is of finite t-character.

Corollary 2.3. *D is of finite t-character if and only if $D[x, y, u, v]$ is of finite t-character.*

Proof. Note that $K[x, y, u, v]$ is a Krull domain. Note also that if Q is a maximal t-ideal of $D[x, y, u, v]$ with $Q \cap D = (0)$, then ht$Q = 1$ by Proposition 2.2(2), and hence $Q_{D-\{0\}}$ is a prime t-ideal of $K[x, y, u, v]$. Hence, each nonzero nonunit of $D[x, y, u, v]$ is contained in only finitely maximal t-ideals Q of $D[x,y,u,v]$ with $Q \cap D = (0)$. Thus, by Corollary 2.2, the result follows. $\square$

Let $A \subseteq B$ be an extension of integral domains. We say that B is *t-linked over A* if $I^{-1} = A$ for a nonzero finitely generated ideal I of A implies $(IB)^{-1} = B$. It is easy to see that B is t-linked over A if and only if $Q \cap A = (0)$ or $Q \cap A \neq (0)$ and $(Q \cap A)_t \subsetneq A$ for all prime t-ideals Q of B [2, Proposition 2.1]; $D[\{X_\alpha\}]$ is t-linked over D [10, Lemma 2.1]; if B is t-linked over A, then the map $\varphi : Cl(A) \to Cl(B)$ given by $\varphi(cl(I)) = cl((IB)_t)$ is a group homomorphism [2, Theorem 2.2]; and if A and B are Krull domains, then B is t-linked over A if and only if $\text{ht}(Q \cap A) \leq 1$ for all maximal t-ideals Q of B, i.e., the condition (PDE) is satisfied (cf. [11, Theorem 6.2]).

Proposition 2.3. *Let $R = D[x,y,u,v]$.*

(1) R is t-linked over D.
(2) The map $\varphi : Cl(D) \to Cl(R)$ given by $\varphi(cl(I)) = cl((IR)_t)$ is a group monomorphism.
(3) The map $\mu : Pic(D) \to Pic(R)$ given by $\mu(cl(I)) = cl(IR)$ is a group monomorphism.

Proof. (1) If I is a nonzero finitely generated ideal of D such that $I^{-1} = D$, then $(IR)^{-1} = I^{-1}R = R$ by Proposition 2.1(1). Thus, R is t-linked over D.

(2) By (1), R is t-linked over D, and thus φ is a group homomorphism. Next, let I be a nonzero t-invertible t-ideal of D such that $(IR)_t = fR$ for some $f \in R$. Then $fD[x,y,\frac{u}{x}][\frac{1}{x}] = fR[\frac{1}{x}] = (IR)_tR[\frac{1}{x}] = (IR[\frac{1}{x}])_t = (ID[x,y,\frac{u}{x}][\frac{1}{x}])_t = I_tD[x,y,\frac{u}{x}][\frac{1}{x}] = ID[x,y,\frac{u}{x}][\frac{1}{x}]$. Note that $fD[x,y,\frac{u}{x}][\frac{1}{x}] = gD[x,y,\frac{u}{x}][\frac{1}{x}]$ for some $g \in D[x,y,\frac{u}{x}]$ with $x \nmid g$ in $D[x,y,\frac{u}{x}]$; hence the previous equality shows that $g \in D$. Thus, $I = ID[x,y,\frac{u}{x}][\frac{1}{x}] \cap D = gD[x,y,\frac{u}{x}][\frac{1}{x}] \cap D = gD$. Hence, φ is injective.

(3) If I is an invertible ideal of D, then IR is invertible by Corollary 2.1, and hence $(IR)_t = IR$. Thus, μ is a well-defined group monomorphism by (2). $\square$

3. Divisibility properties of $D[x,y,u,v]$

As in Section 2, let D denote an integral domain with quotient field K, x, y, u be algebraically independent over D, $v = \frac{yu}{x}$ or $xv = yu$, and $R = D[x,y,u,v]$.

In this section, we study when R is a PvMD, a ring of Krull type, an independent ring of Krull type, a TV-PvMD, a t-SFT PvMD, a t-almost Dedekind domain, an SM domain, a UMT-domain, a weakly Krull domain, or a generalized Krull domain.

Lemma 3.1. *[7, Proposition 10(2)] If R is a PvMD and if A is a t-ideal of R with $A \subseteq R$ and $A \cap D \neq (0)$, then $A \cap D$ is a t-ideal of D and $A = (A \cap D)R$.*

Proof. Since $(x, v)_v = R$ by Lemma 2.3(1), we may assume that $AR[\frac{1}{x}] \subsetneq R[\frac{1}{x}]$. Also, since R is a PvMD, $AR[\frac{1}{x}]$ is a proper t-ideal of $R[\frac{1}{x}]$, and D is integrally closed by Lemma 2.1(3). Let $B = AR[\frac{1}{x}] \cap D[x, y, \frac{u}{x}]$. Then B is a t-ideal of $D[x, y, \frac{u}{x}]$, and since $D[x, y, \frac{u}{x}]$ is a polynomial ring over an integrally closed domain D, $B = (A \cap D)D[x, y, \frac{u}{x}]$ [18, Lemma 4.1] and $A \cap D$ is a t-ideal of D. Hence, $AR[\frac{1}{x}] = (A \cap D)R[\frac{1}{x}]$, and thus $A \subseteq (A \cap D)R[\frac{1}{x}] \cap R = (A \cap D)R$ by Lemma 2.1(5). Clearly, $(A \cap D)R \subseteq A$, and thus $A = (A \cap D)R$. □

We next give the PvMD analogue of [11, Proposition 14.5 and Corollary 14.9] that if D is a Krull domain, then R is a Krull domain and $Cl(R) = Cl(D) \oplus \mathbb{Z}$. Note that if $\{X_\alpha\}$ is a nonempty set of indeterminates over D, then $Cl(D[\{X_\alpha\}]) = Cl(D)$, i.e., there is a group isomorphism from $Cl(D[\{X_\alpha\}])$ onto $Cl(D)$, if and only if D is integrally closed [9, Corollary 2.13]; in particular, $Cl(D[\{X_\alpha\}]) = Cl(D)$ when D is a PvMD.

Theorem 3.1.

(1) [7, Corollary 2.4(1)] D is a PvMD if and only if R is a PvMD.
(2) [7, Theorem 2.6] If D is a PvMD, then $Cl(R) = Cl(D) \oplus \mathbb{Z}$.

Proof. (1) ($\Rightarrow$) Assume that D is a PvMD, and let Q be a maximal t-ideal of R. Then $QR[\frac{1}{x}]$ or $QR[\frac{1}{v}]$ is a maximal t-ideal by Lemma 2.3(2). Note that $R[\frac{1}{x}] = D[x, y, \frac{u}{x}][\frac{1}{x}]$ and $R[\frac{1}{v}] = D[x, v, \frac{u}{v}][\frac{1}{v}]$. Note also that $x, y, \frac{u}{x}$ (resp., $v, y, \frac{u}{v}$) are algebraically independent over D; hence $R[\frac{1}{x}]$ and $R[\frac{1}{v}]$ are PvMDs. Since $R_Q = R[\frac{1}{z}]_{QR[\frac{1}{z}]}$ for $z = x$ or v, R_Q is a valuation domain. Thus, R is a PvMD.

($\Leftarrow$) If R is a PvMD, then $R[\frac{1}{x}]$ is a PvMD. Note that $R[\frac{1}{x}] = D[x, y, \frac{u}{x}][\frac{1}{x}]$ and x is a prime element of $D[x, y, \frac{u}{x}]$; so $D[x, y, \frac{u}{x}]$ is a PvMD. Thus, D is a PvMD.

(2) By Proposition 2.3(2), the map $\varphi : Cl(D) \to Cl(R)$ given by $\varphi(cl(I)) = cl((IR)_t)$ is a group monomorphism. Next, let $S = D - \{0\}$. Then $R_S = K[x, y, u, v]$, and hence R_S is a Krull domain with $Cl(R_S) = \mathbb{Z}$

[11, Proposition 14.8]. Let $\psi : Cl(R) \to Cl(R_S)$ be defined by $\psi(cl(A)) = cl(A_S)$. Then ψ is a group homomorphism, and since R is a PvMD, ψ is a surjection. Note that if I is a nonzero t-invertible t-ideal of D, then $((IR)_t)R_S = (IR_S)_t = R_S$; hence $\psi \circ \varphi = 0$. Let A be a t-invertible t-ideal of R such that $A_S = fR_S$ for some $0 \neq f \in R$. Then $\frac{1}{f}A_S = R_S$, and since A is of finite type, there is an $s \in S$ with $s\frac{1}{f}A \subseteq R$. Note that $s\frac{1}{f}A \cap D \neq (0)$ and $s\frac{1}{f}A$ is a t-ideal; hence by (1) and Lemma 3.1, $s\frac{1}{f}A = (JR)_t$ for some t-ideal J of D. Thus, $cl(A) = cl((JR)_t) = \varphi(cl(J))$, and therefore we have an exact sequence

$$0 \to Cl(D) \to Cl(R) \to Cl(R_S) \to 0.$$

Note that (x, y) is a t-invertible prime t-ideal of R and $Cl(R_S)$ is generated by $cl((x, y)R_S)$ [11, Proof of Proposition 14.8]; so if we let $\theta : Cl(R_S) \to Cl(R)$ be given by $\theta(cl((((x, y)R_S)^n)_t)) = cl(((x, y)^n)_t)$, then θ is a well-defined group homomorphism and $\psi \circ \theta$ is the identity function of $Cl(R_S)$. Hence, the exact sequence above is split, and thus $Cl(R) = Cl(D) \oplus Cl(R_S) = Cl(D) \oplus \mathbb{Z}$. □

Let t-Spec(A) denote the set of prime t-ideals of an integral domain A. Clearly, t-Max$(A) \subseteq t$-Spec(A), and if A is not a field, then t-Max$(A) = t$-Spec(A) if and only if t-dim$(A) = 1$.

Lemma 3.2. *[7, Proposition 2.3(4)] If D is a PvMD, then*

$$t\text{-}Spec(R) = \{PR \mid P \in t\text{-}Spec(D)\} \cup \{Q \in t\text{-}Max(R) \mid Q \cap D = (0)\}.$$

Proof. If P is a prime t-ideal of D, then PR is a prime t-ideal of R by Corollary 2.1. For the reverse containment, let Q be a prime t-ideal of R. If $Q \cap D \neq (0)$, then $Q = (Q \cap D)R$ by Lemma 3.1 and $Q \cap D$ is a prime t-ideal of D by Proposition 2.1(3). Assume $Q \cap D = (0)$, and let M be a maximal t-ideal of R such that $Q \subseteq M$. By Lemma 2.3(2), we may assume that $MR[\frac{1}{x}]$ is a maximal t-ideal. Note that $R[\frac{1}{x}] = D[x, y, u][\frac{1}{x}]$ and $QR[\frac{1}{x}] \cap D[x, y, u]$ is a prime t-ideal with $(QR[\frac{1}{x}] \cap D[x, y, u]) \cap D = (0)$. Since D is a PvMD, $QR[\frac{1}{x}] \cap D[x, y, u]$ is a maximal t-ideal, and thus $QR[\frac{1}{x}] \cap D[x, y, u] = MR[\frac{1}{x}] \cap D[x, y, u]$. Thus, $Q = M$. □

Corollary 3.1. *[7, Corollary 2.4(2)] D is a ring (resp., an independent ring) of Krull type if and only if R is a ring (resp., an independent ring) of Krull type.*

Proof. A ring of Krull type is a PvMD of finite t-character. Thus, by Theorem 3.1(1) and Corollary 2.3, D is a ring of Krull type if and only if

R is a ring of Krull type. Also, note that if P is a prime ideal of D, then $PR \cap D = P$ by Lemma 2.1(2). Thus, by Lemma 3.2, D is an independent ring of Krull type if and only if R is an independent ring of Krull type. □

It is known that D is a TV-PvMD (i.e., TV-domain which is also a PvMD) if and only if D is an independent ring of Krull type whose maximal t-ideals are t-invertible [15, Theorem 3.1].

Corollary 3.2. *[7, Corollary 2.4(5)] D is a TV-PvMD if and only if R is a TV-PvMD.*

Proof. By Corollary 3.1, it suffices to show that each maximal t-ideal is t-invertible. However, note that if Q is a maximal t-ideal of R such that $Q \cap D = (0)$, then Q is t-invertible by Proposition 2.2. Thus, the result follows directly from Corollaries 2.1(3) and 2.2. □

The next result shows that R is not a GCD domain nor a UFD (in fact, (x, y) is a t-invertible t-ideal of R but it is not a principal ideal). Recall that D is a GCD domain if and only if D is a PvMD with $Cl(D) = \{0\}$ [4, Corollary 1.5]. Thus, by Theorem 3.1, we have

Corollary 3.3. *If D is a GCD domain, then R is a PvMD with $Cl(R) = \mathbb{Z}$.*

Let $D[\![\{X_\alpha\}]\!]_1$ be the first type power series ring over D, i.e., $D[\![\{X_\alpha\}]\!]_1 = \bigcup D[\![X_{i_1}, \ldots, X_{i_k}]\!]$, where $\{X_{i_1}, \ldots, X_{i_k}\}$ runs over all finite subsets of $\{X_\alpha\}$. It is known that D is a t-SFT PvMD if and only if $D[\{X_\alpha\}]$ is a t-SFT PvMD, and if D is a t-SFT PvMD, then $D[\![\{X_\alpha\}]\!]_{1D-\{0\}}$ is a Krull domain [6, Theorems 9 and 11].

Corollary 3.4. *D is a t-SFT PvMD if and only if R is a t-SFT PvMD.*

Proof. By Theorem 3.1(1) and [20, Proposition 2.1], it suffices to show that every prime t-ideal is a t-SFT-ideal.

($\Rightarrow$) Let Q be a prime t-ideal of R. If $Q \cap D = (0)$, then Q is t-invertible by Lemma 3.2 and Proposition 2.2, and hence Q is a t-SFT-ideal. Next, assume that $Q \cap D \neq (0)$. Then $Q = (Q \cap D)R$ and $Q \cap D$ is a prime t-ideal of D; hence there exist a nonzero finitely generated ideal $J \subseteq Q \cap D$ and a positive integer k such that $a^k \in J_v$ for all $a \in Q \cap D$. Let $f \in Q$, and let $c(f)$ be the ideal of D generated by the coefficients of f as the polynomial in $D[x, y, \frac{u}{x}]$. Then $c(f) \in Q \cap D$, and if $c(f) = (a_1, \ldots, a_n)$, then $c(f^k)_v = (c(f)^k)_v = (a_1^k, \ldots, a_n^k)_v \subseteq J_v$ [3, Lemma 3.3]. Hence, $f^k \in J_v R = (JR)_v$. Thus, Q is a t-SFT-ideal.

($\Leftarrow$) Let P be a prime t-ideal of D. Then PR is a prime t-ideal of R by Corollary 2.1, and hence there are a nonzero finitely generated ideal $B \subseteq PR$ and a positive integer k such that $f^k \in B_v$ for all $f \in PR$. Clearly, $B_v \cap D \neq (0)$; so $B_v = (B_v \cap D)R$ by Lemma 3.1. Since B_v is of finite type, $B_v \cap D$ is a t-ideal of finite type. Also, since $(B_v \cap D)R \cap K = B_v \cap D$ by Lemma 2.1(2), $a^k \in B_v \cap D$ for all $a \in P$. Thus, P is a t-SFT-ideal. $\square$

Let $\{X_\alpha\}$ be a nonempty set of indeterminates over D, and let $S = \{f \in D[\{X_\alpha\}] \mid c(f) = D\}$. The ring $D[\{X_\alpha\}]_S$ denoted by $D(\{X_\alpha\})$ is called the Nagata ring of D. It is known that D is a Noetherian domain if and only if $D(\{X_\alpha\})$ is a Noetherian domain ([5, Theorem 2.1], [12, Theorem 6]).

Lemma 3.3. *$R_{PR} = D_P(x, y, u)$ for all $P \in t$-$Max(D)$.*

Proof. Clearly, $x \notin PR$, and hence $R_{PR} = R[\frac{1}{x}]_{PR[\frac{1}{x}]}$. Thus, $R_{PR} = D[x,y,u][\frac{1}{x}]_{PD[x,y,u][\frac{1}{x}]} = D[x,y,u]_{PD[x,y,u]} = D_P(x,y,u)$. $\square$

An integral domain D is called an *almost Dedekind domain* (resp., a *t-almost Dedekind domain*) if D_P is a rank-one DVR for all maximal ideals (resp., maximal t-ideals) P of D. Clearly, a t-almost Dedekind domain is a PvMD, and D is an almost Dedekind domain if and only if D is a t-almost Dedekind domain of (Krull) dimension one. Also, D is a t-almost Dedekind domain if and only if $D[X]$ is a t-almost Dedekind domain [18, Theorem 4.2].

Proposition 3.1. *D is a t-almost Dedekind domain if and only if R is a t-almost Dedekind domain.*

Proof. ($\Rightarrow$) Suppose that D is a t-almost Dedekind domain, and let Q be a maximal t-ideal of R. By Lemma 2.3(2), we may assume that $QR[\frac{1}{x}]$ is a maximal t-ideal. Note that $D[x,y,u]$ is a t-almost Dedekind domain; so $R[\frac{1}{x}] = D[x,y,u][\frac{1}{x}]$ is a t-almost Dedekind domain. Thus, $R_Q = R[\frac{1}{x}]_{QR[\frac{1}{x}]}$ is a rank-one DVR.

($\Leftarrow$) Assume that R is a t-almost Dedekind domain, and let P be a maximal t-ideal of D. Then PR is a maximal t-ideal of R by Corollary 2.2, and so R_{PR} is a rank-one DVR. Note that $R_{PR} = D_P(x,y,u)$ by Lemma 3.3 and $D_P(x,y,u) \cap K = D_P$. Thus, D_P is a rank-one DVR. $\square$

Let $N_v = \{f \in D[\{X_\alpha\}] \mid c(f)_v = D\}$. It is known that D is an SM-domain if and only if $D[\{X_\alpha\}]$ is an SM domain, if and only if $D[\{X_\alpha\}]_{N_v}$ is a Noetherian domain [5, Theorem 2.1].

Proposition 3.2. *D is an SM domain if and only if R is an SM domain.*

Proof. By Corollary 2.3, it suffices to show that R is t-locally Noetherian if and only if D is t-locally Noetherian.

($\Rightarrow$) Assume that D is t-locally Noetherian, and let Q be a maximal t-ideal of R. If $Q \cap D = (0)$, then ht$Q = 1$ and Q is t-invertible by Proposition 2.2(2). Hence, R_Q is a rank-one DVR, and thus R_Q is Noeterian. Next, assume that $Q \cap D \neq (0)$. Then $Q \cap D$ is a maximal t-ideal of D and $Q = (Q \cap D)R$ by Proposition 2.2(1). Note that $R_{(Q\cap D)R} = D_{(Q\cap D)}(x, y, u)$ by Lemma 3.3 and $D_{Q\cap D}$ is Noetherian. Thus, $R_{(Q\cap D)R}$ is Noetherian.

($\Leftarrow$) Assume that R is t-locally Noetherian, and let P be a maximal t-ideal of D. Then PR is a maximal t-ideal and $R_{PR} = D_P(x, y, u)$. Hence, $D_P(x, y, u)$ is Noetherian, and thus D_P is Notherian. $\square$

Let $D[X]$ be the polynomial ring over D. A nonzero prime ideal of $D[X]$ is an upper to zero in $D[X]$ if $Q \cap D = (0)$, and we say that D is a UMT-domain if every upper to zero in $D[X]$ is a maximal t-ideal. It is known that D is a UMT-domain if and only if $D[X]$ is a UMT-domain, if and only if the integral closure of D_P is a Prüfer domain for all $P \in t$-Max(D) [10, Theorems 1.5 and 2.4]. Hence, D is a PvMD if and only if D is an integrally closed UMT-domain [16, Proposition 3.2]. We next give the $D[x, y, u, v]$ analogue of [2, Lemma 4.10] that t-dim$(D[X]) = 1$ if and only if D is a UMT-domain and t-dim$(D) = 1$.

Lemma 3.4. *t-dim$(R) = 1$ if and only if D is a UMT-domain with t-dim$(D) = 1$.*

Proof. ($\Rightarrow$) If t-dim$(R) = 1$, then t-dim$(R[\frac{1}{x}]) = 1$, and since $R[\frac{1}{x}] = D[x, y, \frac{u}{x}][\frac{1}{x}]$, t-dim$(D[x, y, \frac{u}{x}]) = 1$. Note that $x, y, \frac{u}{x}$ are algebraically independent over D. Thus, D is a UMT-domain with t-dim$(D) = 1$ [2, Lemma 4.10].

($\Leftarrow$) Assume that D is a UMT-domain with t-dim$(D) = 1$. Then t-dim$(D[x, y, \frac{u}{x}]) = 1$ [2, Lemma 4.10], and hence t-dim$(R[\frac{1}{x}]) = 1$. Let Q be a maximal t-ideal of R. If $Q \cap D = (0)$, then ht$Q = 1$ by Proposition 2.2(2). If $Q \cap D \neq (0)$, then $Q \cap D$ is a maximal t-ideal of D and $Q = (Q \cap D)R$ by Proposition 2.2(1). Note that $(Q \cap D)D[x, y, u]$ is a maximal t-ideal of $D[x, y, u]$ and t-dim$(D[x, y, u]) = 1$. Thus, ht$Q$ = ht$(QR[\frac{1}{x}])$ = ht$((Q \cap D)D[x, y, u, v]) = 1$. $\square$

Proposition 3.3.

(1) D is a UMT-domain if and only if R is a UMT-domain.

(2) D is a weakly Krull UMT-domain if and only if R is a weakly Krull domain.

(3) D is a generalized Krull domain if and only if R is a generalized Krull domain.

Proof. (1) ($\Rightarrow$) Let Q be a maximal t-ideal of R. By Lemma 2.3(2), we may assume that $QR[\frac{1}{x}]$ is a maximal t-ideal. Note that $R[\frac{1}{x}] = D[x, y, u][\frac{1}{x}]$; hence $R[\frac{1}{x}]$ is a UMT-domain because x, y, u are algebraically independent over D. Since $R_Q = R[\frac{1}{x}]_{QR[\frac{1}{x}]}$, the integral closure of R_Q is a Prüfer domain [10, Theorem 1.5]. Thus, R is a UMT-domain.

($\Leftarrow$) If R is a UMT-domain, then $R[\frac{1}{x}]$ is a UMT-domain [10, Proposition 1.2]. Note that $R[\frac{1}{x}] = D[x, y, u][\frac{1}{x}]$ and $xD[x, y, u]$ is a height-one prime ideal of $D[x, y, u]$; hence $D[x, y, u]$ is a UMT-domain. Thus, D is a UMT-domain.

(2) This follows directly from Corollary 2.3 and Lemma 3.4.

(3) Note that a generalized Krull domain is just a weakly Krull PvMD [2, Corollary 4.13]. Thus, the result follows directly from (2) and Theorem 3.1(1). □

It is known that if D is a Dedekind domain in which all but a finite number of prime ideals are principal, then D is a principal ideal domain [11, Lemma 13.9].

Proposition 3.4. *Let D be a Krull domain in which at most a finite number of height-one prime ideals are not principal. Then D is a UFD.*

Proof. Let $P_1, \ldots, P_n \in t\text{-Max}(D)$ be those possible non-principal height-one prime ideals of D. Choose an element $a \in D$ such that $P_1 D_{P_1} = aD_{P_1}$ and $a \notin P_i$ for $i = 2, \ldots, n$. Let Q be the principal ideal qD for other height-one prime ideals Q of D, and let $aD_Q = (QD_Q)^{n_q}$ for an integer $n_q \geq 0$. Hence, if we let $b = a \cdot \prod_{Q \neq P_i} q^{-n_q}$, then $P_1 = bD$, a contradiction. □

A *π-domain* is a Krull domain in which every height-one prime ideal is invertible. Hence, D is a π-domain if and only if D is a Krull domain with $Cl(D) = Pic(D)$.

Corollary 3.5. *Let D be a Krull domain in which at most a finite number of height-one prime ideals are not invertible. Then D is a π-domain.*

Proof. Let M be a maximal ideal of D. Then $Pic(D_M) = \{0\}$, and hence all but a finite number of height-one prime ideals of D_M are principal.

Thus, D_M is a UFD by Proposition 3.4. Hence, D is a π-domain [19, Theorem 5.1]. □

Remark 3.1. (1) Let $T = K[x,y,u,v]$. Then T is a Krull domain with $Cl(T) = \mathbb{Z}$, and hence T is not a UFD. Hence by Proposition 3.4, there are infinitely many non-principal height-one prime ideals of T. Also, note that (x,y) is t-invertible but not invertible (note that $(x,y)^{-1} = (1, \frac{u}{x}, \frac{v}{y})$, and so $(x,y)(x,y)^{-1} = (x,y,u,v)$). Thus, $Cl(T) \neq Pic(T)$.

(2) Let Λ be a nonzero index set, $\mathbb{Z}^{(\Lambda)}$ be the direct sum of Λ-copies of the additive group of integers, $\{X_i, Y_i, U_i, V_i \mid i \in \Lambda\}$ be a set of indeterminates over an integral domain D, and $T = D[\{X_i, Y_i, U_i, V_i\}]/(\{X_iV_i - Y_iU_i\})$. A similar argument of Theorem 3.1 also shows that T is a PvMD if and only if D is a PvMD, and if D is a PvMD, then $Cl(T) = Cl(D) \oplus \mathbb{Z}^{(\Lambda)}$. Precise proofs of these results are given in [7].

(3) Let $X_1, \ldots, X_n, Y_1, \ldots, Y_n$ be indeterminates over D, and let

$$T = D[X_1, \ldots, X_n, Y_1, \ldots, Y_n]/(\{X_iY_j - X_jY_i \mid 1 \leq i < j \leq n\}).$$

It is known that if D is a Krull domain, then T is a Krull domain and $Cl(T) = Cl(D) \oplus \mathbb{Z}$ [11, Proposition 14.11]. Clearly, if $n = 2$, then T is just the ring $D[x,y,u,v]$ with $xv = yu$.

Acknowledgments. This work was supported by Basic Science Research Program through the National Research Foundation of Korea (NRF) funded by the Ministry of Education (2017R1D1A1B06029867).

References

[1] D. D. Anderson and D. F. Anderson, Locally factorial integral domains, *J. Algebra* **90**, 265 (1984).

[2] D. D. Anderson, E. Houston, and M. Zafrullah, t-linked extensions, the t-class group, and Nagata's theorem, *J. Pure Appl. Algebra* **86**, 109 (1993).

[3] D. D. Anderson and M. Zafrullah, Almost Bezout domains, *J. Algebra* **142**, 285 (1991).

[4] A. Bouvier and M. Zafrullah, On some class groups of an integral domain, *Bull. Greek Math. Soc.* **29**, 45 (1988).

[5] G. W. Chang, $*$-Noetherian domains and the ring $D[\mathbf{X}]_{N_*}$, *J. Algebra* **297**, 216 (2006).

[6] G. W. Chang, Power series rings over Prüfer v-multiplication domains, *J. Korean Math. Soc.* **53**, 447 (2016).

[7] G. W. Chang, Every abelian group is the class group of a ring of Krull type, *J. Korean Math. Soc.*, to appear.
[8] L. Claborn, Every abelian group is a class group, *Pacific J. Math.* **18**, 219 (1966).
[9] S. El Baghdadi, L. Izelgue, and S. Kabbaj, On the class group of a graded domain, *J. Pure Appl. Algebra* **171**, 171 (2002).
[10] M. Fontana, S. Gabelli, and E. Houston, UMT-domains and domains with Prüfer integral closure, *Comm. Algebra* **25**, 1017 (1998).
[11] R. M. Fossum, *The Divisor Class Group of a Krull Domain* (Springer-Verlag, New York, 1972).
[12] R. Gilmer and W. Heinzer, The Noetherian property for quotient rings of infinite polynomial rings, *Proc. Amer. Math. Soc.* **76**, 1 (1979).
[13] M. Griffin, Some results on v-multiplication rings, *Canad. J. Math.* **97**, 710 (1967).
[14] J. Hedstrom and E. Houston, Some remarks on star operations, *J. Pure Appl. Algebra* **18**, 37 (1980).
[15] E. Houstan and M. Zafrullah, Integral domains in which each t-ideal is divisorial, *Michigan. Math. J.* **35**, 291 (1988).
[16] E. Houston and M. Zafrullah, On t-invertibility II, *Comm. Algebra* **17**, 1955 (1989).
[17] S. Kabbaj and A. Mimouni, t-class semigroups of integral domains, *J. Reine Angew. Math.* **612**, 213 (2007).
[18] B. G. Kang, Prüfer v-multiplication domains and the ring $R[X]_{N_v}$, *J. Algebra* **123**, 151 (1989).
[19] B. G. Kang, On the converse of a well-known fact about Krull domains, *J. Algebra* **124**, 284 (1989).
[20] B. G. Kang and M. H. Park, A note on t-SFT-rings, *Comm. Algebra* **34**, 3153 (2006).
[21] P. Samuel, *Lectures on Unique Factorization Domains* (Tata Institute of Fundamental Research, Colaba, Bombay 1964).
[22] F. Wang and R. L. McCasland, On strong Mori domains, *J. Pure Appl. Algebra* **135**, 155 (1999).

On the Hochschild cohomology of differential graded categories

B. Keller

UFR de Mathématiques, Université de Paris,
Institut de Mathématiques de Jussieu–PRG, UMR 7586 du CNRS,
75205 Paris Cedex 13, France
E-mail: bernhard.keller@imj-prg.fr
https://webusers.imj-prg.fr/~bernhard.keller/

We survey results on the Hochschild cohomology of differential graded (=dg) categories including recent work on Hochschild cohomology of dg singularity categories and its link to singular Hochschild cohomology.

Keywords: Hochschild cohomology, dg category, derived category.

1. Introduction

Let k be a field. Hochschild cohomology is a classical invariant of an associative k-algebra or, more generally, a k-category. It may be viewed as the derived version of the center of the algebra, which we recall in section 2. It is invariant under Morita equivalence and also under derived equivalence, as we show in section 5 after a reminder on derived categories in section 4. These invariance results suggest that Hochschild cohomology should be defined directly using the derived category. This is not possible in general but we show in section 6 that it may be obtained from the canonical differential graded enhancement of the derived category. Hochschild cohomology is endowed with higher structure: the Hochschild cochain complex is a B_∞-algebra as we recall in section 7, where we also state the corresponding invariance results. In the final section 8, we report on recent results on singular Hochschild cohomology after [32] and [25]. In particular, we sketch a proof of the fact that singular Hochschild cohomology of an algebra is isomorphic, as an algebra, to the classical Hochschild cohomology of its differential graded singularity category.

2. The center of algebras and categories

For simplicity, we work over a field k. Let A be a k-algebra, *i.e.* an associative, possibly non commutative, unital k-algebra. The *center* $Z(A)$ *of* A

is formed by the elements $a \in A$ such that

$$ab = ba$$

for all $b \in A$. Notice that the center is naturally a commutative algebra whereas the trace space is just a vector space. For example, if $n \geq 1$ and $M_n(A)$ denotes the algebra of $n \times n$-matrices with coefficients in A, we have an isomorphism

$$Z(A) \xrightarrow{\sim} Z(M_n(A))$$

taking an element a to the diagonal matrix whose diagonal coefficients all equal a. Observe that by definition, we have a short exact sequence

$$0 \longrightarrow Z(A) \longrightarrow A \longrightarrow \mathrm{Hom}_k(A, A)$$

where the second map is the inclusion and the third one takes $a \in A$ to the commutator $[a, ?]$.

Let $f : A \to B$ be an algebra morphism (it need not preserve the unit). For an element a of the center $Z(A)$, there is no reason for $f(a)$ to lie in $Z(B)$ and there is no induced map between the centers in general. Hence the assignment $A \mapsto Z(A)$ is *not a functor*. However, we will see below how, by passing from algebras to categories, we do gain some functoriality for the center.

Recall [15] that a *k-category* is a category $\mathcal{A}$ whose morphism sets are endowed with k-vector space structures such that the compositions are bilinear. We may (and will) identify a k-algebra A with the k-category with a single object whose endomorphism algebra is A. A general k-category should be viewed as a *k-algebra with several objects* [39]. Recall [15] that a *quiver* is a quadruple $Q = (Q_0, Q_1, s, t)$, where Q_0 is a set of 'vertices', Q_1 a set of 'arrows' and s and t are maps $Q_1 \to Q_0$ taking an arrow to its source respectively its target. By definition, a *morphism of quivers* $f : Q \to Q'$ is a pair of maps $f_0 : Q_0 \to Q'_0$ and $f_1 : Q_1 \to Q'_1$ such that $s'(f_1(\alpha)) = f_0(s(\alpha))$ and $t'(f_1(\alpha)) = f_0(t(\alpha))$ for all arrows $\alpha \in Q_1$. If $\mathcal{C}$ is a small k-category, its *underlying quiver* $Q(\mathcal{C})$ has as vertex set the set of objects of $\mathcal{C}$ and as arrow set the set of all morphisms of $\mathcal{C}$. We define a *morphism* between small k-categories to be a k-linear functor. Then clearly, the assignment $\mathcal{C} \mapsto Q(\mathcal{C})$ underlies a functor from the category of small k-categories to the category of quivers. It is easy to see that this functor admits a left adjoint $Q \mapsto k\text{-cat}(Q)$. Here, the objects of $k\text{-cat}(Q)$ are the vertices of Q and the space of morphisms $x \to y$ is the vector space of formal linear combinations of paths (=formal compositions of arrows) of length ≥ 0 from x to y.

Let $\mathcal{A}$ be a small k-category. We define the *center* $Z(\mathcal{A})$ to be the algebra of endomorphisms of the identity functor $\mathbf{1}_{\mathcal{A}} : \mathcal{A} \to \mathcal{A}$. Thus, an element of the center is a family φX, $X \in \mathcal{A}$, of endomorphisms $\varphi X : X \to X$ of objects of $\mathcal{A}$ such that for each morphism $f : X \to Y$, the square

$$\begin{array}{ccc} X & \xrightarrow{f} & Y \\ {\scriptstyle \varphi X}\downarrow & & \downarrow{\scriptstyle \varphi Y} \\ X & \xrightarrow[f]{} & Y \end{array}$$

commutes. Thus, we have the short exact sequence

$$0 \longrightarrow Z(\mathcal{A}) \longrightarrow \prod_{X \in \mathcal{A}} \mathcal{A}(X,X) \longrightarrow \prod_{X,Y \in \mathcal{A}} \mathrm{Hom}_k(\mathcal{A}(X,Y), \mathcal{A}(X,Y)) \ , \tag{1}$$

where the third map takes a family (φX) to the family of maps taking $f : X \to Y$ to $(\varphi Y) \circ f - f \circ (\varphi X)$.

We may ask whether we can reduce the computation of the center of the k-category $\mathcal{A}$ to that of the center of some k-algebra A. The natural candidate is the *matrix algebra* defined by

$$k[\mathcal{A}] = \bigoplus_{X,Y \in \mathcal{A}} \mathcal{A}(X,Y)$$

and endowed with matrix multiplication. Notice that it is associative but unital only if $\mathcal{A}$ has only finitely many non zero objects. In general, it is still locally unital, i.e. for each finite set of elements a_i, $i \in I$, we may find an idempotent e such that $ea_i = a_i = a_i e$ for all i.

Lemma 2.1. *We have a natural injective algebra morphism $Z(k[\mathcal{A}]) \to Z(\mathcal{A})$ which is an isomorphism iff $\mathcal{A}$ has only finitely many non zero objects.*

The proof is left to the reader as an exercise. We have some functoriality for the assignment $\mathcal{A} \mapsto Z(\mathcal{A})$. Indeed, if $\mathcal{B} \subset \mathcal{A}$ is a full subcategory, then the restriction map

$$(\varphi X)_{X \in \mathcal{A}} \mapsto (\varphi X)_{X \in \mathcal{B}}$$

is an algebra morphism $Z(\mathcal{A}) \to Z(\mathcal{B})$. We see that $\mathcal{A} \mapsto Z(\mathcal{A})$ is *contravariant with respect to fully faithful embeddings*.

Let A be a k-algebra and $\mathsf{Mod}\, A$ the k-category of all right A-modules (by choosing suitable universes, we can dispense with the set-theoretical problem that $\mathsf{Mod}\, A$ is not small). We have a fully faithful embedding $A \subset \mathsf{Mod}\, A$ taking the unique object to the free A-module of rank one.

Lemma 2.2. *The restriction along $A \subset \mathsf{Mod}\, A$ is an isomorphism $Z(\mathsf{Mod}\, A) \xrightarrow{\sim} Z(A)$.*

Proof. Let ρ denote the restriction. We define $\sigma : Z(A) \to Z(\mathsf{Mod}\, A)$ by sending an element $z \in Z(A)$ to the family $\sigma(z)$ whose component at a module M is right multiplication by z. Since z is central, the map $\sigma(z)M$ is indeed an endomorphism of M and since any $f : L \to M$ is A-linear, the family $\sigma(z)$ does lie in $Z(\mathsf{Mod}\, A)$. It is also clear that $\rho(\sigma(z)) = z$ for each $z \in Z(A)$. Thus, the map $\rho : Z(\mathsf{Mod}\, A) \to Z(A)$ is surjective. Suppose that φ is in its kernel. Then $\varphi A = 0$ and in fact $\varphi F = 0$ for each free A-module. If M is an arbitrary A-module, we have an exact sequence

$$F_0 \longrightarrow M \longrightarrow 0$$

where F_0 is free. The fact that φF_0 vanishes then immediately implies that φM vanishes. Thus, the map ρ is also injective. $\checkmark$

Recall that a *generator* (resp. *cogenerator*) of $\mathsf{Mod}\, A$ is a module G such that each module is a quotient of a coproduct of copies of G (resp. a submodule of a product of copies of G). Let $\mathsf{proj}\, A$ denote the category of finitely generated projective modules, $\mathsf{Proj}\, A$ the category of all projective modules and $\mathsf{Inj}\, A$ the category of all injective modules. It is easy to adapt the above proof to obtain the

Lemma 2.3. *Let $\mathcal{B} \subset \mathsf{Mod}\, A$ be a full subcategory containing a generator or a cogenerator of $\mathsf{Mod}\, A$. Then restriction to $\mathcal{B}$ is an isomorphism $Z(\mathsf{Mod}\, A) \xrightarrow{\sim} Z(\mathcal{B})$. In particular, we have isomorphisms*

$$Z(A) \xleftarrow{\sim} Z(\mathsf{proj}\, A) \xleftarrow{\sim} Z(\mathsf{Proj}\, A) \xleftarrow{\sim} Z(\mathsf{Mod}\, A) \xrightarrow{\sim} Z(\mathsf{Inj}\, A).$$

3. Hochschild cohomology of algebras

Let k be a field. We write $\otimes$ for $\otimes_k$. Let A be a k-algebra. The *Hochschild [24] cochain complex of A* is the complex C^*A concentrated in cohomological degrees ≥ 0

$$A \to \mathrm{Hom}_k(A, A) \to \mathrm{Hom}_k(A \otimes A, A) \to \ldots \to \mathrm{Hom}_k(A^{\otimes p}, A) \to \ldots$$

whose differential is given by

$$\begin{aligned}(df)(a_0, \ldots, a_p) = a_0 f(a_1, \ldots, a_p) - \textstyle\sum_{i=0}^{p-1} f(a_0, \ldots, a_i a_{i+1}, \ldots, a_p) \\ + (-1)^p f(a_0, \ldots, a_{p-1}) a_p.\end{aligned}$$

Notice that the first two differentials are given by

$$a \mapsto [a, ?] \quad \text{and} \quad f \mapsto (a \otimes b \mapsto f(a)b - f(ab) + af(b)).$$

We see that in degree 0 we recover the center $Z(A) = HH^0(A)$. We also see that $HH^1(A)$ is equal to the Lie algebra of outer derivations of A (with the bracket induced by the commutator of derivations). Both structures, the commutative multiplication and the Lie bracket, extend to the whole of Hochschild cohomology, as we will see in section 7.

Let $A^e = A^{op} \otimes A$. We identify the category $\mathsf{Mod}(A^e)$ of right A^e-modules with the (isomorphic) category of A-A-bimodules via the rule

$$amb = m(b \otimes a).$$

In particular, we have the *identity bimodule* ${}_AA_A$ given by the algebra A considered as a bimodule over itself.

Theorem 3.1 (Cartan–Eilenberg [6]). *We have a canonical isomorphism*

$$\mathsf{Ext}^*_{A^e}(A, A) \xrightarrow{\sim} HH^*(A).$$

To prove the theorem, one computes the derived functors using the bar resolution of the first argument: Recall that the (augmented) bar resolution is the complex of bimodules

$$0 \longleftarrow {}_AA_A \longleftarrow A \otimes A \longleftarrow \dots \longleftarrow A \otimes A^{\otimes p} \otimes A \longleftarrow \dots$$

where the augmentation is the multiplication of A and the differential is given by

$$d(a_0, \dots, a_{p+1}) = \sum_{i=0}^{p} (-1)^p (a_0, \dots, a_i a_{i+1}, \dots, a_{p+1}).$$

Corollary 3.1. *Hochschild cohomology $HH^*(A)$ carries a natural graded multiplication, the* cup product, *extending that of $Z(A) = HH^0(A)$.*

It is shown in [6] that the cup product is induced by the following associative multiplication on cochains: For $f \in C^pA$ and $g \in C^qA$, define

$$(f \cup g)(a_1, \dots, a_p, a_{p+1}, \dots, a_{p+q}) = (-1)^{pq} f(a_1, \dots, a_p) g(a_{p+1}, \dots, a_{p+q}).$$

4. Reminder on derived categories

We collect basic definitions and results on derived categories. We refer to [30] for more details and references. As before, k is a field and A a k-algebra. The category $\mathcal{C}A = \mathcal{C}\,\mathsf{Mod}\,A$ has as objects the cochain complexes

$$\dots \longrightarrow M^p \xrightarrow{d_M} M^{p+1} \longrightarrow \dots$$

of (right) A-modules. Notice that each such complex has an underlying $\mathbb{Z}$-graded A-module given by the sequence of the M^p, $p \in \mathbb{Z}$. It is endowed with the differential d_M, which is a homogeneous endomorphism of degree $+1$. The *morphisms* of $\mathcal{C}A$ are the morphisms $f : L \to M$ of graded A-modules which are homogeneous of degree 0 and satisfy $d_M \circ f = f \circ d_L$. The *suspension functor* $\Sigma : \mathcal{C}A \to \mathcal{C}A$ takes a complex M to ΣM with components $(\Sigma M)^p = M^{p+1}$ and differential $d_{\Sigma M} = -d_M$. Two morphisms f and $g : L \to M$ of $\mathcal{C}A$ are *homotopic* if there is a homogeneous morphism of graded A-modules $h : L \to M$ of degree -1 such that

$$f - g = d_M \circ h + h \circ d_L .$$

The *homotopy category* $\mathcal{H}A$ has the same objects as $\mathcal{C}A$; its morphisms are homotopy classes of morphisms of $\mathcal{C}A$. A morphism $s : L \to M$ of $\mathcal{C}A$ or $\mathcal{H}A$ is a *quasi-isomorphism* if the induced morphism in homology $H^*(s) : H^*(L) \to H^*(M)$ is invertible. The *derived category* is the localization of $\mathcal{C}A$ (or $\mathcal{H}A$) at the class of all quasi-isomorphisms. Thus, it has the same objects as $\mathcal{C}A$ and its morphisms are equivalence classes of formal compositions of morphisms of $\mathcal{C}A$ (or $\mathcal{H}A$) and formal inverses of quasi-isomorphisms.

The homotopy category $\mathcal{H}A$ is canonically triangulated with suspension functor Σ. Each componentwise split short exact sequence of $\mathcal{C}A$

$$0 \longrightarrow L \longrightarrow M \longrightarrow N \longrightarrow 0$$

yields a canonical triangle

$$L \longrightarrow M \longrightarrow N \longrightarrow \Sigma L .$$

The derived category $\mathcal{D}A$ is triangulated with suspension functor Σ. Each short exact sequence of $\mathcal{C}A$ yields a canonical triangle.

We identify A-modules $M \in \mathsf{Mod}\, A$ with complexes concentrated in degree 0

$$\ldots \longrightarrow 0 \longrightarrow M \longrightarrow 0 \longrightarrow \ldots .$$

Then, for A-modules L and M, we have canonical isomorphisms

$$\mathsf{Ext}^p_A(L, M) \xrightarrow{\sim} \mathrm{Hom}_{\mathcal{D}A}(L, \Sigma^p M)$$

for all $p \in \mathbb{Z}$ (with the convention that Ext^p vanishes for $p < 0$). Moreover, the composition in $\mathcal{D}A$ identifies with the Yoneda product on Ext.

Theorem 4.1. *The projection $\mathcal{H}A \to \mathcal{D}A$ admits a fully faithful left adjoint* $\mathbf{p}$ *and a fully faithful right adjoint* $\mathbf{i}$.

Notice that the analogous statement for the category of complexes $\mathcal{C}A$ instead of the homotopy category $\mathcal{H}A$ is wrong. This is one of the main reasons for introducing $\mathcal{H}A$. The functors $\mathbf{p}$ and $\mathbf{i}$ generalize projective respectively injective resolutions. Indeed, if M is an A-module and $P \to M$ a projective resolution (*i.e.* a quasi-isomorphism where P is right bounded with projective components), then we have $P \xrightarrow{\sim} \mathbf{p}M$ in $\mathcal{H}A$. Similarly, if $M \to I$ is an injective resolution, then $\mathbf{i}M \xrightarrow{\sim} I$ in $\mathcal{H}A$.

Now let B be another algebra and X a complex of A-B-bimodules. For $M \in \mathcal{C}A$, define the $M \otimes_A X \in \mathcal{C}B$ by

$$(M \otimes_A X)^n = \bigoplus_{p+q=n} M^p \otimes_A X^q \quad \text{and} \quad d(m \otimes a) = (dm) \otimes a + (-1)^p m \otimes (dx).$$

For $L \in \mathcal{C}B$, define $\mathrm{Hom}_B(X, L) \in \mathcal{C}A$ as the complex whose nth component is formed by the morphisms $f : X \to L$ of graded B-modules, homogeneous of degree n and whose differential is given by

$$d(f) = d_L \circ f - (-1)^n f \circ d_X.$$

Define objects of the derived categories

$$L \overset{L}{\otimes}_A X = (\mathbf{p}L) \otimes_A X \quad \text{and} \quad \mathrm{RHom}_B(X, L) = \mathrm{Hom}_B(X, ?).$$

Example 4.1. For example, we have

$$HH^*(A) = \mathsf{Ext}^*_{A^e}(A, A) = H^*(\mathrm{RHom}_{A^e}(A, A)).$$

The following lemma is easy to prove.

Lemma 4.1. *We have an adjoint pair*

$$? \overset{L}{\otimes}_A X : \mathcal{D}A \rightleftarrows \mathcal{D}B : \mathrm{RHom}_B(X, ?)\ .$$

It is natural to ask when these adjoints are equivalences. To answer this question, let us observe that $\mathcal{D}A$ has arbitrary coproducts, which are given by coproducts of complexes. An object P of $\mathcal{D}A$ is *compact* if the functor $\mathrm{Hom}(P, ?) : \mathcal{D}A \to \mathsf{Mod}\, k$ commutes with arbitrary coproducts. It is *perfect* if it is quasi-isomorphic to a bounded complex of finitely generated projective modules. For example, the free A-module A_A is compact because

$$\mathrm{Hom}(A, M) \xrightarrow{\sim} H^0 M$$

and of course it is also perfect.

Proposition 4.1. *An object of $\mathcal{D}A$ is compact if and only if it is perfect.*

The *perfect derived category* $\mathsf{per}(A)$ is the full subcategory of $\mathcal{D}A$ formed by the perfect objects. Clearly it is a *thick subcategory, i.e.* a triangulated subcategory stable under taking direct summands.

Proposition 4.2. *The functor* $? \overset{L}{\otimes}_A X : \mathcal{D}A \to \mathcal{D}B$ *is an equivalence if and only if*

a) X_B *is perfect in* $\mathcal{D}B$ *and*
b) X_B *generates* $\mathcal{D}B$ *as a triangulated category with arbitrary coproducts and*
c) *the natural map* $A \to \mathrm{Hom}_{\mathcal{D}B}(X_B, X_B)$ *given by left multiplication is an isomorphism and* $\mathrm{Hom}_{\mathcal{D}B}(X_B, \Sigma^p X_B) = 0$ *for all* $p \neq 0$.

By definition, the bimodule complex X is a *two-sided tilting complex* if these conditions hold. We have the following important class of examples:

Theorem 4.2 (Happel [22]). *If* T *is an* A-B-*bimodule, then* T *is a two-sided tilting complex iff* T_B *is a tilting module and the left action yields an isomorphism* $A \xrightarrow{\sim} \mathsf{End}_B(T)$.

For a proof of the theorem in this form, *cf.* [30].

Theorem 4.3 (Rickard [41, 42]). *There is a triangle equivalence* $\mathcal{D}A \to \mathcal{D}B$ *if and only if there is a 2-sided tilting complex* X.

Notice that the theorem does not claim that a given triangle equivalence $F : \mathcal{D}A \xrightarrow{\sim} \mathcal{D}B$ is isomorphic to a derived functor $? \overset{L}{\otimes}_A X$ for a 2-sided tilting complex X. It is open whether this always holds.

Define $\mathsf{rep}(A, B)$ to be the full subcategory of $\mathcal{D}(A^{op} \otimes B)$ formed by the bimodule complexes X such that X_B is perfect. We think of the objects of $\mathsf{rep}(A, B)$ as 'representations up to homotopy' of A in $\mathsf{per}(B)$. Notice that a bimodule complex X belongs to $\mathsf{rep}(A, B)$ if and only if the functor $? \overset{L}{\otimes}_A X$ takes $\mathsf{per}(A)$ to $\mathsf{per}(B)$. For $X \in \mathsf{rep}(A, B)$, put

$$X^{\vee} = \mathrm{RHom}_B(X, B)$$

and notice that this is naturally an object of $\mathcal{D}(B^{op} \otimes A)$.

Lemma 4.2. *We have a canonical isomorphism*

$$? \overset{L}{\otimes}_B X^{\vee} \xrightarrow{\sim} \mathrm{RHom}_B(X, ?).$$

Thus, the functor $? \overset{L}{\otimes}_B X^\vee$ is right adjoint to $? \overset{L}{\otimes}_A X$. The adjunction morphisms are produced by the *action morphism*

$$\alpha : A \to \mathrm{RHom}_B(X, X) \xleftarrow{\sim} X \overset{L}{\otimes}_B X^\vee \text{ in } \mathcal{D}(A^e)$$

and the *evaluation morphism*

$$\varepsilon : X^\vee \overset{L}{\otimes}_A X = \mathrm{RHom}_B(X, B) \overset{L}{\otimes}_A X \to B \text{ in } \mathcal{D}(B^e).$$

Notice that $? \overset{L}{\otimes}_A X$ is fully faithful if and only if the action morphism $A \to X \overset{L}{\otimes}_B X^\vee$ is invertible.

5. Invariance theorems

Let A and B be k-algebras. Let $X \in \mathsf{rep}(A, B)$ and

$$X^\vee = \mathrm{RHom}_B(X, B) \in \mathcal{D}(B^{op} \otimes A).$$

Note that in general, $X^\vee$ is *not perfect over* A and so does not belong to $\mathsf{rep}(B, A)$. Recall the canonical action and evaluation morphisms

$$\alpha : A \to X \overset{L}{\otimes}_B X^\vee \text{ in } \mathcal{D}(A^e) \quad \text{and} \quad \varepsilon : X^\vee \overset{L}{\otimes}_A X \to B \text{ in } \mathcal{D}(B^e).$$

Now recall that a fully faithful functor $F : \mathcal{A} \to \mathcal{B}$ between k-categories yields a restriction morphism $F^* : Z(\mathcal{B}) \to Z(\mathcal{A})$. Since Hochschild cohomology may be viewed as a 'derived center', the following theorem is quite natural.

Theorem 5.1. *Suppose we have $X \in \mathsf{rep}(A, B)$ such that the functor $? \overset{L}{\otimes}_A X : \mathcal{D}A \to \mathcal{D}B$ is fully faithful. Then we have a canonical 'restriction' morphism*

$$HH^*(X) : HH^*(B) \to HH^*(A).$$

It is an isomorphism if $? \overset{L}{\otimes}_A X : \mathcal{D}A \to \mathcal{D}B$ is an equivalence.

Below, we will show using other methods that even if X_B is not perfect but $? \overset{L}{\otimes}_A X : \mathcal{D}A \to \mathcal{D}B$ is fully faithful, we still have such a restriction morphism.

Sketch of proof. It is not hard to show, without any hypothesis on $X \in \mathsf{rep}(A, B)$, that we have an adjoint pair

$$X^\vee \overset{L}{\otimes}_A ? \overset{L}{\otimes}_A X : \mathcal{D}A^e \rightleftarrows \mathcal{D}B^e : X \overset{L}{\otimes}_B ? \overset{L}{\otimes}_B X^\vee \,.$$

Now we construct a map from

$$HH^pB = \mathrm{Hom}_{\mathcal{D}B^e}(B, \Sigma^p B)$$

to HH^pA as follows:

$$\begin{array}{ccc} \mathrm{Hom}_{\mathcal{D}B^e}(B, \Sigma^p B) & \longrightarrow & \mathrm{Hom}_{\mathcal{D}B^e}(X^\vee \overset{L}{\otimes}_A X, \Sigma^p B) \\ & & \downarrow{\scriptstyle adj} \\ \mathrm{Hom}_{\mathcal{D}A^e}(A, \Sigma^p A) & \xrightarrow{\sim} & \mathrm{Hom}_{\mathcal{D}A^e}(A, \Sigma^p X \overset{L}{\otimes}_B X^\vee). \end{array}$$

√

6. Differential graded categories

Recall from Lemma 2.2, that for an algebra A, the restriction along the inclusion $A \subset \mathsf{Mod}\, A$ is an isomorphism

$$Z(\mathsf{Mod}\, A) \xrightarrow{\sim} Z(A).$$

It is natural to ask what the derived version of this fact is. In the derived version, Hochschild cohomology should replace the center and the derived category should replace the module category. So we would like to know how to recover Hochschild cohomology from the derived category $\mathcal{D}A$. Unfortunately, this seems to be an ill-posed question nobody knows to how to answer. However, it is easy to recover Hochschild cohomology from the differential graded (=dg) version of $\mathcal{D}A$, namely the dg category $\mathcal{D}_{dg}A$. Our first aim in this section is to define the dg category $\mathcal{D}_{dg}A$ in an intrinsic way, via a universal property.

6.1. *Dg categories and their derived categories*

Recall that the category of complexes $\mathcal{C}k$ is *monoidal*, *i.e.* it is endowed with the bifunctor $(L, M) \mapsto L \otimes M$ given by

$$(L \otimes M)^n = \bigoplus_{p+q=n} L^p \otimes M^q, \quad d(l \otimes m) = (dl) \otimes m + (-1)^{|l|} l \otimes dm$$

enjoying a number of desirable properties. A *dg category* is a category $\mathcal{A}$ enriched in the monoidal category of complexes. Thus, the morphism spaces $\mathcal{A}(X, Y)$ are complexes and the compositions

$$\mathcal{A}(Y, Z) \otimes \mathcal{A}(X, Y) \to \mathcal{A}(X, Z)$$

and units $k \to \mathcal{A}(X, X)$ are morphisms of complexes.

For example, if B is an algebra, the dg category $\mathcal{C}_{dg}B$ has the same objects as $\mathcal{C}B$ and its morphism complexes are defined by

$$(\mathcal{C}_{dg}B)(L, M) = \mathrm{Hom}_B(L, M) ,$$

cf. section 4. As in the case of k-categories, we identify dg categories with one object with dg algebras.

If $\mathcal{A}$ is a dg category, the *category* $H^0\mathcal{A}$ has the same objects as $\mathcal{A}$ and the morphism spaces $H^0(\mathcal{A}(X, Y))$ with the natural compositions. For example, this yields another viewpoint on the homotopy category $\mathcal{H}B$ via the equality of categories

$$H^0(\mathcal{C}_{dg}B) = \mathcal{H}B.$$

A *dg functor* $F : \mathcal{A} \to \mathcal{B}$ is a functor such that

$$F : \mathcal{A}(X, Y) \to \mathcal{B}(FX, FY)$$

is a morphism of complexes for all $X, Y \in \mathcal{A}$. It is a *quasi-equivalence* if $F : \mathcal{A}(X, Y) \to \mathcal{B}(FX, FY)$ is a quasi-isomorphism for all $X, Y \in \mathcal{A}$ and the induced functor $H^0F : H^0\mathcal{A} \to H^0\mathcal{B}$ is an equivalence. The *category* Hqe is the localization of the category dgcat of small dg categories at the class of all quasi-equivalences. For example, if $f : A \to B$ is a quasi-isomorphism between dg algebras, it may be viewed as a quasi-equivalence between dg categories with one object.

Theorem 6.1 (Tabuada [43]). *The category* dgcat *carries a (cofibrantly generated) Quillen model structure whose weak equivalences are the quasi-equivalences. In particular, the morphisms* $\mathcal{A} \to \mathcal{B}$ *in* Hqe *form a* set *for all small dg categories* $\mathcal{A}$, $\mathcal{B}$.

Thanks to the theorem, we can speak about representable functors on the category Hqe without having to 'enlarge the universe'.

Theorem 6.2 (1999 [28]). *Let* $\mathcal{A}$ *be a dg category and* $\mathcal{N} \subseteq \mathcal{A}$ *a full dg subcategory. Then there is a morphism* $Q : \mathcal{A} \to \mathcal{A}/\mathcal{N}$ *of* Hqe *which kills* $\mathcal{N}$ *(*i.e. *we have* $\mathbf{1}_{QN} = 0$ *in* $H^0(\mathcal{A}/\mathcal{N})$ *for all* $N \in \mathcal{N}$*) and which is universal in* Hqe *among the morphisms killing* $\mathcal{N}$.

We define the *dg quotient of* $\mathcal{A}$ *by* $\mathcal{N}$ to be the dg category $\mathcal{A}/\mathcal{N}$ of the theorem. It is thus unique up to unique isomorphism in Hqe.

Theorem 6.3 (Drinfeld 2004 [11]). *The dg category* $\mathcal{A}/\mathcal{N}$ *is obtained from* $\mathcal{A}$ *by adjoining a contracting homotopy* h_N *for each object* N *on* $\mathcal{N}$ *(*i.e. h_N *is of degree* -1 *and* $d(h_N) = \mathbf{1}_N$*).*

For an algebra B, we define the *dg derived category of B* to be the dg quotient

$$\mathcal{D}_{dg}B = \mathcal{C}_{dg}B/\mathcal{A}c_{dg}B\,,$$

where $\mathcal{A}c_{dg}B$ is the full dg subcategory of $\mathcal{C}_{dg}B$ whose objects are the acyclic complexes.

Theorem 6.4 ([11, 28]). *We have a canonical equivalence $H^0(\mathcal{D}_{dg}B) \xrightarrow{\sim} \mathcal{D}B$.*

Our next aim is to define the derived category $\mathcal{D}\mathcal{A}$ (as well as its dg version $\mathcal{D}_{dg}\mathcal{A}$) of a dg category $\mathcal{A}$. Define the *opposite dg category $\mathcal{A}^{op}$* to be the dg category with the same objects, with the morphism complexes $\mathcal{A}^{op}(X,Y) = \mathcal{A}(Y,X)$ and the compositions given by

$$f \circ_{\mathcal{A}^{op}} g = (-1)^{|f||g|} g \circ f$$

for all homogeneous $f \in \mathcal{A}^{op}(Y,Z)$ and $g \in \mathcal{A}^{op}(X,Y)$. For two dg functors $F, G : \mathcal{A} \to \mathcal{B}$, define the complex $\mathrm{Hom}(F,G)$ to be the subcomplex of

$$\prod_{X \in \mathcal{A}} \mathcal{B}(FX, GX)$$

formed by the families (φX) such that

$$(Gf) \circ (\varphi X) = (-1)^{|\varphi||f|} (\varphi Y) \circ (Ff)$$

for all $X \in \mathcal{A}$. In this way, the category $\mathrm{Fun}_{dg}(\mathcal{A}, \mathcal{B})$ of dg functors from $\mathcal{A}$ to $\mathcal{B}$ becomes a dg category. We define the dg category of dg right $\mathcal{A}$-modules to be

$$\mathcal{C}_{dg}\mathcal{A} = \mathrm{Fun}_{dg}(\mathcal{A}^{op}, \mathcal{C}_{dg}k)$$

and the category of right $\mathcal{A}$-modules to be $Z^0\mathcal{C}_{dg}\mathcal{A}$ (same objects as $\mathcal{C}_{dg}\mathcal{A}$ and morphism spaces $Z^0(\mathcal{C}_{dg}\mathcal{A})(L,M)$). We define the *homotopy category of dg $\mathcal{A}$-modules* as

$$\mathcal{H}\mathcal{A} = H^0\mathcal{C}_{dg}\mathcal{A}.$$

For example, for each object $X \in \mathcal{A}$, we have the *representable dg module*

$$X^\wedge = \mathcal{A}(?, X) : \mathcal{A}^{op} \to \mathcal{C}_{dg}k.$$

Whence the *dg Yoneda functor*

$$\mathcal{A} \to \mathcal{C}_{dg}\mathcal{A}\,,\ X \mapsto X^\wedge.$$

As an exercise, the reader may want to prove the dg Yoneda lemma:

Lemma 6.1. *For X in $\mathcal{A}$ and M in $\mathcal{CA}$, we have a natural isomorphism*

$$\mathrm{Hom}_{\mathcal{A}}(X^{\wedge}, M) \xrightarrow{\sim} MX\ ,\ f \mapsto f(\mathbf{1}_X).$$

Notice that when B is an ordinary algebra and $\mathcal{A}$ the dg category whose endomorphism algebra is B (concentrated in degree 0), then $\mathcal{C}_{dg}\mathcal{A} = \mathcal{C}_{dg}B$

Let $\mathcal{A}$ be a small dg category. A morphism $s : L \to M$ of dg $\mathcal{A}$-modules is a *quasi-isomorphism* if

$$sX : LX \to MX$$

is a quasi-isomorphism for each $X \in \mathcal{A}$. We define the *derived category* $\mathcal{DA}$ to be the localization of $\mathcal{CA}$ (respectively $\mathcal{HA}$) at the class of quasi-isomorphisms and the *dg derived category* $\mathcal{D}_{dg}\mathcal{A}$ to be the dg quotient

$$\mathcal{C}_{dg}(\mathcal{A})/\mathcal{A}c_{dg}(\mathcal{A}).$$

As in Theorem 4.1, the quotient functor $\mathcal{HA} \to \mathcal{DA}$ admits a left adjoint $\mathbf{p}$ and a right adjoint $\mathbf{i}$, *cf.* [27], and the construction of the derived functors generalizes.

Let us give an example where we have a beautiful description of the derived category of a non trivial dg category: Let A be a right noetherian algebra (concentrated in degree 0) and $\mathsf{mod}\, A$ the abelian category of finitely generated (right) A-modules. Let $\mathcal{C}^b_{dg}(\mathsf{mod}\, A) \subset \mathcal{C}_{dg}A$ be the full dg subcategory of bounded complexes over $\mathsf{mod}\, A$ and

$$\mathcal{D}^b_{dg}(\mathsf{mod}\, A) = \mathcal{C}^b_{dg}(\mathsf{mod}\, A)/\mathcal{A}c^b_{dg}(\mathsf{mod}\, A).$$

Let $\mathsf{Inj}\, A$ denote the category of all injective modules and $\mathcal{H}\,\mathsf{Inj}\, A$ the homotopy category of (unbounded) complexes of injective modules.

Theorem 6.5 (Krause [33]). *We have a canonical triangle equivalence*

$$\mathcal{H}\,\mathsf{Inj}\, A \xrightarrow{\sim} \mathcal{D}(\mathcal{D}^b_{dg}(\mathsf{mod}\, A)).$$

6.2. *Hochschild cohomology of dg categories*

Let $\mathcal{A}$ be a small dg category. We have the following generalization of Proposition 4.1. Recall that a *thick subcategory* of a triangulated category is a full triangulated subcategory stable under taking direct factors.

Proposition 6.1. *An object $P \in \mathcal{DA}$ is compact if and only if it is* perfect, i.e. *contained in the thick subcategory generated by the representable modules $X^{\wedge}$, $X \in \mathcal{A}$.*

Let $\mathcal{B}$ be another dg category. The *tensor product* $\mathcal{A} \otimes \mathcal{B}$ is the dg category whose objects are the pairs (X, Y), $X \in \mathcal{A}$, $Y \in \mathcal{B}$, and whose morphisms are given by

$$(\mathcal{A} \otimes \mathcal{B})((X,Y),(X',Y')) = \mathcal{A}(X,X') \otimes \mathcal{B}(Y,Y').$$

We define $\mathsf{rep}(\mathcal{A},\mathcal{B})$ as the full subcategory of $\mathcal{D}(\mathcal{B} \otimes \mathcal{A}^{op})$ formed by the dg bimodules X whose restriction to $\mathcal{B}$ is perfect.

Let $\mathcal{A}^e = \mathcal{A} \otimes \mathcal{A}^{op}$. The *identity bimodule* $I_{\mathcal{A}}$ sends (X, Y) to $\mathcal{A}(X, Y)$, $X, Y \in \mathcal{A}$. We put

$$HH^*(\mathcal{A}) = H^* \mathrm{RHom}_{\mathcal{A}^e}(I_{\mathcal{A}}, I_{\mathcal{A}}).$$

This may also be computed as the homology of the complex $C^*\mathcal{A}$ constructed as follows: It is the product total complex of the bicomplex

$$\prod \mathcal{A}(X_0, X_0) \longrightarrow \prod \mathrm{Hom}_k(\mathcal{A}(X_0, X_1), \mathcal{A}(X_0, X_1)) \longrightarrow \ldots$$

whose pth column ($p \geq 0$) is

$$\prod \mathrm{Hom}_k(\mathcal{A}(X_{p-1}, X_p) \otimes \ldots \otimes \mathcal{A}(X_0, X_1), \mathcal{A}(X_0, X_p))$$

where the product is taken over all sequences of objects $X_0, \ldots, X_p$ of $\mathcal{A}$ and whose horizontal differential is given by formula (1) adjusted following the Koszul sign rule.

Theorem 6.6 ([29]). *Let $\mathcal{A}$ and $\mathcal{B}$ be dg categories and $X \in \mathcal{D}(\mathcal{B} \otimes \mathcal{A}^{op})$. Suppose that the functor $? \overset{L}{\otimes}_A X : \mathcal{D}\mathcal{A} \to \mathcal{D}\mathcal{B}$ is fully faithful (but X is not necessarily right perfect). Then there is a canonical restriction morphism*

$$\mathrm{res}_X : HH^*\mathcal{B} \to HH^*\mathcal{A}.$$

It is an isomorphism if the functor $X \overset{L}{\otimes}_B ? : \mathcal{D}(\mathcal{B}^{op}) \to \mathcal{D}(\mathcal{A}^{op})$ is also fully faithful.

We will sketch a proof after Theorem 7.1.

Corollary 6.1 (Lowen–Van den Bergh [36]). *Let $\mathcal{A}$ be a dg category. The restriction along the Yoneda functor $\mathcal{A} \to \mathcal{D}_{dg}\mathcal{A}$ induces an isomorphism*

$$HH^*(\mathcal{D}_{dg}\mathcal{A}) \xrightarrow{\sim} HH^*\mathcal{A}.$$

A similar result was obtained by Toën in [44]. It should be viewed as the derived version of the isomorphism

$$Z(\mathsf{Mod}\, A) \xrightarrow{\sim} Z(A)$$

of Lemma 2.2 for an algebra A.

7. Higher structure on the Hochschild complex

Let A be an algebra (for simplicity) and C^*A its Hochschild cochain complex, *cf.* section 3. If c and u, v, ..., w are Hochschild cochains, *i.e.* homogeneous elements of C^*A, one defines the *brace operation* $c\{u, v, \ldots, w\}$ by substituting the cochains u, v, ..., w for some of the arguments of c, inserting suitable signs and summing over all possibilities of doing this, cf. Figure 1. We refer to [26] or section 5.1 of [16] for the details and in particular the exact choice of signs.

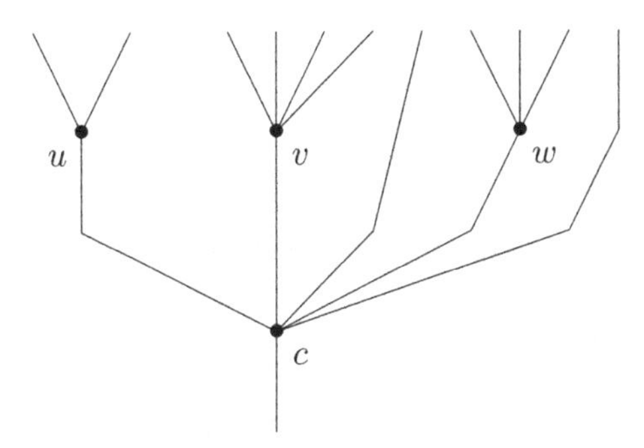

Fig. 1. A term of the brace operation $c\{u, v, w\}$

The complex C^*A together with the cup product $\cup$ and the brace operations is an example of a B_∞-algebra in the sense of section 5.2 of Getzler–Jones' [16], *i.e.* a graded vector space V such that the tensor coalgebra $T^c(\Sigma V)$ (with the deconcatenation coproduct) is endowed with a dg bialgebra structure whose comultiplication is deconcatenation. Here the letter 'B' stands for 'Baues', in honour of Hans–Joachim Baues, who showed [3] that the singular cochain complex with integer coefficients of any topological space carries a natural B_∞-algebra structure. By definition, a *morphism of B_∞-algebras* $f : V \to V'$ is a morphism of the corresponding dg bialgebras. It is given by a family of homogeneous morphisms of graded vector spaces $f_n : V^{\otimes n} \to V$, $n \geq 0$. It is a *quasi-isomorphism* if the component $f_1 : V \to V'$ is a quasi-isomorphism of complexes. The *homotopy category of B_∞-algebras* is the localization of the category of B_∞-algebras at the class of all quasi-isomorphisms.

Notice that B_∞-structures are closely related to monoidal structures (*cf.* [35] and the references given there). In the case of the Hochschild cochain complex, the monoidal category is the derived category $\mathcal{D}(A^e)$ with the derived tensor product $\overset{L}{\otimes}_A$ over A; in the case of the singular cochain complex on a topological space X, it is the derived category of sheaves of abelian groups on X with the derived tensor product. The B_∞-algebra structure on C^*A contains in particular the information on the Gerstenhaber bracket, which may be recovered via

$$[c, u] = c\{u\} \mp u\{c\}.$$

By the *homotopy category of B_∞-algebras* we mean the localization of the category of B_∞-algebras with respect to all morphisms of B_∞-algebras inducing isomorphisms in homology.

Let B be another algebra and $X \in \mathcal{D}(B \otimes A^{op})$ a complex of A-B-bimodules.

Theorem 7.1 ([29]). *Suppose that the functor $? \overset{L}{\otimes}_A X : \mathcal{D}A \to \mathcal{D}B$ is fully faithful (but X_B is not necessarily perfect). Then there is a canonical restriction morphism*

$$\mathrm{res}_X : C^*B \to C^*A$$

in the homotopy category of B_∞-algebras. It is invertible if the functor $X \overset{L}{\otimes}_B ? : \mathcal{D}(B^{op}) \to \mathcal{D}(A^{op})$ is also fully faithful (for example if $? \overset{L}{\otimes}_A X$ is an equivalence).

We sketch the proof: Let G be the 'glued' dg category with two objects 1 and 2 such that $G(1,1) = B$, $G(2,2) = A$, $G(1,2) = X$ and $G(2,1) = 0$. We have obvious forgetful (or 'restriction') maps

$$C^*B \xleftarrow{\mathrm{res}_B} C^*G \xrightarrow{\mathrm{res}_A} C^*A$$

which clearly respect the B_∞-structure. It is a classical fact, *cf.* [4, 8, 9, 17, 18, 20, 23, 38], that we have a homotopy bicartesian square

$$\begin{array}{ccc} C^*G & \xrightarrow{\mathrm{res}_A} & C^*A \\ {\scriptstyle \mathrm{res}_B}\downarrow & & \downarrow{\scriptstyle \alpha} \\ C^*B & \longrightarrow & \mathrm{RHom}_{B \otimes A^{op}}(X, X). \end{array}$$

We claim that α and hence res_B is invertible in $\mathcal{D}k$. Indeed, we have a commutative square in $\mathcal{D}k$

$$\begin{array}{ccc} C^*A & \xrightarrow{\sim} & \mathrm{RHom}_{A^e}(A,A) \\ \downarrow{\scriptstyle\alpha} & & \downarrow{\scriptstyle\beta} \\ \mathrm{RHom}_{B\otimes A^{op}}(X,X) & \xrightarrow{\sim} & \mathrm{RHom}_{A^e}(A,\mathrm{RHom}_B(X,X)), \end{array}$$

where β is induced by the action morphism

$$A \to \mathrm{RHom}_B(X,X).$$

This is invertible by our assumption that the functor $?\overset{L}{\otimes}_A X$ is fully faithful. We put $\mathrm{res}_X = \mathrm{res}_A \circ \mathrm{res}_B^{-1}$.

Corollary 7.1. *For an algebra A, the isomorphism*

$$HH^*(\mathcal{D}_{dg}A) \xrightarrow{\sim} HH^*(A)$$

of Corollary 6.1 lifts to an isomorphism in the homotopy category of B_∞-algebras.

8. Tate–Hochschild cohomology

Let A be a right noetherian algebra and $\mathsf{mod}\,A$ the abelian category of finitely generated (right) A-modules. Let $\mathcal{D}^b(\mathsf{mod}\,A)$ be its bounded derived category. The perfect derived category $\mathsf{per}(A)$ is a thick subcategory of $\mathcal{D}^b(\mathsf{mod}\,A)$. They coincide if A is of finite global dimension. We define the *singularity category* $\mathsf{sg}(A)$ *to be the quotient* $\mathcal{D}^b(\mathsf{mod}\,A)/\,\mathsf{per}(A)$. This category was first considered by Buchweitz [5] in 1986 and rediscovered by Orlov [40] in 2003 in a geometric setting. It measures the non regularity of the algebra A.

As an example, let $S = k[x_1, \ldots, x_n]$ and let $A = S/(f)$ for a non zero $f \in S$. It follows from the work of Eisenbud [13] that the singularity category $\mathsf{sg}(A)$ is triangle equivalent to the homotopy category of matrix factorizations of f. By definition, this category equals $H^0(\mathsf{mf}_{dg}(f))$, where $\mathsf{mf}_{dg}(f)$ is the *dg category of matrix factorizations of* f: its objects are the 2-periodic diagrams (not complexes!)

$$\cdots \xrightarrow{d_0} P_1 \xrightarrow{d_1} P_0 \xrightarrow{d_0} P_1 \xrightarrow{d_1} P_0 \xrightarrow{d_0} \cdots$$

where the P_i are finitely generated projective S-modules and d^2 is the multiplication with f. For two such objects P and Q, the *morphism complex* $\mathrm{Hom}(P,Q)$ has as its nth component the space of homogeneous S-linear maps $g : P \to Q$ of degree n. The differential is given by $d(g) = d \circ g - (-1)^n g \circ d$. We leave it as an exercise for the reader to check that $d^2(g) = 0$. It is an important point that the complexes $\mathrm{Hom}(P, Q)$ are also 2-periodic, so that we may also view $\mathsf{mf}_{dg}(f)$ as a differential $\mathbb{Z}/2\mathbb{Z}$-graded category!

Now suppose that $A^e = A \otimes A^{op}$ is also (right) noetherian. We define *Tate–Hochschild cohomology of A* to be the Yoneda algebra

$$HH^*_{sg}(A) = \mathsf{Ext}_{\mathsf{sg}(A^e)}(A, A)$$

of the identity bimodule in the singularity category of the enveloping algebra. By definition, it is an algebra and it is not hard to check directly that it is (graded) commutative. However, the singularity category $\mathsf{sg}(A^e)$ is *not monoidal* in any natural way. We may nevertheless ask whether Tate–Hochschild cohomology carries the same rich structure as classical Hochschild cohomology. This problem was open for some time and finally solved in the thesis of Zhengfang Wang:

Theorem 8.1 (Zhengfang Wang). *a) $HH^*_{sg}(A)$ carries a natural (but intricate!) Gerstenhaber bracket [47].*

b) There is a natural B_∞-algebra $C^*_{sg}A$ computing $HH^*_{sg}A$ with its Gerstenhaber bracket [46].

Thus, we see that there is a complete structural analogy between Tate–Hochschild cohomology and classical Hochschild cohomology. It is therefore natural to ask whether Tate–Hochschild cohomology is not an instance of classical Hochschild cohomology, i.e. whether the Tate–Hochschild cohomology of A is classical Hochschild cohomology of some more complicated object associated with A in analogy with Corollary 6.1. This object is the *dg singularity category* defined as the dg quotient

$$\mathsf{sg}_{dg}(A) = \mathcal{D}^b_{dg}(\mathsf{mod}\, A)/\,\mathsf{per}_{dg}(A)$$

of the bounded dg derived category of A by its full dg subcategory on the perfect complexes, cf. section 6.1. Recall that a dg category $\mathcal{A}$ is *smooth* if the identity bimodule $I_\mathcal{A} : (X, Y) \mapsto \mathcal{A}(X, Y)$ is perfect in the derived category $\mathcal{D}(\mathcal{A}^e)$ of bimodules.

Theorem 8.2 ([32]). *There is a canonical morphism of graded algebras*

$$HH^*_{sg}A \to HH^*(\mathsf{sg}_{dg}(A)).$$

It is an isomorphism if the dg category $\mathcal{D}^b_{dg}(\mathsf{mod}\, A)$ is smooth.

According to Theorem A of Elagin–Lunts–Schnürer's [14], the dg category $\mathcal{D}^b_{dg}(\mathsf{mod}\, A)$ is smooth if A is a finite-dimensional algebra over any field k such that $A/\mathsf{rad}(A)$ is separable over k (which is automatic if k is perfect). By Theorem B of [loc. cit.], it also holds if the algebra A is right noetherian and finitely generated over its center and the center is a finitely generated algebra over a perfect field k.

Conjecture 8.1. *The morphism of the theorem lifts to a morphism in the homotopy category of B_∞-algebras.*

Note that this morphism will be an isomorphism if the bounded dg derived category $\mathcal{D}^b_{dg}(\mathsf{mod}\, A)$ is smooth. In particular, this should hold for each finite-dimensional algebra defined by a quiver with an admissible ideal of relations. The following theorem confirms the conjecture for radical square 0 algebras.

Theorem 8.3 (Chen–Li–Wang [7]). *The conjecture holds if $A = kQ/(Q_1)^2$, where Q is a finite quiver without sinks or sources and $(Q_1)^2$ the square of the ideal of the path algebra kQ generated by the arrows.*

To show why the conjecture is probably not easy to prove, let us sketch the construction of the isomorphism in Theorem 8.2. Let $\mathcal{M} = \mathcal{D}^b_{dg}(\mathsf{mod}\, A)$ and $\mathcal{S} = \mathsf{sg}_{dg}(A)$. We have natural dg functors

$$A \xrightarrow{\ i\ } \mathcal{M} \xrightarrow{\ p\ } \mathcal{S}$$

whose composition vanishes in the homotopy category of dg categories. We construct the following square (commutative up to isomorphism)

$$\begin{array}{ccccc}
\mathcal{D}^b(A \otimes A^{op}) & \xrightarrow{(1\otimes i)^*} & \mathcal{D}(A \otimes \mathcal{M}^{op}) & \xrightarrow{(i\otimes 1)^!} & \mathcal{D}(\mathcal{M} \otimes \mathcal{M}^{op}) \\
\downarrow & & & & \downarrow {\scriptstyle (p\otimes p)^*} \\
\mathsf{sg}(A \otimes A^{op}) & - - - - - - - - - - - - & & \dashrightarrow & \mathcal{D}(\mathcal{S} \otimes \mathcal{S}^{op})
\end{array}$$

Here, for a dg functor $f : \mathcal{A}_1 \to \mathcal{A}_2$, we denote by f^* the left adjoint and by $f^!$ the right adjoint of the restriction functor $f_* : \mathcal{D}\mathcal{A}_2 \to \mathcal{D}\mathcal{A}_1$. One checks that the dashed triangle functor exists, takes the identity bimodule A to the identity bimodule $\mathcal{S}(?, -)$ and induces an isomorphism between the Yoneda algebras of these objects. Since the functor is induced by the composition of a right derived with a left derived functor, it is hard to compute it explicitly and that is why the conjecture is not obvious.

9. Application: Two reconstruction theorems

We apply the results of the preceding section to the reconstruction of singularities.

9.1. *Isolated hypersurface singularities*

Theorem 9.1 (Hua-K [25]). *Let $S = \mathbb{C}[[x_1, \dots, x_n]]$ and suppose that $R = S/(f)$ is an isolated singularity. Then R is determined up to isomorphism by its dimension and the dg singularity category $\mathsf{sg}_{dg}(R)$.*

Notice that because of Knörrer periodicity, the dg singularity category alone does not determine R.

Sketch of proof. We consider the center

$$Z(\mathsf{sg}_{dg}(R)) = HH^0(\mathsf{sg}_{dg}(R)).$$

By Theorem 8.2, it is isomorphic to $HH^0_{sg}(R)$. Now since R is a hypersurface, the dg singularity category may be described by matrix factorizations and is therefore 2-periodic. Hence its Hochschild cohomology is 2-periodic and, again by Theorem 8.2, so is $HH^*_{sg}(R)$. Thus, we have an isomorphism

$$HH^0_{sg}(R) \xrightarrow{\sim} HH^{2r}_{sg}(R).$$

Now by a theorem of Buchweitz [5], for Gorenstein algebras, in sufficiently high degrees, Tate–Hochschild cohomology agrees with classical Hochschild cohomology and we get an isomorphism

$$HH^{2r}_{sg}(R) \xrightarrow{\sim} HH^{2r}(R).$$

Thus, we are reduced to the computation of Hochschild cohomology of a hypersurface. Thanks to the results of [21], we find that $HH^{2r}(R)$ and $HH^0(R)$ are isomorphic to the Tyurina algebra

$$S/(f, \frac{\partial f}{\partial x_1}, \dots, \frac{\partial f}{\partial x_n}).$$

Now the Tyurina algebra together with the dimension determine R by the Mather–Yau theorem [37], more precisely its formal series version proved in [19]. Notice that in this sketch, we have neglected the technical problems arising from the fact that R is a topological algebra. $\checkmark$

Notice that in the above computation, we have considered the dg singularity category $\mathsf{sg}_{dg}(R)$ as a differential $\mathbb{Z}$-graded category. If one considers it as a differential $\mathbb{Z}/2$-graded category, one obtains a different result for the center, namely the Milnor algebra

$$Z_{\mathbb{Z}/2}(\mathsf{sg}_{dg}(R)) = S/(\frac{\partial f}{\partial x_1}, \dots, \frac{\partial f}{\partial x_n})$$

as shown by Dyckerhoff [12].

9.2. *Compound Du Val singularities*

Let R be a complete local isolated compound Du Val singularity (thus, it is 3-dimensional, normal and a generic hyperplane section through the origin is a Kleinian surface singularity). Let

$$f : Y \to X = \mathrm{Spec}(R)$$

be a small crepant resolution (thus, it is birational, an isomorphism in codimension 1, an isomorphism outside the exceptional fibre and $f^*(\omega_X) \cong \omega_Y$). Then the reduced exceptional fibre $\mathcal{F}$ of f is a tree of rational curves $\mathbb{P}^1$. The morphism f contracts this tree to a point. Associated to this situation is the *contraction algebra* Λ introduced by Donovan–Wemyss [10]. It is a finite-dimensional algebra which represents the deformations with non commutative base of the exceptional fibre of f. It is known that numerous invariants of the singularity can be computed from the algebra Λ. This has lead Donovan–Wemyss to conjecture [2] that the derived equivalence class of Λ determines R up to isomorphism. We show a weakened version: Thanks to work of Van den Bergh and de Thanhoffer de Voelcsey [45], it is known that Λ is the Jacobian algebra of a quiver Q with potential W. By definition, the potential is an element of $HH_0(\mathbb{C}Q)$. Let $\overline{W}$ be its image in $HH_0(\Lambda)$.

Theorem 9.2 (Hua–K [25]). *The derived equivalence class of the pair* $(\Lambda, \overline{W})$ *determines* R.

The proof uses, among other things, Theorem 9.1 and silting theory.

Let us point out the link to cluster theory: It turns out that the singularity category $\mathsf{sg}(R)$ is triangle equivalent to a generalized cluster category in the sense of Amiot [1], namely the generalized cluster category $\mathcal{C}_{Q,W}$ associated with the quiver Q with potential W. This category has therefore the same main properties as the categories appearing in the (additive) categorification of Fomin–Zelevinsky cluster algebras, *cf.* [31]. However, the quivers that appear are quite different: Whereas the quivers in Donovan–Wemyss' theory have many loops and 2-cycles, the quivers appearing in cluster theory never have loops or 2-cycles.

Acknowledgment

I thank the organizers of the Eighth China–Japan–Korea International Symposium on Ring Theory, and in particular Hideto Asashiba, for the wonderful job they have done and for their exquisite hospitality. I am grateful to the referees for their careful reading and helpful suggestions.

References

[1] Claire Amiot, *Cluster categories for algebras of global dimension* 2 *and quivers with potential*, Annales de l'institut Fourier **59** (2009), no. 6, 2525–2590.

[2] Jenny August, *On the finiteness of the derived equivalence classes of some stable endomorphism rings*, arXiv:1801.05687 [math.RT].

[3] H. J. Baues, *The double bar and cobar constructions*, Compositio Math. **43** (1981), no. 3, 331–341.

[4] Belkacem Bendiffalah and Daniel Guin, *Cohomologie de l'algèbre triangulaire et applications*, Algebra Montpellier Announcements **01-2003** (2003), 1–5, electronic.

[5] Ragnar-Olaf Buchweitz, *Maximal Cohen–Macaulay modules and Tate cohomology over Gorenstein rings*, `http:hdl.handle.net/1807/16682` (1986), 155 pp.

[6] Henri Cartan and Samuel Eilenberg, *Homological algebra*, Princeton University Press, Princeton, N. J., 1956.

[7] Xiao-Wu Chen, Huanhuan Li, and Zhengfang Wang, *On Tate–Hochschild cohomology of radical square zero algebras*, in preparation.

[8] Claude Cibils, *Tensor Hochschild homology and cohomology*, Interactions between ring theory and representations of algebras (Murcia), Lecture Notes in Pure and Appl. Math., vol. 210, Dekker, New York, 2000, pp. 35–51.

[9] Claude Cibils, Eduardo Marcos, María Julia Redondo, and Andrea Solotar, *Cohomology of split algebras and of trivial extensions*, arXiv:math.KT/0102194.

[10] Will Donovan and Michael Wemyss, *Noncommutative deformations and flops*, Duke Math. J. **165** (2016), no. 8, 1397–1474.

[11] Vladimir Drinfeld, *DG quotients of DG categories*, J. Algebra **272** (2004), no. 2, 643–691.

[12] Tobias Dyckerhoff, *Compact generators in categories of matrix factorizations*, Duke Math. J. **159** (2011), no. 2, 223–274.

[13] David Eisenbud, *Homological algebra on a complete intersection, with*

an application to group representations, Trans. Amer. Math. Soc. **260** (1980), no. 1, 35–64.

[14] Alexey Elagin, Valery A. Lunts, and Olaf M. Schnürer, *Smoothness of derived categories of algebras*, arXiv:1810.07626 [math.AG].

[15] P. Gabriel and A.V. Roiter, *Representations of finite-dimensional algebras*, Encyclopaedia Math. Sci., vol. 73, Springer–Verlag, 1992.

[16] Ezra Getzler and J. D. S. Jones, *Operads, homotopy algebra, and iterated integrals for double loop spaces*, hep-th/9403055.

[17] Edward L. Green, Eduardo N. Marcos, and Nicole Snashall, *The Hochschild cohomology ring of a one point extension*, Comm. Algebra **31** (2003), no. 1, 357–379.

[18] Edward L. Green and Øyvind Solberg, *Hochschild cohomology rings and triangular rings*, Proceedings of the Ninth International Conference, Beijing, 2000, vol. II, Beijing Normal University Press, Beijing, 2002, eds. D. Happel and Y. B. Zhang, pp. 192–200.

[19] Gert-Martin Greuel and Thuy Huong Pham, *Mather-Yau theorem in positive characteristic*, J. Algebraic Geom. **26** (2017), no. 2, 347–355.

[20] Jorge A. Guccione and Juan J. Guccione, *Hochschild cohomology of triangular matrix algebras*, arXiv:math.KT/0104068.

[21] Jorge Alberto Guccione, Jose Guccione, Maria Julia Redondo, and Orlando Eugenio Villamayor, *Hochschild and cyclic homology of hypersurfaces*, Adv. Math. **95** (1992), no. 1, 18–60.

[22] Dieter Happel, *On the derived category of a finite-dimensional algebra*, Comment. Math. Helv. **62** (1987), no. 3, 339–389.

[23] Dieter Happel, *Hochschild cohomology of finite-dimensional algebras*, Séminaire d'Algèbre Paul Dubreil et Marie-Paul Malliavin, 39ème Année (Paris, 1987/1988), Lecture Notes in Math., vol. 1404, Springer, Berlin, 1989, pp. 108–126.

[24] G. Hochschild, *On the cohomology groups of an associative algebra*, Ann. of Math. (2) **46** (1945), 58–67.

[25] Zheng Hua and Bernhard Keller, *Cluster categories and rational curves*, arXiv:1810.00749 [math.AG].

[26] T. V. Kadeishvili, *A_∞-algebra structure in cohomology and the rational homotopy type*, Preprint 37, Forschungsschwerpunkt Geometrie, Universität Heidelberg, Mathematisches Institut, 1988.

[27] ———, *Deriving DG categories*, Ann. Sci. École Norm. Sup. (4) **27** (1994), no. 1, 63–102.

[28] ———, *On the cyclic homology of exact categories*, J. Pure Appl. Algebra **136** (1999), no. 1, 1–56.

[29] Bernhard Keller, *Derived invariance of higher structures on the Hochschild complex*, preprint, 2003, available at the author's home page.
[30] ———, *Derived categories and tilting*, Handbook of Tilting Theory, LMS Lecture Note Series, vol. 332, Cambridge Univ. Press, Cambridge, 2007, pp. 49–104.
[31] ———, *Cluster algebras and derived categories*, Derived categories in algebraic geometry, EMS Ser. Congr. Rep., Eur. Math. Soc., Zürich, 2012, pp. 123–183.
[32] ———, *Singular Hochschild cohomology via the singularity category*, arXiv:1809.05121 [math.RT].
[33] Henning Krause, *The stable derived category of a Noetherian scheme*, Compos. Math. **141** (2005), no. 5, 1128–1162.
[34] Jean-Louis Loday, *Cyclic homology*, second ed., Grundlehren der Mathematischen Wissenschaften [Fundamental Principles of Mathematical Sciences], vol. 301, Springer-Verlag, Berlin, 1998, Appendix E by María O. Ronco, Chapter 13 by the author in collaboration with Teimuraz Pirashvili.
[35] Wendy Lowen and Michel Van den Bergh, *The B_∞-structure on the derived endomorphism algebra of the unit in a monoidal category*, arXiv:1907.06026 [math.KT].
[36] ———, *Hochschild cohomology of abelian categories and ringed spaces*, Adv. Math. **198** (2005), no. 1, 172–221.
[37] John N. Mather and Stephen S. T. Yau, *Classification of isolated hypersurface singularities by their moduli algebras*, Invent. Math. **69** (1982), no. 2, 243–251.
[38] Sandra Michelena and María Inés Platzeck, *Hochschild cohomology of triangular matrix algebras*, J. Algebra **233** (2000), no. 2, 502–525.
[39] Barry Mitchell, *Rings with several objects*, Advances in Math. **8** (1972), 1–161.
[40] D. O. Orlov, *Triangulated categories of singularities and D-branes in Landau-Ginzburg models*, Tr. Mat. Inst. Steklova **246** (2004), no. Algebr. Geom. Metody, Svyazi i Prilozh., 240–262.
[41] Jeremy Rickard, *Morita theory for derived categories*, J. London Math. Soc. **39** (1989), 436–456.
[42] ———, *Derived equivalences as derived functors*, J. London Math. Soc. (2) **43** (1991), no. 1, 37–48.

[43] Gonçalo Tabuada, *Une structure de catégorie de modèles de Quillen sur la catégorie des dg-catégories*, C. R. Math. Acad. Sci. Paris **340** (2005), no. 1, 15–19.

[44] Bertrand Toën, *The homotopy theory of dg-categories and derived Morita theory*, Invent. Math. **167** (2007), no. 3, 615–667.

[45] Michel Van den Bergh and Louis de Thanhoffer de Völcsey, *Explicit models for some stable categories of maximal Cohen–Macaulay modules*, arXiv:1006.2021 [math.RA].

[46] Zhengfang Wang, *Gerstenhaber algebra and Deligne's conjecture on Tate–Hochschild cohomology*, arXiv:1801.07990.

[47] ———, *Singular Hochschild cohomology and Gerstenhaber algebra structure*, arXiv:1508.00190 [math.RT].

Color Hom-Lie bialgebras

Ibrahima Bakayoko* and Sei-Qwon Oh†

**Département de Mathématiques, UJNK/Centre Universitaire de N'Zérékoré, BP : 50, N'Zérékoré, Guinea*
E-mail: ibrahimabakayoko27@gmail.com

†Department of Mathematics, Chungnam National University, 99 Daehak-ro, Yuseong-gu, Daejeon 34134, Korea
E-mail: sqoh@cnu.ac.kr

Here we give a definition of a color Hom-Lie bialgebra, which is a generalization of the Hom-Lie bialgebra, and a technique to get a color Hom-Lie bialgebra from a color Lie bialgebra with an endomorphism.

Keywords: Color Hom-Lie algebra, color Hom-Lie coalgebra, Lie bialgebra, color Hom-Lie bialgebra.

1. Introduction

Lie algebra and Lie bialgebra have played a prominent role in Mathematics and Mathematical Physics. The Lie algebra was generalized to a color Lie algebra by Scheunert [14] and to a color Hom-Lie algebra by Abdaoui, F. Ammar, A. Makhlouf [1, 2], Saadaoui [3], Yuan [18], Silvestrov [13], etc. On the other hand, Lie bialgebra was generalized to a Hom-Lie bialgebra by Yau [16]. Their cohomology and representation theories were studied by many people, for example [1, 3, 6, 9, 12, 15, 17], and examples of color (Hom)-Lie algebras were provided in [6, 10, 18]. A main purpose of this paper is to give a definition of a color Hom-Lie bialgebra, which is a generalization of the Hom-Lie bialgebra.

The paper is organized as follows. In the section 2, we recall graded vector spaces, skew-symmetric bicharacters and some basic properties of color Hom-Lie algebras. Then we introduce a color Hom-Lie coalgebra as a dual concept of the color Hom-Lie algebra, see Lemma 2.5 and Definition 2.3. We also give a technique to obtain a color Hom-Lie coalgebra from a color Lie coalgebra with an endomorphism, see Lemma 2.6.

The section 3 contains the principal results. Here we introduce an adjoint map and a cochain $C^0 \longrightarrow C^1 \longrightarrow C^2$ for color Hom-Lie algebras.

Then we give a definition of a color Hom-Lie bialgebra which is a generalization of the Hom-Lie bialgebra, see Definition 3.2, and give a technique to get a color Hom-Lie bialgebra from a color Lie bialgebra with an endomorphism, see Theorem 3.1. For $i = 1, 2$, let $\mathfrak{g}_i$ be a color Lie bialgebra with an endomorphism α_i and let $\mathfrak{g}_{\alpha_i}$ be the color Hom-Lie bialgebra obtained from $(\mathfrak{g}_i, \alpha_i)$. We prove that if $\mathfrak{g}_{\alpha_1}$ is isomorphic to $\mathfrak{g}_{\alpha_2}$ then $\mathfrak{g}_1$ is also isomorphic to $\mathfrak{g}_2$. In particular, we introduce a coboundary color Hom-Lie bialgebra which is a generalization of the coboundary Lie bialgebra.

Assume throughout the paper that $\mathbf{k}$ denotes a field of characteristic zero, all vector spaces are over $\mathbf{k}$ and id is the identity map on a set. We also denote by

- G an abelian group,
- e the identity element of G,
- ε a skew-symmetric bicharacter on G, the definition of which will be given in section 2.1,
- $\mathcal{H}(V)$ the set of homogeneous elements of the G-graded vector space V.

Moreover we assume that all algebras have unity unless stated otherwise.

2. Preliminaries

2.1. *Color Hom-Lie algebra*

A map $\varepsilon : G \times G \longrightarrow \mathbf{k}^* = \mathbf{k} \backslash \{0\}$ is said to be a *skew-symmetric bicharacter* if ε satisfies the following conditions: For all $a, b, c \in G$,

(i) $\varepsilon(a, b) = \varepsilon(b, a)^{-1}$,
(ii) $\varepsilon(a + b, c) = \varepsilon(a, c)\varepsilon(b, c)$.

Note that

$$\varepsilon(e, a) = 1 = \varepsilon(a, e), \ \varepsilon(a, a) = 1 \text{ or } -1 \tag{1}$$

for all $a \in G$ and that

$$\varepsilon^{-1} : G \times G \longrightarrow \mathbf{k}^*, \quad \varepsilon^{-1}(a, b) = \varepsilon(a, b)^{-1}$$

is also a skew-symmetric bicharacter.

A vector space V is said to be *G-graded* if there exists a family $(V_a)_{a \in G}$ of vector subspaces of V such that

$$V = \oplus_{a \in G} V_a.$$

An element $x \in V$ is said to be *homogeneous of degree* $a \in G$ if $x \in V_a$. In this case, we write

$$a = |x|.$$

Let $V = \oplus_{a\in G} V_a$ and $W = \oplus_{a\in G} W_a$ be two G-graded vector spaces. A linear map $f : V \to W$ is said to be *homogeneous of degree* b if

$$f(V_a) \subseteq W_{a+b}$$

for all elements $a \in G$. If f is homogeneous of degree zero, i.e. $f(V_a) \subseteq W_a$ holds for all $a \in G$, then f is said to be *even*. The vector space $V \otimes W$ is a G-graded vector space with the canonical grading

$$(V \otimes W)_r = \sum_{a+b=r} V_a \otimes V_b, \quad r \in G.$$

Definition 2.1. (See [18].) A *color Hom-Lie algebra* is a quadruple $\mathfrak{g} = (\mathfrak{g}, [\cdot,\cdot], \varepsilon, \alpha)$ consisting of a G-graded vector space $\mathfrak{g} = \bigoplus_{a\in G} \mathfrak{g}_a$, an even bilinear map

$$[\cdot,\cdot] : \mathfrak{g} \times \mathfrak{g} \longrightarrow \mathfrak{g}, \quad (x, y) \mapsto [x, y],$$

a skew-symmetric bicharacter ε on G and an even linear map $\alpha : \mathfrak{g} \longrightarrow \mathfrak{g}$ satisfying the following conditions: For all $x \in \mathfrak{g}_a, y \in \mathfrak{g}_b, z \in \mathfrak{g}_c$,

- (i) $[\mathfrak{g}_a, \mathfrak{g}_b] \subseteq \mathfrak{g}_{a+b}$, ($G$-gradation)
- (ii) $[x, y] = -\varepsilon(a, b)[y, x]$, ($\varepsilon$-skew symmetry)
- (iii) $\varepsilon(c, a)[\alpha(x), [y, z]] + \varepsilon(a, b)[\alpha(y), [z, x]] + \varepsilon(b, c)[\alpha(z), [x, y]] = 0$, ($\varepsilon$-Jacobi identity)
- (iv) $\alpha([x, y]) = [\alpha(x), \alpha(y)]$, (multiplicativity).

If $\alpha = \mathrm{id}$ then $\mathfrak{g}$ is said to be a *color Lie algebra* and simply written by $(\mathfrak{g}, [\cdot,\cdot], \varepsilon)$.

In a color Hom-Lie algebra $\mathfrak{g} = (\mathfrak{g}, [\cdot,\cdot], \varepsilon, \alpha)$, observe, by ε-Jacobi identity, that

$$[[x, y], \alpha(z)] = [\alpha(x), [y, z]] - \varepsilon(|x|, |y|)[\alpha(y), [x, z]] \tag{2}$$

for $x, y, z \in \mathcal{H}(\mathfrak{g})$.

Definition 2.2. (See [18].) Let $(\mathfrak{g}, [\cdot,\cdot], \varepsilon, \alpha)$ and $(\mathfrak{g}', [\cdot,\cdot]', \varepsilon', \alpha')$ be two color Hom-Lie algebras. An even linear map $f : \mathfrak{g} \to \mathfrak{g}'$ is said to be a *morphism of color Hom-Lie algebras* if

$$f\alpha = \alpha' f \quad \text{and} \quad f[\cdot,\cdot] = [\cdot,\cdot]'(f \otimes f).$$

Lemma 2.1. *(See [18].) Let* $(\mathfrak{g}, [\cdot,\cdot], \varepsilon)$ *be a color Lie algebra and let* $\alpha : \mathfrak{g} \to \mathfrak{g}$ *be a morphism of color Lie algebras. Then* $(\mathfrak{g}, [\cdot,\cdot]_\alpha, \varepsilon, \alpha)$ *is a color Hom-Lie algebra, where* $[\cdot,\cdot]_\alpha = \alpha[\cdot,\cdot]$.

Moreover, suppose that $(\mathfrak{g}', [\cdot,\cdot]', \varepsilon')$ *is another color Lie algebra and that* $\alpha' : \mathfrak{g}' \to \mathfrak{g}'$ *is a morphism of color Lie algebras. If* $f : \mathfrak{g} \to \mathfrak{g}'$ *is a color Lie algebra morphism such that* $\alpha' f = f\alpha$ *then*

$$f : (\mathfrak{g}, [\cdot,\cdot]_\alpha, \varepsilon, \alpha) \to (\mathfrak{g}', [\cdot,\cdot]'_{\alpha'}, \varepsilon', \alpha')$$

is a morphism of color Hom-Lie algebras.

An algebra A is said to be *G-graded* if its underlying vector space A is G-graded, i.e. $A = \oplus_{a\in G} A_a$, and

$$A_a A_b \subseteq A_{a+b}$$

for all $a, b \in G$. Hence the homogeneous space A_e is a subalgebra and the unity 1 is a homogeneous element of degree e. Let B be another G-graded algebra. An even linear map $f : A \to B$ is said to be a *morphism of G-graded algebras* if it is also an algebra morphism. The following well-known fact gives us many examples of the color Hom-Lie algebra.

Lemma 2.2. *Let* A *be a G-graded algebra. Define a* **k***-bilinear map* $[\cdot,\cdot] : A \times A \longrightarrow A$ *by*

$$[x, y] = xy - \varepsilon(|x|, |y|)yx$$

for all $x, y \in \mathcal{H}(A)$. *Then* $(A, [\cdot,\cdot], \varepsilon)$ *is a color Lie algebra such that*

$$[x, yz] = \varepsilon(|x|, |y|)y[x, z] + [x, y]z \tag{3}$$

for all $x, y, z \in \mathcal{H}(A)$.

Moreover if $\alpha : A \longrightarrow A$ *is a morphism of G-graded algebras then* α *is an endomorphism of* $(A, [\cdot,\cdot], \varepsilon)$ *and* $(A, [\cdot,\cdot]_\alpha, \varepsilon, \alpha)$ *is a color Hom-Lie algebra.*

Proof. We give a proof for completion. By [18, Proposition 3.13], $(A, [\cdot,\cdot], \varepsilon)$ is a color Lie algebra. Moreover (3) is obtained by the following observation

$$\begin{aligned}[x, yz] &= xyz - \varepsilon(|x|, |y| + |z|)yzx \\ &= (xy - \varepsilon(|x|, |y|)yx)z + \varepsilon(|x|, |y|)y(xz - \varepsilon(|x|, |z|)zx) \\ &= [x, y]z + \varepsilon(|x|, |y|)y[x, z]\end{aligned}$$

for all $x, y, z \in \mathcal{H}(A)$.

If $\alpha : A \longrightarrow A$ is a morphism of G-graded algebras then α is clearly a morphism of color Lie algebras. Hence $(A, [\cdot,\cdot]_\alpha, \varepsilon, \alpha)$ is a color Hom-Lie algebra by Lemma 2.1. □

2.2. *Color Hom-Lie coalgebra*

Here we define a color Hom-Lie coalgebra which is a dual concept of the color Hom-Lie algebra.

Let $V = \oplus_{a\in G} V_a$ be a G-graded vector space. Fix a basis $\mathfrak{B} = \{v_i | i \in I\}$ of V such that all v_i are homogeneous elements and denote by v_i^* the dual element of v_i for all $i \in I$ and let W be the subspace of V^* spanned by all v_i^*. (If V is finite dimensional then $W = V^*$.) Give a degree on v_i^* by

$$|v_i^*| = |v_i|$$

for each $i \in I$. Then W becomes a G-graded vector space such that the homogeneous subspace W_a of W is the subspace spanned by all v_i^* with $|v_i| = a$.

Lemma 2.3. *Let $\alpha : V \longrightarrow V$ be an even linear map such that $\alpha^*(W) \subseteq W$, where α^* is the dual map of α. Then α^* is an even linear map from W into itself.*

Proof. Let $\alpha(v_k) = \sum_i a_{ki} v_i$ for each $k \in I$, where $a_{ki} \in \mathbf{k}$. Note that $|v_k| = |v_i|$ for all i such that $a_{ki} \neq 0$ since α is even. Since $\alpha^*(v_\ell^*)(v_k) = v_\ell^* \alpha(v_k) = a_{k\ell}$, we have

$$\alpha^*(v_\ell^*) = \sum_k a_{k\ell} v_k^*$$

and

$$|v_\ell^*| = |v_\ell| = |v_k| = |v_k^*|$$

for all k such that $a_{k\ell} \neq 0$. Thus α^* is even. □

Lemma 2.4. *Let $\omega : G \times G \longrightarrow \mathbf{k}$ be a map. Then*

$$\omega(|f|, |g|) f(x) g(y) = \omega(|x|, |y|) f(x) g(y)$$

for $f, g \in \mathcal{H}(W)$, $x, y \in \mathcal{H}(V)$.

Proof. Let $|f| = a, |g| = b$. Then $f(V_c) = 0$ for all $c \neq a$, where V_c is the homogeneous subspace of V with degree $c \in G$. Similarly, $g(V_c) = 0$ for all $c \neq b$. Hence if $|x| \neq a$ or $|y| \neq b$ then

$$\omega(|f|, |g|) f(x) g(y) = 0 = \omega(|x|, |y|) f(x) g(y).$$

If $|x| = a$ and $|y| = b$ then $|f| = |x|, |g| = |y|$ and thus $\omega(|f|, |g|) = \omega(|x|, |y|)$ and

$$\omega(|f|, |g|)f(x)g(y) = \omega(|x|, |y|)f(x)g(y).$$

□

Lemma 2.5. *Let $\gamma : V \longrightarrow V \otimes V$ be a linear map and let $[\cdot, \cdot]$ be the dual map of γ on W, that is,*

$$\langle [f, g], x\rangle = \langle f \otimes g, \gamma(x)\rangle \tag{4}$$

where $\langle \cdot, \cdot\rangle : W \otimes V \longrightarrow \mathbf{k}$ is the canonical map defined by

$$\langle f, x\rangle = f(x).$$

Let $\alpha : V \longrightarrow V$ be an even linear map such that $\alpha^*(W) \subseteq W$.

(i) $[\cdot, \cdot]$ is even if and only if γ is even.

(ii) Suppose that both γ and $[\cdot, \cdot]$ are even. Then $[f, g] = -\varepsilon^{-1}(|f|, |g|)[g, f]$ for all $f, g \in \mathcal{H}(W)$ if and only if $\gamma(x) = -\sum \varepsilon(|x_1|, |x_2|)x_2 \otimes x_1$ for all $x \in \mathcal{H}(V)$.

(iii) $\alpha^*([f, g]) = [\alpha^*(f), \alpha^*(g)]$ for all $f, g \in W$ if and only if $\gamma\alpha = (\alpha \otimes \alpha)\gamma$.

(iv) Suppose that both γ and $[\cdot, \cdot]$ are even. Then

$$\varepsilon^{-1}(|h|, |f|)[\alpha^*(f), [g, h]] + \varepsilon^{-1}(|f|, |g|)[\alpha^*(g), [h, f]] + \varepsilon^{-1}(|g|, |h|)[\alpha^*(h), [f, g]] = 0$$

for $f, g, h \in \mathcal{H}(W)$ if and only if

$$\sum \varepsilon(|x_1|, |x_3|)(\alpha(x_1) \otimes x_2 \otimes x_3 + x_2 \otimes x_3 \otimes \alpha(x_1) + x_3 \otimes \alpha(x_1) \otimes x_2) = 0$$

for all $x \in \mathcal{H}(V)$, where $(\alpha \otimes \gamma)\gamma(x) = \sum \alpha(x_1) \otimes x_2 \otimes x_3$.

Proof. (i) Note that, for each $v_i \in \mathfrak{B}$, $\gamma(v_i) = \sum_{j\in J} \alpha_j v'_j \otimes v''_j$ for some $\alpha_j \in \mathbf{k}^*$ and $\{v'_j\}, \{v''_j\} \subseteq \mathfrak{B}$. Suppose that γ is even. For $f, g \in \mathcal{H}(W)$, let $[f, g] = \sum_{k\in K} \beta_k v^*_k$, where $\{v_k\} \subseteq \mathfrak{B}$ and $\beta_k \in \mathbf{k}^*$. Since γ is even, we have that $|v_i| = |v'_j| + |v''_j|$ for each j. Moreover, for each $i \in K$,

$$\sum_j \alpha_j f(v'_j)g(v''_j) = \langle f \otimes g, \gamma(v_i)\rangle = \langle [f, g], v_i\rangle = \sum_k \beta_k v^*_k(v_i) = \beta_i \neq 0.$$

Thus $f(v'_j)g(v''_j) \neq 0$ for some j. It follows that $|f| + |g| = |v'_j| + |v''_j| = |v_i| = |v^*_i|$ for each $i \in K$ and thus $[\cdot, \cdot]$ is even. Conversely, suppose that $[\cdot, \cdot]$ is even. Then

$$\langle [v'^*_j, v''^*_j], v_i\rangle = \langle v'^*_j \otimes v''^*_j, \gamma(v_i)\rangle = \alpha_j \neq 0$$

and thus $|v_j'|+|v_j''| = |v_j'^*|+|v_j''^*| = |[v_j'^*, v_j''^*]| = |v_i|$ for each $j \in J$. Hence γ is even. It follows that γ is even if and only if $[\cdot,\cdot]$ is even.

(ii) Let $x \in \mathcal{H}(V)$ and $\gamma(x) = \sum x_1 \otimes x_2$. Suppose that $\gamma(x) = -\sum \varepsilon(|x_1|,|x_2|)x_2 \otimes x_1$. Then, for any $f, g \in \mathcal{H}(W)$,

$$\begin{aligned}
\langle [f,g], x\rangle &= \langle f \otimes g, \gamma(x)\rangle = \langle f \otimes g, -\sum \varepsilon(|x_1|,|x_2|)x_2 \otimes x_1\rangle \\
&= -\sum \varepsilon(|x_1|,|x_2|)f(x_2)g(x_1) \\
&= -\sum \varepsilon(|g|,|f|)g(x_1)f(x_2) \qquad \text{(by Lemma 2.4)} \\
&= -\varepsilon^{-1}(|f|,|g|)\langle g \otimes f, \gamma(x)\rangle \\
&= -\varepsilon^{-1}(|f|,|g|)\langle [g,f], x\rangle.
\end{aligned}$$

Hence $[f,g] = -\varepsilon^{-1}(|f|,|g|)[g,f]$. Conversely, if $[f,g] = -\varepsilon^{-1}(|f|,|g|)[g,f]$ then

$$\begin{aligned}
\langle f \otimes g, \gamma(x)\rangle &= \langle [f,g], x\rangle = -\varepsilon^{-1}(|f|,|g|)\langle [g,f], x\rangle \\
&= -\varepsilon^{-1}(|f|,|g|)\langle g \otimes f, \gamma(x)\rangle \\
&= -\sum \varepsilon^{-1}(|f|,|g|)g(x_1)f(x_2) \\
&= -\sum \varepsilon^{-1}(|x_2|,|x_1|)f(x_2)g(x_1) \qquad \text{(by Lemma 2.4)} \\
&= \langle f \otimes g, -\sum \varepsilon(|x_1|,|x_2|)x_2 \otimes x_1\rangle.
\end{aligned}$$

Hence $\gamma(x) = -\sum \varepsilon(|x_1|,|x_2|)x_2 \otimes x_1$.

(iii) Observe that

$$\begin{aligned}
\langle \alpha^*([f,g]), x\rangle &= \langle f \otimes g, \gamma\alpha(x)\rangle, \\
\langle [\alpha^*(f), \alpha^*(g)], x\rangle &= \langle \alpha^*(f) \otimes \alpha^*(g), \gamma(x)\rangle \\
&= \langle f \otimes g, (\alpha \otimes \alpha)\gamma(x)\rangle
\end{aligned}$$

for any $x \in V$. Hence the result holds.

(iv) Note that α^* is even by Lemma 2.3. Let $f, g, h \in \mathcal{H}(W)$, $x \in \mathcal{H}(V)$ and let $(\mathrm{id} \otimes \gamma)\gamma(x) = \sum x_1 \otimes x_2 \otimes x_3$. Observe that

$$\begin{aligned}
\langle \varepsilon^{-1}(|h|,|f|)[\alpha^*(f), [g,h]], x\rangle &= \varepsilon(|f|,|h|)\langle f \otimes g \otimes h, (\alpha \otimes \gamma)\gamma(x)\rangle \\
&= \sum \varepsilon(|f|,|h|)f(\alpha(x_1))g(x_2)h(x_3) \\
&= \langle f \otimes g \otimes h, \sum \varepsilon(|x_1|,|x_3|)\alpha(x_1) \otimes x_2 \otimes x_3\rangle,
\end{aligned} \tag{5}$$

$$\begin{aligned}\langle\varepsilon^{-1}(|f|,|g|)[\alpha^*(g),[h,f]],x\rangle &= \varepsilon(|g|,|f|)\langle g\otimes h\otimes f,(\alpha\otimes\gamma)\gamma(x)\rangle\\ &= \sum\varepsilon(|g|,|f|)g(\alpha(x_1))h(x_2)f(x_3)\\ &= \langle f\otimes g\otimes h,\sum\varepsilon(|x_1|,|x_3|)x_3\otimes\alpha(x_1)\otimes x_2\rangle,\end{aligned} \tag{6}$$

$$\begin{aligned}\langle\varepsilon^{-1}(|g|,|h|)[\alpha^*(h),[f,g]],x\rangle &= \varepsilon(|h|,|g|)\langle h\otimes f\otimes g,(\alpha\otimes\gamma)\gamma(x)\rangle\\ &= \sum\varepsilon(|h|,|g|)h(\alpha(x_1))f(x_2)g(x_3)\\ &= \langle f\otimes g\otimes h,\sum\varepsilon(|x_1|,|x_3|)x_2\otimes x_3\otimes\alpha(x_1)\rangle.\end{aligned} \tag{7}$$

by Lemma 2.4. Hence the result follows immediately by (5), (6) and (7). □

The above lemma gives us the definition of color Hom-Lie coalgebra which is a dual concept of color Hom-Lie algebra.

Definition 2.3. A *color Hom-Lie coalgebra* is a quadruple $\mathfrak{g}=(\mathfrak{g},\gamma,\varepsilon,\alpha)$ consisting of a G-graded vector space $\mathfrak{g}=\oplus_{a\in G}\mathfrak{g}_a$, an even linear map $\gamma:\mathfrak{g}\to\mathfrak{g}\otimes\mathfrak{g}$, a skew-symmetric bicharacter ε on G and an even linear map $\alpha:\mathfrak{g}\to\mathfrak{g}$ satisfying the following conditions:

(i) $\gamma(x)=-\sum\varepsilon(|x_1|,|x_2|)x_2\otimes x_1$, ($\varepsilon$-skew-cocommutation)
(ii) $\gamma\alpha=(\alpha\otimes\alpha)\gamma$, (comultiplication)
(iii) $\sum\varepsilon(|x_1|,|x_3|)(\alpha(x_1)\otimes x_2\otimes x_3+x_2\otimes x_3\otimes\alpha(x_1)+x_3\otimes\alpha(x_1)\otimes x_2)=0$, ($\varepsilon$-co-Jacobi identity)

where $x\in\mathcal{H}(\mathfrak{g})$, $\gamma(x)=\sum x_1\otimes x_2$ and $(\alpha\otimes\gamma)\gamma(x)=\sum\alpha(x_1)\otimes x_2\otimes x_3$.

If $\alpha=\mathrm{id}$ then $\mathfrak{g}$ is called a *color Lie coalgebra* and simply written by $(\mathfrak{g},\gamma,\varepsilon)$.

Corollary 2.1. *Let $\mathfrak{g}$ be a finite dimensional G-graded vector space, let $\gamma:\mathfrak{g}\longrightarrow\mathfrak{g}\otimes\mathfrak{g}$ be a linear map and let $\alpha:\mathfrak{g}\longrightarrow\mathfrak{g}$ be an even linear map. Let $[\cdot,\cdot]$ be the dual map of γ, that is,*

$$\langle[f,g],x\rangle=\langle f\otimes g,\gamma(x)\rangle$$

for $f,g\in\mathfrak{g}^*$ and $x\in\mathfrak{g}$. Then $(\mathfrak{g},\gamma,\varepsilon,\alpha)$ is a color Hom-Lie coalgebra if and only if $(\mathfrak{g}^*,[\cdot,\cdot],\varepsilon^{-1},\alpha^*)$ is a color Hom-Lie algebra.

Proof. It follows immediately by Lemma 2.5. □

Definition 2.4. Let $(\mathfrak{g},\gamma,\varepsilon,\alpha)$ and $(\mathfrak{g}',\gamma',\varepsilon',\alpha')$ be two color Hom-Lie coalgebras. An even linear map $f:\mathfrak{g}\to\mathfrak{g}'$ is said to be a *morphism of color*

Hom-Lie coalgebras if

$$f\alpha = \alpha' f \quad \text{and} \quad \gamma' f = (f \otimes f)\gamma.$$

The following lemma is a color Lie coalgebra version of Lemma 2.1.

Lemma 2.6. *Let $(\mathfrak{g}, \gamma, \varepsilon)$ be a color Lie coalgebra and let α be a morphism of color Lie coalgebra $\mathfrak{g}$ into itself. Then $(\mathfrak{g}, \gamma_\alpha = \gamma\alpha, \varepsilon, \alpha)$ is a color Hom-Lie coalgebra.*

Moreover suppose that $(\mathfrak{g}', \gamma', \varepsilon')$ is another color Lie coalgebra and that $\alpha' : \mathfrak{g}' \to \mathfrak{g}'$ is a morphism of color Lie coalgebras. If $f : \mathfrak{g} \to \mathfrak{g}'$ is a morphism of color Lie coalgebras such that $f\alpha = \alpha' f$ then $f : (\mathfrak{g}, \gamma_\alpha, \varepsilon, \alpha) \to (\mathfrak{g}', \gamma'_{\alpha'}, \varepsilon', \alpha')$ is a morphism of color Hom-Lie coalgebras.

Proof. Let us show that $(\mathfrak{g}, \gamma_\alpha = \gamma\alpha, \varepsilon, \alpha)$ is a color Hom-Lie coalgebra. Since α is a morphism of color Lie colagebras,

$$\gamma_\alpha \alpha = \gamma\alpha\alpha = (\alpha \otimes \alpha)\gamma\alpha = (\alpha \otimes \alpha)\gamma_\alpha.$$

Hence α satisfies the comultiplication. Let $x \in \mathcal{H}(\mathfrak{g})$ and set $\gamma(x) = \sum x_1 \otimes x_2$. Since $\gamma_\alpha = \gamma\alpha = (\alpha \otimes \alpha)\gamma$, we have $\gamma_\alpha(x) = \sum \alpha(x_1) \otimes \alpha(x_2)$ and

$$\begin{aligned} \gamma_\alpha(x) &= (\alpha \otimes \alpha)\gamma(x) = (\alpha \otimes \alpha)(-\sum \varepsilon(|x_1|, |x_2|) x_2 \otimes x_1) \\ &= -\sum \varepsilon(|x_1|, |x_2|) \alpha(x_2) \otimes \alpha(x_1). \end{aligned}$$

Hence γ_α satisfies the ε-skew-cocommutation. Observe that

$$\begin{aligned} (\alpha \otimes \gamma_\alpha)\gamma_\alpha(x) &= (\alpha \otimes \gamma_\alpha)(\alpha \otimes \alpha)\gamma(x) \\ &= (\alpha^2 \otimes (\gamma\alpha^2))\gamma(x) \\ &= (\alpha^2 \otimes \alpha^2 \otimes \alpha^2)(\mathrm{id} \otimes \gamma)\gamma(x) \\ &= (\alpha^2 \otimes \alpha^2 \otimes \alpha^2)(\mathrm{id} \otimes \gamma)\gamma(x). \quad (\because \alpha \text{ is even}) \end{aligned}$$

Hence γ_α satisfies the color Hom-co-Jacobi identity since γ satisfies the color co-Jacobi identity. It follows that $(\mathfrak{g}, \gamma_\alpha = \gamma\alpha, \varepsilon, \alpha)$ is a color Hom-Lie coalgebra.

Note that $(\mathfrak{g}', \gamma'_{\alpha'}, \varepsilon', \alpha')$ is a color Hom-Lie coalgebra by the above statement. Since

$$\gamma'_{\alpha'} f = \gamma'\alpha' f = \gamma' f\alpha = (f \otimes f)\gamma\alpha = (f \otimes f)\gamma_\alpha,$$

$f : (\mathfrak{g}, \gamma_\alpha, \varepsilon, \alpha) \to (\mathfrak{g}', \gamma'_{\alpha'}, \varepsilon', \alpha')$ is a morphism of color Hom-Lie coalgebras. □

3. Color Hom-Lie bialgebras

3.1. *Color Hom-Lie bialgebra*

Let $\mathfrak{g} = (\mathfrak{g}, [\cdot,\cdot], \varepsilon, \alpha)$ be a color Hom-Lie algebra. Let us recall the definition of $\mathfrak{g}$-module in [1, §3].

Definition 3.1. Let $\mathfrak{g} = (\mathfrak{g}, [\cdot,\cdot], \varepsilon, \alpha)$ be a color Hom-Lie algebra. A pair (M, β) consisting of a G-graded vector space $M = \oplus_{a\in G} M_a$ and an even linear map $\beta : M \longrightarrow M$ is said to be a $\mathfrak{g}$-*module* if there exists a bilinear map $\mathfrak{g} \times M \longrightarrow M$, $(x, m) \mapsto x \cdot m$ subject to the following conditions:

(i) $\mathfrak{g}_a \cdot M_b \subseteq M_{a+b}$,
(ii) $\beta(x \cdot m) = \alpha(x) \cdot \beta(m)$,
(iii) $[x, y] \cdot \beta(m) = \alpha(x) \cdot (y \cdot m) - \varepsilon(|x|, |y|)\alpha(y) \cdot (x \cdot m)$

for all $a, b \in G, x, y \in \mathcal{H}(\mathfrak{g}), m \in M$.

Note that if $\beta = \mathrm{id}_M$ and $\alpha = \mathrm{id}_{\mathfrak{g}}$ then $\mathfrak{g} = (\mathfrak{g}, [\cdot,\cdot], \varepsilon)$ is a color Lie algebra and M is a graded $\mathfrak{g}$-module.

Lemma 3.1. *Let $\mathfrak{g} = (\mathfrak{g}, [\cdot,\cdot], \varepsilon, \alpha)$ be a color Hom-Lie algebra, let (M, β_M) and (N, β_N) be $\mathfrak{g}$-modules and let k be a nonnegative integer.*

(i) $(\mathfrak{g}, \alpha)$ is a $\mathfrak{g}$-module with action

$$x \cdot m = [\alpha^k(x), m]$$

for $x, m \in \mathfrak{g}$.

(ii) $(M \otimes N, \beta_M \otimes \beta_N)$ is a $\mathfrak{g}$-module with action

$$x \cdot (m \otimes n) = (\alpha^k(x) \cdot m) \otimes \beta_N(n) + \varepsilon(|x|, |m|)\beta_M(m) \otimes (\alpha^k(x) \cdot n) \quad (8)$$

for $x \in \mathcal{H}(\mathfrak{g}), m \in \mathcal{H}(M), n \in \mathcal{H}(N)$.

(iii) $(\mathfrak{g} \otimes \mathfrak{g}, \alpha \otimes \alpha)$ is a $\mathfrak{g}$-module with action

$$x \cdot (y_1 \otimes y_2) = [\alpha(x), y_1] \otimes \alpha(y_2) + \varepsilon(|x|, |y_1|)\alpha(y_1) \otimes [\alpha(x), y_2] \quad (9)$$

for $x, y_1, y_2 \in \mathcal{H}(\mathfrak{g})$.

(iv) For $x \in \mathcal{H}(\mathfrak{g})$, define $ad_x : \mathfrak{g} \otimes \mathfrak{g} \longrightarrow \mathfrak{g} \otimes \mathfrak{g}$ by

$$ad_x(y_1 \otimes y_2) = [x, y_1] \otimes \alpha(y_2) + \varepsilon(|x|, |y_1|)\alpha(y_1) \otimes [x, y_2] \quad (10)$$

for $y_1, y_2 \in \mathcal{H}(\mathfrak{g})$. Then $(\mathfrak{g} \otimes \mathfrak{g}, \alpha \otimes \alpha)$ is a $\mathfrak{g}$-module with action

$$x \cdot (y_1 \otimes y_2) = ad_x(y_1 \otimes y_2).$$

The map ad_x is called the adjoint action on $\mathfrak{g} \otimes \mathfrak{g}$ by x.

Proof. (i) Let $x, y, z \in \mathcal{H}(\mathfrak{g})$. Since α is an even endomorphism, we have that

$$\mathfrak{g}_a \cdot \mathfrak{g}_b = [\alpha^k(\mathfrak{g}_a), \mathfrak{g}_b] \subseteq \mathfrak{g}_{a+b},$$

$$\alpha(x \cdot y) = \alpha([\alpha^k(x), y]) = [\alpha^{k+1}(x), \alpha(y)] = \alpha(x) \cdot \alpha(y)$$

and

$$\begin{aligned}
[x, y] \cdot \alpha(z) &= [\alpha^k([x, y]), \alpha(z)] = [[\alpha^k(x), \alpha^k(y)], \alpha(z)] \\
&= [\alpha^{k+1}(x), [\alpha^k(y), z]] - \varepsilon(|x|, |y|)[\alpha^{k+1}(y), [\alpha^k(x), z]] \\
&= \alpha(x) \cdot y \cdot z - \varepsilon(|x|, |y|)\alpha(y) \cdot x \cdot z
\end{aligned}$$

by (2). Hence $(\mathfrak{g}, \alpha)$ is a $\mathfrak{g}$-module.

(ii) For homogeneous subspaces $\mathfrak{g}_a \subseteq \mathfrak{g}, M_b \subseteq M, N_c \subseteq N$,

$$\begin{aligned}
\mathfrak{g}_a \cdot (M_b \otimes N_c) &\subseteq (\alpha^k(\mathfrak{g}_a) \cdot M_b) \otimes \beta_N(N_c) + \beta_M(M_b) \otimes (\alpha^k(\mathfrak{g}_a) \cdot N_c) \\
&\subseteq M_{a+b} \otimes N_c + M_b \otimes N_{a+c} \\
&\subseteq (M \otimes N)_{a+b+c}.
\end{aligned}$$

For $x, y \in \mathcal{H}(\mathfrak{g}), m \in \mathcal{H}(M), n \in \mathcal{H}(N)$,

$$\begin{aligned}
(\beta_M \otimes \beta_N)(x \cdot (m \otimes n)) &= (\beta_M \otimes \beta_N)((\alpha^k(x) \cdot m) \otimes \beta_N(n) \\
&\qquad + \varepsilon(|x|, |m|)\beta_M(m) \otimes (\alpha^k(x) \cdot n)) \\
&= (\alpha^{k+1}(x) \cdot \beta_M(m)) \otimes \beta_N^2(n) \\
&\qquad + \varepsilon(|x|, |m|)\beta_M^2(m) \otimes (\alpha^{k+1}(x) \cdot \beta_N(n)) \\
&= \alpha(x) \cdot (\beta_M \otimes \beta_N)(m \otimes n)
\end{aligned}$$

and

$$\begin{aligned}
\alpha(x) \cdot y \cdot &(m \otimes n) - \varepsilon(|x|, |y|)\alpha(y) \cdot x \cdot (m \otimes n) \\
&= (\alpha^{k+1}(x) \cdot \alpha^k(y) \cdot m - \varepsilon(|x|, |y|)\alpha^{k+1}(y) \cdot \alpha^k(x) \cdot m) \otimes \beta_N^2(n) \\
&\quad + \varepsilon(|x| + |y|, |m|)\beta_M^2(m) \otimes (\alpha^{k+1}(x) \cdot \alpha^k(y) \cdot n \\
&\qquad - \varepsilon(|x|, |y|)\alpha^{k+1}(y) \cdot \alpha^k(x) \cdot n) \\
&= ([\alpha^k(x), \alpha^k(y)] \cdot \beta_M(m)) \otimes \beta_N^2(n) \\
&\quad + \varepsilon(|x| + |y|, |m|)\beta_M^2(m) \otimes ([\alpha^k(x), \alpha^k(y)] \cdot \beta_N(n)) \\
&= [x, y] \cdot (\beta_M \otimes \beta_N)(m \otimes n).
\end{aligned}$$

Hence $(M \otimes N, \beta_M \otimes \beta_N)$ is a $\mathfrak{g}$-module with action (8).

(iii) It follows by (i) and (ii) by setting $k = 1$ and $(M, \beta_M) = (N, \beta_N) = (\mathfrak{g}, \alpha)$.

(iv) It follows by (i) and (ii) by setting $k = 0$ and $(M, \beta_M) = (N, \beta_N) = (\mathfrak{g}, \alpha)$. □

Let $\mathfrak{g} = (\mathfrak{g}, [\cdot,\cdot], \varepsilon, \alpha)$ be a color Hom-Lie algebra. A homogeneous linear map $f : \mathfrak{g} \otimes \mathfrak{g} \longrightarrow \mathfrak{g} \otimes \mathfrak{g}$ is said to be *ε-alternating* if

$$f(x_1 \otimes x_2) = -\varepsilon(|x_1|, |x_2|) f(x_2 \otimes x_1)$$

for any $x_1, x_2 \in \mathcal{H}(\mathfrak{g})$. Set

$$\begin{aligned} C^0 &= \text{the subspace of } \mathfrak{g} \otimes \mathfrak{g} \text{ spanned by } \{z \in \mathcal{H}(\mathfrak{g} \otimes \mathfrak{g}) | (\alpha \otimes \alpha)(z) = z\}, \\ C^1 &= \text{the subspace of } \mathrm{Hom}_{\mathbf{k}}(\mathfrak{g}, \mathfrak{g} \otimes \mathfrak{g}) \text{ spanned by} \\ &\qquad \{f \in \mathrm{Hom}_{\mathbf{k}}(\mathfrak{g}, \mathfrak{g} \otimes \mathfrak{g}) | f \text{ is homogeneous, } f\alpha = (\alpha \otimes \alpha) f\}, \\ C^2 &= \text{the subspace of } \mathrm{Hom}_{\mathbf{k}}(\mathfrak{g} \otimes \mathfrak{g}, \mathfrak{g} \otimes \mathfrak{g}) \text{ spanned by} \\ &\qquad \{f \in \mathrm{Hom}_{\mathbf{k}}(\mathfrak{g} \otimes \mathfrak{g}, \mathfrak{g} \otimes \mathfrak{g}) | f \text{ is homogeneous, } \varepsilon\text{-alternating,} \\ &\qquad\qquad f(\alpha \otimes \alpha) = (\alpha \otimes \alpha) f\}. \end{aligned}$$

Note that if $\alpha = \mathrm{id}$ then $C^0 = \mathfrak{g} \otimes \mathfrak{g}$.

Define a $\mathbf{k}$-linear map $d^0 : C^0 \longrightarrow C^1$ by

$$d^0(z)(x) = \mathrm{ad}_x(z)$$

for $z \in C^0$ and $x \in \mathfrak{g}$. Suppose that z and x be homogeneous elements with degree r and s respectively. By (10), the degree of $\mathrm{ad}_x(z)$ is $r + s$ and thus $d^0(z)$ is homogeneous of degree r. Let $z = \sum z_1 \otimes z_2$. Then $(\alpha \otimes \alpha)(z) = \sum \alpha(z_1) \otimes \alpha(z_2) = z$ and thus

$$\begin{aligned} &(\alpha \otimes \alpha)\mathrm{ad}_x(z) \\ &= (\alpha \otimes \alpha^2)(\textstyle\sum [x, z_1] \otimes z_2) + \varepsilon(|x|, |z_1|)(\alpha^2 \otimes \alpha)(\textstyle\sum z_1 \otimes [x, z_2]) \\ &= (\mathrm{id} \otimes \alpha)(\textstyle\sum [\alpha(x), z_1] \otimes z_2) + \varepsilon(|x|, |z_1|)(\alpha \otimes \mathrm{id})(\textstyle\sum z_1 \otimes [\alpha(x), z_2]) \\ &= \mathrm{ad}_{\alpha(x)}(z). \end{aligned}$$

It follows that $d^0(z)\alpha = (\alpha \otimes \alpha) d^0(z)$ and hence $d^0(z) \in C^1$.

Note that the module action (9) is represented by the adjoint action. That is,

$$x \cdot (y_1 \otimes y_2) = [\alpha(x), y_1] \otimes \alpha(y_2) + \varepsilon(|x|, |y_1|)\alpha(y_1) \otimes [\alpha(x), y_2] = \mathrm{ad}_{\alpha(x)}(y_1 \otimes y_2)$$

for $x, y_1, y_2 \in \mathcal{H}(\mathfrak{g})$. Let $f \in C^1$ be homogeneous of degree r and $x, y \in \mathcal{H}(\mathfrak{g})$. Define a linear map $d^1 : C^1 \longrightarrow C^2$ by

$$d^1(f)(x \otimes y) = \mathrm{ad}_{\alpha(x)} f(y) - \varepsilon(|x|, |y|)\mathrm{ad}_{\alpha(y)} f(x) - f([x, y]). \tag{11}$$

(The equation (11) should be compared with [1, (3.17)].) Then it is easy to check that $d^1(f)$ is homogeneous of degree r and ε-alternating. Set $f(x) = \sum x_1 \otimes x_2$ and $f(y) = \sum y_1 \otimes y_2$. Then

$$f\alpha(x) = \sum \alpha(x_1) \otimes \alpha(x_2), \quad f\alpha(y) = \sum \alpha(y_1) \otimes \alpha(y_2)$$

since $(\alpha \otimes \alpha)f = f\alpha$. Hence

$$\begin{aligned}
d^1(f)(\alpha \otimes \alpha)(x \otimes y) &= \mathrm{ad}_{\alpha^2(x)} f\alpha(y) - \varepsilon(|x|,|y|)\mathrm{ad}_{\alpha^2(y)} f\alpha(x) - f\alpha([x,y]) \\
&= \sum \mathrm{ad}_{\alpha^2(x)}(\alpha(y_1) \otimes \alpha(y_2)) \\
&\quad - \sum \varepsilon(|x|,|y|)\mathrm{ad}_{\alpha^2(y)}(\alpha(x_1) \otimes \alpha(x_2)) - (\alpha \otimes \alpha)f([x,y]), \\
(\alpha\otimes\alpha)d^1(f)(x \otimes y) &= (\alpha \otimes \alpha)(\mathrm{ad}_{\alpha(x)} f(y) - \varepsilon(|x|,|y|)\mathrm{ad}_{\alpha(y)} f(x) - f([x,y])) \\
&= \sum \mathrm{ad}_{\alpha^2(x)}(\alpha(y_1) \otimes \alpha(y_2)) \\
&\quad - \sum \varepsilon(|x|,|y|)\mathrm{ad}_{\alpha^2(y)}(\alpha(x_1) \otimes \alpha(x_2)) - (\alpha \otimes \alpha)f([x,y])
\end{aligned}$$

and thus we have $d^1(f)(\alpha \otimes \alpha) = (\alpha \otimes \alpha)d^1(f)$. It follows that $d^1(f) \in C^2$.

Lemma 3.2. $d^1 d^0 = 0$. *That is,*

$$\{0\} \longrightarrow C^0 \xrightarrow{d^0} C^1 \xrightarrow{d^1} C^2$$

is a cochain.

Proof. Observe that

$$\begin{aligned}
d^1 d^0(z)(x \otimes y) &= \mathrm{ad}_{\alpha(x)} d^0(z)(y) - \varepsilon(|x|,|y|)\mathrm{ad}_{\alpha(y)} d^0(z)(x) - d^0(z)([x,y]) \\
&= \mathrm{ad}_{\alpha(x)}\mathrm{ad}_y(z) - \varepsilon(|x|,|y|)\mathrm{ad}_{\alpha(y)}\mathrm{ad}_x(z) - \mathrm{ad}_{[x,y]}(z) \\
&= 0
\end{aligned}$$

since $\mathrm{ad}_{[x,y]}(z) = \mathrm{ad}_{[x,y]}(\alpha \otimes \alpha)(z)$ and the adjoint action makes the pair $(\mathfrak{g} \otimes \mathfrak{g}, \alpha \otimes \alpha)$ a $\mathfrak{g}$-module by Lemma 3.1(iv). Hence $d^1 d^0 = 0$. $\square$

An element $f \in C^1$ is said to be a 1-*cocycle* if $d^1(f) = 0$. That is, a linear map $f : \mathfrak{g} \longrightarrow \mathfrak{g} \otimes \mathfrak{g}$ is said to be a 1-cocycle if

$$f([x,y]) = \mathrm{ad}_{\alpha(x)} f(y) - \varepsilon(|x|,|y|)\mathrm{ad}_{\alpha(y)} f(x), \quad (\alpha \otimes \alpha)f = f\alpha.$$

Definition 3.2. A quintuple $(\mathfrak{g}, [\cdot,\cdot], \gamma, \varepsilon, \alpha)$ is said to be a *color Hom-Lie bialgebra* if

(i) $(\mathfrak{g}, [\cdot,\cdot], \varepsilon, \alpha)$ is a color Hom-Lie algebra,
(ii) $(\mathfrak{g}, \gamma, \varepsilon, \alpha)$ is a color Hom-Lie coalgebra,
(iii) γ is a 1-cocycle.

If $\alpha = \text{id}$ then $(\mathfrak{g}, [\cdot,\cdot], \gamma, \varepsilon, \alpha)$ is said to be a *color Lie bialgebra* and simply written by $(\mathfrak{g}, [\cdot,\cdot], \gamma, \varepsilon)$.

Note that $d^0(z)$ is a 1-cocycle for all $z \in C^0$ by Lemma 3.2. A color Hom-Lie bialgebra $(\mathfrak{g}, [\cdot,\cdot], \gamma, \varepsilon, \alpha)$ is said to be *coboundary* if $\gamma = d^0(z)$ for some $z \in C^0$, that is, there exists $z \in \mathfrak{g} \otimes \mathfrak{g}$ such that

$$\gamma(x) = \text{ad}_x(z), \quad (\alpha \otimes \alpha)(z) = z.$$

Definition 3.3. Let $(\mathfrak{g}, [\cdot,\cdot], \gamma, \varepsilon, \alpha)$ and $(\mathfrak{g}', [\cdot,\cdot]', \gamma', \varepsilon', \alpha')$ be two color Hom-Lie bialgebras. An even linear map $f : \mathfrak{g} \to \mathfrak{g}'$ is said to be a *morphism of color Hom-Lie bialgebras* if it is both a morphism of color Hom-Lie algebras and a morphism of color Hom-Lie coalgebras, i.e.

$$f\alpha = \alpha' f, \tag{12}$$

$$f[\cdot,\cdot] = [\cdot,\cdot]'(f \otimes f), \tag{13}$$

$$\gamma' f = (f \otimes f)\gamma. \tag{14}$$

3.2. *Construction theorem*

Here we give a method to get color Hom-Lie bialgebras from color Lie bialgebras with endomorphisms.

Theorem 3.1. *Let $(\mathfrak{g}, [\cdot,\cdot], \gamma, \varepsilon)$ be a color Lie bialgebra and let α be an endomorphism of color Lie bialgebra $\mathfrak{g}$. Then $\mathfrak{g}_\alpha = (\mathfrak{g}, [\cdot,\cdot]_\alpha = \alpha[\cdot,\cdot], \gamma_\alpha = \gamma\alpha, \varepsilon, \alpha)$ is a color Hom-Lie bialgebra.*

Moreover suppose that $(\mathfrak{g}', [\cdot,\cdot]', \gamma', \varepsilon')$ is another color Lie bialgebra and $\alpha' : \mathfrak{g}' \to \mathfrak{g}'$ is a morphism of color Lie bialgebras. If $f : \mathfrak{g} \to \mathfrak{g}'$ is a color Lie bialgebra morphism such that $f\alpha = \alpha' f$ then $f : \mathfrak{g}_\alpha = (\mathfrak{g}, [\cdot,\cdot]_\alpha, \gamma_\alpha, \varepsilon, \alpha) \to \mathfrak{g}'_{\alpha'} = (\mathfrak{g}', [\cdot,\cdot]'_{\alpha'}, \gamma'_{\alpha'}, \varepsilon', \alpha')$ is a morphism of color Hom-Lie bialgebras.

Proof. By Lemma 2.1, $(\mathfrak{g}, [\cdot,\cdot]_\alpha, \varepsilon, \alpha)$ is a color Hom-Lie algebra and, by Lemma 2.6, $(\mathfrak{g}, \gamma_\alpha, \varepsilon, \alpha)$ is a color Hom-Lie coalgebra. Let us show that γ_α is a 1-cocycle. Clearly $(\alpha \otimes \alpha)\gamma_\alpha = \gamma_\alpha \alpha$ since $(\alpha \otimes \alpha)\gamma = \gamma\alpha$. For any $x, y \in \mathcal{H}(\mathfrak{g})$,

$$
\begin{aligned}
&\gamma_\alpha([x,y]_\alpha) - \mathrm{ad}_{\alpha(x)}\gamma_\alpha(y) + \varepsilon(|x|,|y|)\mathrm{ad}_{\alpha(y)}\gamma_\alpha(x)\\
&\quad = \gamma\alpha^2([x,y]) - \mathrm{ad}_{\alpha(x)}\gamma\alpha(y) + \varepsilon(|x|,|y|)\mathrm{ad}_{\alpha(y)}\gamma\alpha(x)\\
&\quad = (\alpha^2\otimes\alpha^2)\gamma([x,y]) - \mathrm{ad}_{\alpha(x)}(\alpha\otimes\alpha)\gamma(y) + \varepsilon(|x|,|y|)\mathrm{ad}_{\alpha(y)}(\alpha\otimes\alpha)\gamma(x)\\
&\quad = (\alpha^2\otimes\alpha^2)\gamma([x,y]) - \sum \mathrm{ad}_{\alpha(x)}(\alpha(y_1)\otimes\alpha(y_2))\\
&\qquad\qquad + \sum \varepsilon(|x|,|y|)\mathrm{ad}_{\alpha(y)}(\alpha(x_1)\otimes\alpha(x_2))\\
&\quad = (\alpha^2\otimes\alpha^2)\gamma([x,y]) - \sum [\alpha(x),\alpha(y_1)]_\alpha\otimes\alpha^2(y_2)\\
&\qquad\qquad - \sum \varepsilon(|x|,|y_1|)\alpha^2(y_1)\otimes[\alpha(x),\alpha(y_2)]_\alpha\\
&\qquad + \sum \varepsilon(|x|,|y|)([\alpha(y),\alpha(x_1)]_\alpha\otimes\alpha^2(x_2)\\
&\qquad\qquad + \sum \varepsilon(|y|,|x_1|)\alpha^2(x_1)\otimes[\alpha(y),\alpha(x_2)]_\alpha)\\
&\quad = (\alpha^2\otimes\alpha^2)(\gamma([x,y]) - \sum [x,y_1]\otimes y_2 - \sum \varepsilon(|x|,|y_1|)y_1\otimes[x,y_2])\\
&\qquad + \varepsilon(|x|,|y|)(\alpha^2\otimes\alpha^2)(\sum [y,x_1]\otimes x_2 + \sum \varepsilon(|y|,|x_1|)x_1\otimes[y,x_2])\\
&\quad = (\alpha^2\otimes\alpha^2)(\gamma([x,y] - x\cdot\gamma(y) + \varepsilon(|x|,|y|)y\cdot\gamma(x))\\
&\quad = 0.
\end{aligned}
$$

Thus the first part of the theorem holds.

For proving the second part, we have to prove (13) and (14). Observe that

$$f[\cdot,\cdot]_\alpha = f\alpha[\cdot,\cdot] = \alpha' f[\cdot,\cdot] = \alpha'[\cdot,\cdot]'(f\otimes f) = [\cdot,\cdot]'_{\alpha'}(f\otimes f), \tag{15}$$

$$\gamma'_{\alpha'} f = \gamma'\alpha' f = \gamma' f\alpha = (f\otimes f)\gamma\alpha = (f\otimes f)\gamma_\alpha. \tag{16}$$

Hence (13) and (14) hold. □

Remark

We recover [16, Theorem 3.5] in the case when $\varepsilon = 1$.

The following corollary gives the converse of Theorem 3.1 in the case when f is an isomorphism.

Corollary 3.1. *Under the conditions of Theorem 3.1, let $f : \mathfrak{g}_\alpha \to \mathfrak{g}'_{\alpha'}$ be an isomorphism of color Hom-Lie bialgebras such that α' and $\alpha'\otimes\alpha'$ are injective. Then the color Lie bialgebras $\mathfrak{g}$ and $\mathfrak{g}'$ are isomorphic.*

Proof. In (15) and (16), we have

$$\alpha' f[\cdot,\cdot] = \alpha'[\cdot,\cdot]'(f\otimes f), \quad (\alpha'\otimes\alpha')\gamma' f = \gamma' f\alpha = (f\otimes f)\gamma\alpha = (\alpha'\otimes\alpha')(f\otimes f)\gamma.$$

Hence $f[\cdot,\cdot] = [\cdot,\cdot]'(f \otimes f)$ and $\gamma' f = (f \otimes f)\gamma$. Since the morphism f is an isomorphism, we have the result. □

Acknowledgments The second author is supported by National Research Foundation of Korea, NRF-2017R1A2B4008388.

References

[1] K. Abdaoui, F. Ammar, and A. Makhlouf, *Constructions and cohomology of color Hom-Lie algebras*, arXiv: 13072612v1 (2013).
[2] F. Ammar and A. Makhlouf, *Hom-Lie superalgebras and Hom-Lie admissible superalgebras*, Journal of Algebra **324** (2010), 1513–1528.
[3] F. Ammar, A. Makhlouf, and N. Saadaoui, *Cohomology of Hom-Lie superalgebras and q-deformed Witt superalgebra*, arXiv:1204.6244v1 (2011).
[4] I. Bakayoko, *Modules over color Hom-Poisson algebras*, J. Gen. Lie Theory Appl. **8:1** (2014).
[5] V. Chari and A. Pressley, *A guide to quantum groups*, Cambridge University Press, Providence, 1994.
[6] X.-W. Chen, S. D. Silvestrov, and F. Van Oystaeyen, *Representations and cocycle twists of color Lie algebras*, Algebr. Represent. Theor. **9** (2006), 633–650.
[7] E.-H. Cho and S.-Q. Oh, *Lie bialgebras arising from Poisson bialgebras*, J. Korean Math. Soc. **47** (2010), 705–718.
[8] J. Dixmier, *Enveloping algebras*, The 1996 printing of the 1977 English translation Graduate Studies in Mathematics, vol. 11, American Mathematical Society, Providence, 1996.
[9] M. A. Farinati and P. Jancsa, *Trivial central extensions of Lie bialgebras*, arXiv :1110.1072v1 (2011).
[10] Y. Félex, S. Halperin, and J.-C. Thomas, *Rational homotopy theory*, Graduate Texts in Mathematics, vol. 205, Springer-Verlag, New York, Inc., 2001.
[11] I. N. Herstein, *Noncommutative rings*, The Carus Mathematical Monographs, vol. 15, The Mathematical Association of America.
[12] J. H. Lu, *Lie bialgebras and Lie algebra cohomology*, non publié (1996).
[13] A. Makhlouf and S. Silvestrov, *Hom-algebra structures*, J. Gen. Lie Theory Appl. **2** (2008), 51–64.
[14] M. Scheunert, *The theory of Lie superalgebras*, Lecture Notes in Mathematics, vol. 716, Springer-Verlag, Berlin Heidelberg New York, 1979.
[15] M. Scheunert and R. B. Zhang, *Cohomology of Lie superalgebras*

and their generalizations, Journal of Mathematical Physics **39** (1998), 5024.

[16] D. Yau, *The classical Hom-Yang-Baxter equation and Hom-Lie bialgebras*, ArXiv:0905.1890v1 (2009).

[17] ———, *Hom-algebras and homology*, J. Lie Theory **19** (2009), 409–421.

[18] L. Yuan, *Hom-Lie color algebras*, Comm. in algebra **40** (2012), 575–592.

Introduction to Auslander-Bridger theory for unbounded projective complexes over commutative Noetherian rings*

Yuji Yoshino

Mathematics Department, Okayama University,
Okayama, 700-8530 Japan
E-mail: yoshino@math.okayama-u.ac.jp

This is an expository presentation that aims at surveying my recent work on unbounded projective complexes over a commutative Noetherian ring. We develop the theory for such complexes in the homotopy category that mimics the stable module theory of Auslander-Bridger. We can apply this theory to totally reflexive modules.

Keywords: Stable module theory, unbounded projective complexes, totally reflexive modules.

1. Introduction

Throughout the present article R always denotes a commutative Noetherian ring. We are interested in chain complexes consisting of finitely generated projective modules over R. The main question considered in this article is to find how we can determine if the acyclicity property of such complexes is preserved under taking R-dual. The answer to this question is almost obvious if the complex is bounded in either end, so unbounded complexes are brought into question in this article.

Historically such problem was first considered by M. Auslander and M. Bridger for complete resolutions of R-modules. In fact they established so-called *'stable module theory'* in 1969. (It was exactly half a century ago!) Their purpose was to study torsion-free modules, reflexive modules and G-dimension. Actually they considered them in relation with syzygies. It was a landmark study on module theory that they have introduced the stable module category for their purpose, which is described as the factor category $\underline{\mathrm{mod}}(R) = \mathrm{mod}(R)/\mathrm{add}(R)$, where $\mathrm{add}(R)$ is the full subcategory

*All the results presented here are taken from the paper [4] that has been put on arXiv. The detailed version of this paper will be submitted for publication elsewhere.

of projective R-modules in the category $\text{mod}(R)$ of finitely generated R-modules.

In my paper [4], I have given a parallel argument to the stable module theory, not for the module category, but for the homotopy category of projective complexes, which I call *stable complex theory* or *Auslander-Bridger theory for projective complexes.* The aim of the present article is to give a concise introduction to the stable complex theory. So I will try to give an outline of the theory in this article, in which I had to omit the detailed proofs for several theorems. The reader should refer to my paper [4] for further details.

In this section, first we settle the notation, and later we shall present the main theorem of this article.

Let $\mathscr{K}(R) = K(\text{proj}(R))$ be the homotopy category of all complexes of finitely generated projective modules over R, so that its objects are complexes of the form;

$$X = \left[\cdots \longrightarrow X^{i-1} \xrightarrow{d_X^{i-1}} X^i \xrightarrow{d_X^i} X^{i+1} \longrightarrow \cdots \right],$$

with each $X^i \in \text{proj}(R)$, which means each X^i is finitely generated projective, while the morphism set is defined as

$$\text{Hom}_{\mathscr{K}(R)}(X, Y) = \{\text{chain maps} : X \to Y\}/\{\text{chain homotopy}\}.$$

Note that an object X of $\mathscr{K}(R)$ is not necessarily bounded in either end, hence it may be an unbounded complex. Note also that each cohomology module $H^i(X)$ $(i \in \mathbb{Z})$ is a finite generated R-module for $X \in \mathscr{K}(R)$.

It is well-known that $\mathscr{K}(R)$ has a structure of triangulated category. In fact, $X[n]$ for $n \in \mathbb{Z}$ is a shifted complex of X in ordinary sense, and a sequence $X \to Y \to Z \to X[1]$ is a triangle in $\mathscr{K}(R)$ if and only if there is a short exact sequence of complexes $0 \to X \to Y \oplus N \to Z \to 0$ where N is a null complex.[a]

The point is that $\mathscr{K}(R)$ has a duality: For $X \in \mathscr{K}(R)$ we define the R-dual complex as

$$X^* = \text{Hom}_R(X, R) = \left[\cdots \longrightarrow X^{i+1*} \xrightarrow{d_X^{i\,*}} X^{i*} \xrightarrow{d_X^{i-1\,*}} X^{i-1*} \longrightarrow \cdots \right].$$

[a] A complex is called a null complex if it is not only acyclic but also split as a long exact sequence.

Then the functor $(-)^* : \mathscr{K}(R)^{op} \to \mathscr{K}(R)$ gives a duality as $X^{**} \cong X$ for $X \in \mathscr{K}(R)$.[b]

The main theorem of [4] is the following. See [4, Theorem 1.1].

Theorem 1.1 (Main Thorem). *Assume that R is a generically Gorenstein ring. Then, $X \in \mathscr{K}(R)$ is acyclic if and only if X^* is acyclic.*

Recall that R is called a generically Gorenstein ring if the total ring of quotients is a Gorenstein ring. For example, integral domains, more generally reduced rings, are generically Gorenstein. The Main Theorem 1.1 is a consequence of the stable complex theory that we are explaining in this article from the next section below.

The main theorem includes the following form of Tachikawa conjecture that has been proved by Avramov, Buchweitz and Şega in [2] in 2005. See [4, Corollary 1.5].

Corollary 1.1 ([2]). *Let R be a Cohen-Macaulay ring with canonical module ω. Furthermore assume that R is a generically Gorenstein ring. If $\mathrm{Ext}^i_R(\omega, R) = 0$ for all $i > 0$, then R is Gorenstein.*

Moreover Theorem 1.1 gives the dependence of total reflexivity conditions as in the following corollary. See also [4, Corollary 1.3]. Recall that a finitely generated R-module M is called totally reflexive if $M \cong M^{**}$ and $\mathrm{Ext}^i_R(M, R) = \mathrm{Ext}^i_R(M^*, R) = 0$ for all $i > 0$. Note also that this is equivalent to saying that there is a complex $X \in \mathscr{K}(R)$ with $H(X) = H(X^*) = 0$ and $X \cong \mathrm{Ker}(d^0_X)$. The complex X is called the complete resolution of M.

Corollary 1.2. *Assume that R is a generically Gorenstein ring. Then the following conditions are equivalent for a finitely generated R-module M.*

(1) M is a totally reflexive R-module.
(2) $\mathrm{Ext}^i_R(M, R) = 0$ for all $i > 0$.
(3) M is an infinite syzygy, i.e. there is an exact sequence of infinite length of the form $0 \longrightarrow M \longrightarrow P_0 \longrightarrow P_1 \longrightarrow P_2 \longrightarrow \cdots$, where each P_i is a finitely generated projective R-module.

A proof of (2) $\Rightarrow$ (1) in Corollary 1.2 easily goes as follows: Take a

[b]Note that since each component X^i is finitely generated projective, one has $X^i \cong (X^i)^{**}$ for all $i \in \mathbb{Z}$, which yields this duality.

projective resolution of M and M^* as

$$\begin{cases} \cdots \xrightarrow{f_2} F_1 \xrightarrow{f_1} F_0 \longrightarrow M \longrightarrow 0, \\ \cdots \xrightarrow{g_2} G_1 \xrightarrow{g_1} G_0 \longrightarrow M^* \longrightarrow 0. \end{cases}$$

Since $0 \longrightarrow M^* \longrightarrow F_0^* \xrightarrow{f_1^*} F_1^* \xrightarrow{f_2^*} F_2^* \longrightarrow \cdots$ is an exact sequence by the assumption (2), just combining the sequences, we have an acyclic complex

$$\cdots \longrightarrow G_1 \longrightarrow G_0 \longrightarrow F_0^* \longrightarrow F_1^* \longrightarrow F_2^* \longrightarrow \cdots,$$

which we denote by X. Then X belongs to $\mathscr{K}(R)$. Thus Theorem 1.1 implies X^* is also acyclic, which shows that M appears in its complete resolution X, hence M is totally reflexive. $\square$

Jorgensen and Şega [3] gave an example of a module M over a non-Gorenstein Artinian ring that disproves the implication (2) $\Rightarrow$ (1) in Corollary. Hence the generic Gorenstein assumption is necessary in the main theorem.

We are able to give a proof of Corollary 1.1 directly from Corollary 1.2 as follows: If the canonical module ω of a Cohen-Macaulay, generically Gorenstein local ring R satisfies that $\mathrm{Ext}_R^i(\omega, R) = 0$ for $i > 0$, then ω is totally reflexive by Corollary 1.2. Then it satisfies the condition (3) in Corollary 1.2, so there is an exact sequence $0 \longrightarrow \omega \longrightarrow P_0 \longrightarrow P_1 \longrightarrow P_2 \longrightarrow \cdots$, where each P_i is a finitely generated projective R-module. However since it is known that ω is a relatively injective object in the category of maximal Cohen-Macaulay modules, the sequence must split, and hence ω is a free R-module. This shows that R is Gorenstein.

Furthermore we note that the following corollary is also easily proved from Theorem 1.1. See [4, Corollary 1.4] and [4, Theorem 12.3] for the proof.

Corollary 1.3. *Under the assumption that R is a generically Gorenstein ring, we have the equality of G-dimension;*

$$\text{G-dim}_R M = \sup\{n \in \mathbb{Z} \mid \mathrm{Ext}_R^n(M, R) \neq 0\ \},$$

for any finitely generated R-module M.

2. *torsion-free and *reflexive complexes

Let $X \in \mathscr{K}(R)$ and $i \in \mathbb{Z}$. By the definition of the morphism sets in $\mathscr{K}(R)$, $\mathrm{Hom}_{\mathscr{K}(R)}(X[i], R)$ equals to the $(-i)$th cohomology of the complex X^*, i.e.

$$H^{-i}(X^*) = \mathrm{Hom}_{\mathscr{K}(R)}(X[i], R).$$

Thus any element $f \in H^{-i}(X^*)$ is represented by the homotopy class of the chain map $f : X[i] \to R$, and hence the cohomology mapping $H^0(f) : H^i(X) \to R$ is uniquely determined. In such a way we have a natural mapping[c]

$$\rho_X^i : H^{-i}(X^*) \longrightarrow H^i(X)^* \quad ; \quad f \mapsto H^0(f)$$

for all $X \in \mathscr{K}(R)$ and $i \in \mathbb{Z}$. The following definition is a starting point of the stable complex theory.

Definition 2.1. Let $X \in \mathscr{K}(R)$ as above.

(1) X is called **torsion-free* if ρ_X^i are injective for all $i \in \mathbb{Z}$.
(2) X is called **reflexive* if ρ_X^i are bijective for all $i \in \mathbb{Z}$.

It is easy to see from the definition that X is *torsion-free iff it satisfies the following condition:

($*$) For any chain map $f : X \to R[-i]$ with $i \in \mathbb{Z}$, if $H(f) = 0$ then $f = 0$ as a morphism in $\mathscr{K}(R)$.

Assume that X satisfies the condition ($*$). Then X is *reflexive iff it satisfies the following condition:

($**$) If $a : H^i(X) \to R$ is an R-module homomorphism with $i \in \mathbb{Z}$, then there is a chain map $f : X \to R[-i]$ such that $H^i(f) = a$.

It is also easy to see that if R is a Gorenstein ring of dimension zero, then every complex $X \in \mathscr{K}(R)$ is *reflexive (and hence *torsion-free), since R is an injective R-module, and taking cohomology H commutes with taking $(-)^*$.

To see if a given complex is *torsion-free or not, the following exact sequence is useful. See [4, Theorem 2.3] for the proof.

Theorem 2.1. *There is an exact sequence of R-modules*

$$0 \to \mathrm{Ext}_R^1(C^{i+1}(X), R) \to H^{-i}(X^*) \xrightarrow{\rho_X^i} H^i(X)^* \to \mathrm{Ext}_R^2(C^{i+1}(X), R)$$

[c]This map is called the Künneth map.

for all $X \in \mathscr{K}(R)$ *and* $i \in \mathbb{Z}$*, where* $C^i(X)$ *denotes* $\mathrm{Coker}(d_X^i)$ *for the complex* X*.*

Thus the following theorem is easily proved by Theorem 2.1. See [4, Theorem 3.4].

Theorem 2.2. *Let* $C(X) = \bigoplus_{i\in\mathbb{Z}} C^i(X)$ *for* $X \in \mathscr{K}(R)$*.*

(1) *Then* X *is *torsion-free iff* $\mathrm{Ext}_R^1(C(X), R) = 0$*.*

(2) *If* $\mathrm{Ext}_R^1(C(X), R) = \mathrm{Ext}_R^2(C(X), R) = 0$*, then* X *is *reflexive.*

Theorem 2.1 is in fact regarded as a generalization of the Auslander-Bridger sequence [1]. To see this, let M be a finitely generated R-module with a finite projective presentation as

$$P_1 \xrightarrow{f} P_0 \longrightarrow M \longrightarrow 0\,.$$

Recall the transpose $\mathrm{Tr}(M)$ is the cokernel of $f^* = \mathrm{Hom}_R(f, R)$. Set $X = \left[0 \longrightarrow P_0^* \xrightarrow{f^*} P_1^* \longrightarrow 0\right] \in \mathscr{K}(R)$. Then $C^1(X) = \mathrm{Tr}(M)$ and $X^* = \left[0 \longrightarrow P_1 \xrightarrow{f} P_0 \longrightarrow 0\right]$. Therefore $H^0(X^*) = M$ and $H^0(X)^* = M^{**}$ in this case, and the mapping ρ_X^0 is the natural mapping $M \to M^{**}$. Thus the theorem shows an exact sequence

$$0 \longrightarrow \mathrm{Ext}^1(\mathrm{Tr}(M), R) \longrightarrow M \longrightarrow M^{**} \longrightarrow \mathrm{Ext}^2(\mathrm{Tr}(M), R).$$

This is the Auslander-Bridger sequence [1, Theorem (2.8)].

We give some examples of *torsion-free and *reflexive complexes below from [4, Example 3.6].

Example 2.1. Let M be a finitely generated R-module and let

$$\cdots \longrightarrow P^{-2} \longrightarrow P^{-1} \longrightarrow P^0 \longrightarrow M \longrightarrow 0$$

be a projective resolution of M with P^{-i} being finitely generated projective for all $i \geq 0$.

(1) Setting

$$X = \left[\cdots \longrightarrow P^{-2} \longrightarrow P^{-1} \longrightarrow P^0 \longrightarrow 0\right] \in \mathscr{K}(R),$$

we can see the following three conditions are equivalent for X:

(*i*) X is *torsion-free,

(*ii*) X is *reflexive,

(*iii*) $\mathrm{Ext}_R^i(M, R) = 0$ for all $i > 0$.

For the proof, notice that $C^{-i-1}(X) = \Omega_R^i M$ for $i \geq 0$, where the RHS denotes the ith syzygy module of M, and $C^j(X) = 0$ for $j \geq 0$. Hence X is *torsion-free iff $\mathrm{Ext}_R^1(\Omega_R^{i-1} M, R) = 0$ for all $i > 0$, by Theorem 2.2. This shows the equivalence $(i) \Leftrightarrow (iii)$. Notice, in this case, that X is *reflexive, since $\mathrm{Ext}_R^2(C^{i+1}(X), R) = \mathrm{Ext}_R^1(C^i(X), R) = 0$ for all $i \in \mathbb{Z}$.

(2) Let $n > 0$ be an integer. Considering the truncation of X, we set

$$X_{(n)} = \left[0 \longrightarrow P^{-n} \longrightarrow \cdots \longrightarrow P^{-1} \longrightarrow P^0 \longrightarrow 0 \right] \in \mathscr{K}(R).$$

Then we can see the equivalences:

(i) $X_{(n)}$ is *torsion-free $\Leftrightarrow$ $\mathrm{Ext}_R^i(M, R) = 0$ for $1 \leq i \leq n$,

(ii) $X_{(n)}$ is *reflexive $\Leftrightarrow$ $\mathrm{Ext}_R^i(M, R) = 0$ for $1 \leq i \leq n+1$.

To prove this, notice that

$$C^{-i}(X_{(n)}) = \begin{cases} \Omega_R^{i-1} M & (0 < i \leq n), \\ P^{-n} & (i = n+1), \\ 0 & (\text{otherwise}). \end{cases}$$

Thus the equivalence (i) follows from Theorem 2.2. Note that $H^0(X_{(n)}) = M$, $H^{-n}(X_{(n)}) = \Omega_R^{n+1} M$ and $H^i(X_{(n)}) = 0$ for $i \neq 0, -n$. Then it is easy to see that the condition $(**)$ after Definition 2.1 is equivalent to that any R-linear map $\Omega_R^{n+1} M \to R$ lifts to a chain map $X_{(n)} \to R[n]$. This holds if and only if $\mathrm{Ext}_R^1(\Omega_R^n M, R) = 0$, which is equivalent to that $\mathrm{Ext}_R^{n+1}(M, R) = 0$. □

Recall that a finitely generated R-module M is called *torsionless* if it satisfies one of the following equivalent conditions:

(1) M is a submodule of a free R-module.
(2) The natural mapping $M \to M^{**}$ is injective.
(3) $\mathrm{Ext}_R^1(\mathrm{Tr} M, R) = 0$.

On the other hand an R-module M is said to be *torsion-free* if the natural mapping $M \to S^{-1}M$ is injective, where S is the multiplicatively closed subset $R \setminus \bigcup_{\mathfrak{p} \in \mathrm{Ass}(R)} \mathfrak{p}$ consisting of all non-zero divisors of R. Note that every torsionless module is torsion-free.

Recall that a Noetherian commutative ring R is said to be *generically Gorenstein* if every localization $R_{\mathfrak{p}}$ for $\mathfrak{p} \in \mathrm{Ass}(R)$ is a Gorenstein local ring, or equivalently the total quotient ring of R is a Gorenstein ring of dimension zero. The following theorem explains why we call a complex *torsion-free. See [4, Theorem 4.2] for the proof.

Theorem 2.3. *Let R be a generically Gorenstein ring. Then a complex $X \in \mathscr{K}(R)$ is *torsion-free iff each cohomology module $H^i(X^*)$ is a torsion-free R-module for $i \in \mathbb{Z}$.*

Note that a finitely generated module M over a commutative Noetherian ring R is said to be reflexive if the natural mapping $M \to M^{**}$ is an isomorphism. The following theorem holds, and that gives of course a reason for the name '*reflexive'. See [4, Theorem 4.5].

Theorem 2.4. *Suppose that the ring R satisfies the condition that any localization $R_\mathfrak{p}$ is Gorenstein for $\mathfrak{p} \in \mathrm{Spec}R$ if $\mathrm{depth}R_\mathfrak{p} \le 1$. (For example, R is a normal domain.) Then a complex $X \in \mathscr{K}(R)$ is *reflexive iff each cohomology module $H^i(X^*)$ is a reflexive R-module for $i \in \mathbb{Z}$.*

3. The subcategory Add(R)

Note that the category $\mathscr{K}(R)$ admits finite direct sums, and moreover some kind of infinite direct sums can be possibly taken inside $\mathscr{K}(R)$. The direct sum $\coprod_{i\in\mathbb{Z}} R[i]$ is one of such typical examples of infinite direct sums, actually it is a complex of the form $\left[\cdots \xrightarrow{0} R \xrightarrow{0} R \xrightarrow{0} R \xrightarrow{0} \cdots\right]$ that belongs to $\mathscr{K}(R)$.

Definition 3.1. We define Add(R) as the smallest additive subcategory of $\mathscr{K}(R)$ containing R that is closed under the shift functor and admits possibly infinite coproducts. Equivalently Add(R) is the intersection of all the full subcategories $\mathcal{U}$ satisfying the following conditions:

(i) $\mathcal{U}$ is closed under isomorphism and $R \in \mathcal{U}$.
(ii) If $Y \in \mathcal{U}$ then $Y[i] \in \mathcal{U}$ for all $i \in \mathbb{Z}$.
(iii) If Z is a direct summand of $Y \in \mathcal{U}$ then $Z \in \mathcal{U}$.
(iv) Let $\{Y_j \mid j \in J\}$ be a set of objects in $\mathcal{U}$ and assume that the coproduct $\coprod_{j\in J} Y_j$ in $\mathscr{K}(R)$ exists. Then $\coprod_{j\in J} Y_j \in \mathcal{U}$.

The following theorem gives sufficiently satisfactory conditions for a complex belonging to Add(R). See [4, Theorem 5.8].

Theorem 3.1. *The following conditions are equivalent for $X \in \mathscr{K}(R)$.*

(1) X belongs to Add(R).

(2) X is a split complex, i.e. there is a graded R-module homomorphisms $s : X \to X[-1]$ satisfying $d_X s d_X = d_X$.

(3) The natural mapping

$$H : \mathrm{Hom}_{\mathscr{K}(R)}(X, Y) \longrightarrow \mathrm{Hom}_{\mathrm{graded}R-\mathrm{mod}}(H(X), H(Y))$$

which sends f to $H(f)$ is bijective for all $Y \in \mathscr{K}(R)$.

As a result of Theorem 3.1, every complex F in $\mathrm{Add}(R)$ is decomposed as $F = \coprod_{j \in \mathbb{Z}} H^j(F)[-j]$ where $H^j(F) \in \mathrm{proj}(R)$ for all $j \in \mathbb{Z}$. Note that $F^* = \coprod_{j \in \mathbb{Z}} H^j(F)^*[j]$ in this case. Moreover it is easy to see from the condition (3) in Theorem 3.1 that every complex in $\mathrm{Add}(R)$ is *reflexive.

On the other hand the following theorem holds. See [4, Proposition 5.9].

Proposition 3.1. *Let $X, F \in \mathscr{K}(R)$. Assume that F belongs to* $\mathrm{Add}(R)$ *and that X is *torsion-free (resp. *reflexive). Then the mapping*

$$H \ : \ \mathrm{Hom}_{\mathscr{K}(R)}(X, F) \longrightarrow \mathrm{Hom}_{\mathrm{graded}R-\mathrm{mod}}(H(X), H(F)) \ ; \ f \mapsto H(f)$$

is injective (resp. bijective).

To prove this proposition, note that if $F = R$ then the proposition is obviously true from the definition of *torsion-free and *reflexive complexes. Let $\mathcal{U}$ be the object class of F for which Proposition 3.1 holds true. Then it is not difficult to see that $\mathcal{U}$ satisfies the conditions (i)–(iv) in Definition 3.1. Hence the Proposition 3.1 is true for all the objects F of $\mathrm{Add}(R)$.

The following theorem is one of the crucial results on *torsion-free complexes, on which the proof of Theorem 1.1 will deeply rely. The proof of the theorem will be reduced to the proposition above. See [4, Theorem 5.10].

Theorem 3.2. *Assume that $X \in \mathscr{K}(R)$ is *torsion-free and that $F \in$* $\mathrm{Add}(R)$. *Let $f \in \mathrm{Hom}_{\mathscr{K}(R)}(X, F)$. Setting $S = R \backslash \bigcup_{\mathfrak{p} \in \mathrm{Ass}(R)} \mathfrak{p}$, if $S^{-1}f = 0$ as a morphism $S^{-1}X \to S^{-1}F$ in $\mathscr{K}(S^{-1}R)$, then we have that $f = 0$ as a morphism in $\mathscr{K}(R)$.*

Proof. If $S^{-1}f = 0$ then $H(S^{-1}f) = 0$ as an $S^{-1}R$-module homomorphism $H(S^{-1}X) \to H(S^{-1}F)$. Thus we see that $S^{-1}H(f) = 0$ as a mapping $S^{-1}H(X) \to S^{-1}H(F)$. Since $H(F)$ is a projective R-module, any elements of S act on $H(F)$ as non zero divisors. It thus follows that $H(f) = 0$ as a mapping $H(X) \to H(F)$. Then from Proposition 3.1 we have $f = 0$. □

4. The stable category of $\mathscr{K}(R)$

It is almost easy to verify the following theorem holds. See [4, Lemmas 7.2, 7.5].

Theorem 4.1. Add(R) *is a functorially finite subcategory of* $\mathscr{K}(R)$.

In fact, let $X \in \mathscr{K}(R)$ *and* $F \in \text{Add}(R)$. *Then a morphism* $p : F \to X$ *in* $\mathscr{K}(R)$ *is a right* Add(R)*-approximation iff the cohomology mapping* $H(p)$ *is surjective. In particular, there always exists a right* Add(R)*-approximation for any* $X \in \mathscr{K}(R)$.

Furthermore, if we take a right Add(R)*-approximation* $p' : F' \to X^*$ *of the* R*-dual* X^*, *then* $(p')^* : X \to (F')^*$ *is a left* Add(R)*-approximation of* X.

The main objective of the stable complex theory is to consider the nature of complexes in $\mathscr{K}(R)$ up to Add(R)-summands. For this reason we introduce the notion of the stable category of $\mathscr{K}(R)$.

Definition 4.1. We denote by $\underline{\mathscr{K}(R)}$ the factor category $\mathscr{K}(R)$ modulo the subcategory Add(R);

$$\underline{\mathscr{K}(R)} = \mathscr{K}(R)/\text{Add}(R).$$

We call $\underline{\mathscr{K}(R)}$ the *stable category* of $\mathscr{K}(R)$.

The objects of $\underline{\mathscr{K}(R)}$ are the same as $\mathscr{K}(R)$, while the morphism set is given by

$$\text{Hom}_{\underline{\mathscr{K}(R)}}(X, Y) = \text{Hom}_{\mathscr{K}(R)}(X, Y)/\text{Add}(R)(X, Y),$$

for $X, Y \in \underline{\mathscr{K}(R)}$, where Add$(R)(X, Y)$ is the R-submodule of $\text{Hom}_{\mathscr{K}(R)}(X, Y)$ consisting of all morphisms factoring through objects of Add(R).

We can define the syzygy functor and the cosyzygy functor as follows: For any $X \in \mathscr{K}(R)$ we take a right Add(R)-approximation $p_X : F_X \to X$ and we define $\Omega(X)$ by the triangle in $\mathscr{K}(R)$:

$$\Omega(X) \longrightarrow F_X \xrightarrow{p_X} X \longrightarrow \Omega(X)[1]$$

Similarly, taking a left Add(R)-approximation $q_X : X \to G_X$, we define $\Sigma(X)$ by the triangle:

$$\Sigma(X)[-1] \longrightarrow X \xrightarrow{q_X} G_X \longrightarrow \Sigma(X)$$

$\Omega(X)$ and $\Sigma(X)$ are not necessarily unique in the category $\mathscr{K}(R)$, however in the stable category $\underline{\mathscr{K}(R)}$ they are determined uniquely up to isomorphisms. Actually Ω and Σ define the functors $\underline{\mathscr{K}(R)} \to \underline{\mathscr{K}(R)}$. See [4, Definitions 7.3, 7.8]. By Theorem 4.1 it follows that $\Sigma(X) = \Omega(X^*)^*$. Once we define the functors Ω and Σ, it is easy to see that they are adjoint. See [4, Theorem 7.11].

Theorem 4.2. *As functors from $\underline{\mathscr{K}(R)}$ to itself, (Σ, Ω) is an adjoint pair, i.e. there are functorial isomorphisms*

$$\mathrm{Hom}_{\underline{\mathscr{K}(R)}}(\Sigma X, Y) \cong \mathrm{Hom}_{\underline{\mathscr{K}(R)}}(X, \Omega Y),$$

for all $X, Y \in \underline{\mathscr{K}(R)}$.

The proof of the theorem is clear from the following commutative diagram with the rows being triangles:

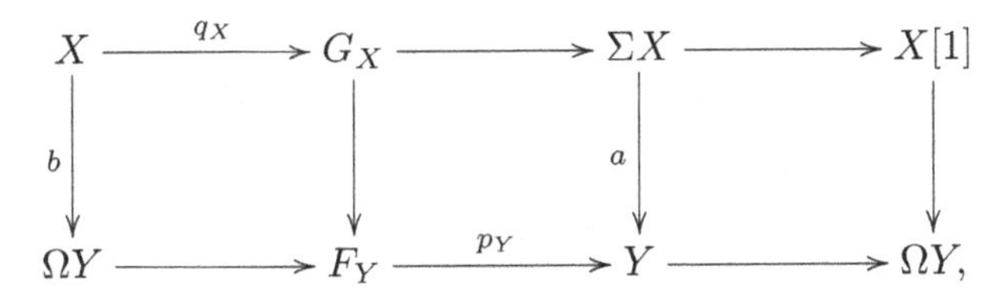

where p_Y (resp. q_X) is a right (resp. left) $\mathrm{Add}(R)$-approximation. Then the correspondence $a \leftrightarrow b$ gives the desired isomorphism in the theorem.

The following example is taken from [4, Example 7.12].

Example 4.1. Let M be a finitely generated R-module and let

$$\cdots \longrightarrow P_n \xrightarrow{u_n} P_{n-1} \longrightarrow \cdots \longrightarrow P_1 \xrightarrow{u_1} P_0 \xrightarrow{u_0} M \longrightarrow 0$$

be a projective resolution of M with $P_i \in \mathrm{proj}(R)$ for $i \in \mathbb{N}$. Now, set the complex X to be $\left[\ 0 \longrightarrow P_1 \xrightarrow{u_1} P_0 \longrightarrow 0\ \right]$. Then one can easily see that

$$\Omega X = \left[\ 0 \longrightarrow P_2 \xrightarrow{u_2} P_1 \longrightarrow 0\ \right].$$

In fact, the following chain map p_0;

$$\begin{array}{ccccccc} 0 & \longrightarrow & P_2 & \xrightarrow{0} & P_0 & \longrightarrow & 0 \\ & & \Big\downarrow{\scriptstyle u_2} & & \Big\downarrow{\scriptstyle 1} & & \\ 0 & \longrightarrow & P_1 & \xrightarrow{u_1} & P_0 & \longrightarrow & 0 \end{array}$$

gives a surjection of cohomology modules. Hence, if we set F_0 to be the complex of the first row, i.e. $F_0 = P_0 \oplus P_2[1]$, we have a right $\mathrm{Add}(R)$-approximation $p_0 : F_0 \to X$ by Theorem 4.1. And $\Omega X[1]$ is, by definition, the mapping cone of p_0.

More generally we have

$$\Omega^n X (= \Omega(\Omega^{n-1}X)) = \left[0 \longrightarrow P_{n+1} \xrightarrow{u_{n+1}} P_n \longrightarrow 0 \right],$$

for $n > 0$.

On the other hand let

$$\cdots \longrightarrow Q_n \xrightarrow{v_n} Q_{n-1} \longrightarrow \cdots \longrightarrow Q_1 \xrightarrow{v_1} Q_0 \xrightarrow{v_0} M^* \longrightarrow 0$$

be a projective resolution of M^*. Then one can see that

$$\Sigma X = \left[0 \longrightarrow P_0 \xrightarrow{w} Q_0^* \longrightarrow 0 \right],$$

where w is the composition $P_0 \xrightarrow{u_0} M \xrightarrow{\text{natural}} M^{**} \xrightarrow{v_0^*} Q_0^*$.

In fact, the following chain map q_0;

$$\begin{array}{ccccccc} 0 & \longrightarrow & P_1 & \xrightarrow{u_1} & P_0 & \longrightarrow & 0 \\ & & \downarrow{\scriptstyle 1} & & \downarrow{\scriptstyle w} & & \\ 0 & \longrightarrow & P_1 & \xrightarrow{0} & Q_0^* & \longrightarrow & 0 \end{array}$$

is a left $\mathrm{Add}(R)$-approximation, since q_0^* gives a surjection of cohomology modules.

For $n > 1$, it can be easily seen that

$$\Sigma^n X = \left[0 \longrightarrow Q_{n-2}^* \xrightarrow{v_{n-1}^*} Q_{n-1}^* \longrightarrow 0 \right].$$

The following theorems give characterizations of *torsion-free complexes and *reflexive complexes in terms of syzygy and cosyzygy functors. In fact they are part of main results of the paper [4]. See [4, Theorem 7.14, Theorem 7.17, Corollary 7.18] for each of the following results.

Theorem 4.3. *The following conditions are equivalent for $X \in \mathscr{K}(R)$.*

*(1) X is *torsion-free.*

(2) There are complexes $Y \in \mathscr{K}(R)$ and $F \in \mathrm{Add}(R)$ such that X is a direct summand of $\Sigma Y \oplus F$ in $\mathscr{K}(R)$.

(3) There is an isomorphism $X \cong \Sigma\Omega X$ in $\underline{\mathscr{K}(R)}$.

Theorem 4.4. *Under the assumption that R is generically Gorenstein, $\Sigma^2 X = \Sigma(\Sigma X)$ is always *reflexive for any $X \in \mathscr{K}(R)$.*

Theorem 4.5. *Assume that R is a generically Gorenstein ring. Then the following two conditions for $X \in \mathscr{K}(R)$ are equivalent:*
*(1) X is *reflexive.*
(2) $\Sigma^2\Omega^2 X \cong X$ in $\underline{\mathscr{K}(R)}$.

5. Add(R)-resolution

We say that a finite sequence of morphisms in $\mathscr{K}(R)$;

$$0 \longrightarrow X_n \xrightarrow{q_n} F_{n-1} \xrightarrow{f_{n-1}} F_{n-2} \longrightarrow \cdots \longrightarrow F_1 \xrightarrow{f_1} F_0 \xrightarrow{p_0} X_0 \longrightarrow 0$$

is $\mathscr{K}(R)$-*exact* if there are triangles $X_{i+1} \xrightarrow{q_{i+1}} F_i \xrightarrow{p_i} X_i \xrightarrow{\omega_i} X_{i+1}[1]$ and equalities $f_i = q_i p_i$ for $0 \le i \le n-1$.

The $\mathscr{K}(R)$-exact sequence is described in a single diagram as

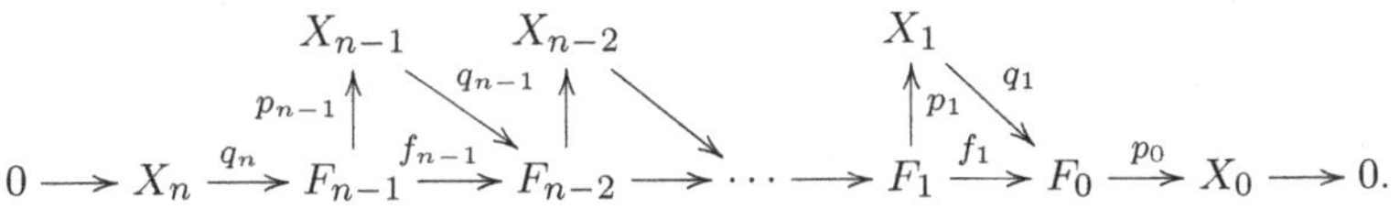

$$\begin{array}{ccccccccccccc} & & X_{n-1} & & X_{n-2} & & & & X_1 & & & & \\ & & {\scriptstyle p_{n-1}}\uparrow & \searrow^{q_{n-1}} & \uparrow & \searrow & & & \uparrow{\scriptstyle p_1} & \searrow^{q_1} & & & \\ 0 \longrightarrow & X_n & \xrightarrow{q_n} F_{n-1} & \xrightarrow{f_{n-1}} & F_{n-2} & \longrightarrow & \cdots & \longrightarrow & F_1 & \xrightarrow{f_1} & F_0 & \xrightarrow{p_0} X_0 & \longrightarrow 0. \end{array} \tag{1}$$

Call the $\mathscr{K}(R)$-exact sequence a *partial* Add(R)-*resolution of* X_0 if $F_i \in$ Add(R) for all $0 \le i < n$. We say the sequence (1) is an Add(R)-*resolution of* X_0 *of length* $n-1$ if it is a partial Add(R)-resolution with $X_n = 0$.

When we are given a partial Add(R)-resolution (1) where $d_{F_i} = 0$ for $0 \le i \le n-1$. Then we can construct a complex $\widetilde{F}$ with a triangle

$$X_n[n-1] \xrightarrow{\psi_n} \widetilde{F} \xrightarrow{\varphi_n} X_0 \xrightarrow{\widetilde{\omega_n}} X_n[n],$$

such that

(*i*) $\widetilde{F} = F_{n-1}[n-1] \oplus F_{n-2}[n-2] \oplus \cdots \oplus F_1[1] \oplus F_0$, as graded R-module.

(*ii*) Let $in_i : F_i[i] \to \widetilde{F}$, $pr_i : \widetilde{F} \to F_i[i]$ be natural maps. Then $pr_j \cdot d_{\widetilde{F}} \cdot in_i = 0$ for $0 \le i \le j \le n-1$.

As an object of $\mathscr{K}(R)$, such a complex $\widetilde{F}$ is unique up to isomomorphism. We call $\widetilde{F}$ the *contraction* of the partial Add(R)-resolution (1).

The construction of contraction is somewhat technical but not difficult, and it just requires to use the mapping cone construction successively. See [4, Theorem and Definition 8.2] for the detail.

Now recall from Theorem 4.2 that there is a bijection

$$\mathrm{Hom}_{\underline{\mathscr{K}(R)}}(\Sigma^n X, Y) \cong \mathrm{Hom}_{\underline{\mathscr{K}(R)}}(X, \Omega^n Y),$$

for all $X, Y \in \mathscr{K}(R)$ and $n > 0$, which of course means that (Σ^n, Ω^n) is an adjoint pair. Then we can take a counit morphism associated with the adjoint pair;

$$\pi_X^n : \Sigma^n \Omega^n X \to X.$$

Recall from Theorem 4.3 that π_X^1 is isomorphic iff X is *torsion-free. And under the assumption that R is generically Gorenstein, π_X^2 is isomorphic iff X is *reflexive, by Theorem 4.5.

Now we define $\Delta^n(X) \in \mathscr{K}(R)$ by the following triangle in $\mathscr{K}(R)$;

$$\Delta^n(X) \longrightarrow \Sigma^n \Omega^n X \xrightarrow{\pi_X^n} X \longrightarrow \Delta^n(X)[1].$$

Adding the condition that $H(\pi_X^n)$ is surjective, we can prove that $\Delta^n(X)$ is unique up to isomorphism as an object $\underline{\mathscr{K}(R)}$.

Remark 5.1. As we have observed above, we have an equivalence;

$$X \text{ is *torsion-free} \quad \Leftrightarrow \quad \Delta^1(X) = 0 \text{ as an object in } \underline{\mathscr{K}(R)}.$$

Moreover, if R is a generically Gorenstein ring, then

$$X \text{ is *reflexive} \quad \Leftrightarrow \quad \Delta^2(X) = 0 \text{ as an object in } \underline{\mathscr{K}(R)}.$$

One of the main result of the stable complex theory is the following theorem. See [4, Theorem 10.2].

Theorem 5.1. *$\Delta^n(X)$ has a finite* Add(R)*-resolution of length at most $n-1$.*

Note that if R is generically Gorenstein, then the Add(R)-resolution of $\Delta^n(X)$ in the theorem is generically split. (I.e. All the triangles involved in the Add(R)-resolution are split after localizing at $\mathfrak{p}$ for all $\mathfrak{p} \in \mathrm{Ass}(R)$.)

The proof of the theorem is done by the following commutative diagram:

$$\begin{array}{ccccccccccc}
0 \to & \Omega^n X & \xrightarrow{q_n} & F_{n-1} & \xrightarrow{f_{n-1}} \cdots \xrightarrow{f_1} & F_0 & \xrightarrow{p_0} & X & \to 0, \\
& \| & & \uparrow{\scriptstyle a_{n-1}} & & \uparrow{\scriptstyle a_0} & & \uparrow{\scriptstyle \pi_X^n} & \\
0 \to & \Omega^n X & \xrightarrow{q^0} & G_{n-1} & \xrightarrow{g^{n-1}} \cdots \xrightarrow{g^1} & G_0 & \xrightarrow{p^0} & \Sigma^n \Omega^n X & \to 0 \\
& & & \uparrow{\scriptstyle b^{n-1}} & & \uparrow{\scriptstyle b^0} & & \uparrow & \\
& 0 & \longrightarrow & L_{n-1} & \xrightarrow{\ell^{n-1}} \cdots \xrightarrow{\ell^1} & L_0 & \longrightarrow & \Delta^n(X) & \to 0
\end{array}$$

where the first row is a partial Add(R)-resolution constructed by taking right Add(R)-approximations successively, and the second by taking left Add(R)-approximations successively. The sequence in the third row is obtained by successive use of octahedral axiom. One can take each L_i as an object of Add(R) and one can show that the third row is again a $\mathscr{K}(R)$-exact sequence. This is what we have claimed in Theorem 5.1. See [4, §10] for the detail.

6. Complexes with $H(X^*) = 0$

Now in this section we shall give an outline of the proof of Main Theorem 1.1 after [4, §11].

One of the major consequences of the stable complex theory is the following theorem that is [4, Theorem 11.1].

Theorem 6.1. *Let R be a generically Gorenstein ring, and let $X \in \mathscr{K}(R)$. If $H(X^*) = 0$, then $\Omega^r X$ is *torsion-free for $r \geq 0$.*

Since the natural mapping $H(X^*) = 0 \to H(X)^*$ is injective, X itself is clearly *torsion-free. By induction on r and the argument using the contractions, one can prove the theorem for all $r \geq 0$. But this is somewhat technically difficult part of the proof. See [4, Theorem 11.1] for the detail.

Once we have proved Theorem 6.1 above, we can strengthen it as follows: (See [4, Theorem 11.4].)

Theorem 6.2. *Let R be a generically Gorenstein ring, and let $X \in \mathscr{K}(R)$. If $H(X^*) = 0$, then $\Omega^r X$ is *reflexive for $r \geq 0$.*

Proof. There is a triangle of the following form:

$$\Omega^{r+1}X \xrightarrow{\quad q \quad} F_r \xrightarrow{\quad p \quad} \Omega^r X \xrightarrow{\quad \omega \quad} \Omega^{r+1}X[1],$$

where p is a right Add(R)-approximation, hence $H(\omega) = 0$ by Theorem 4.1. Thus the sequence of graded R-modules;

$$0 \longrightarrow H(\Omega^{r+1}X) \xrightarrow{H(q)} H(F_r) \xrightarrow{H(p)} H(\Omega^r X) \longrightarrow 0$$

is exact. Given a graded R-module homomorphism $\alpha : H(\Omega^r X) \to R[i]$ for some $i \in \mathbb{Z}$, we find a morphism $b : F_r \to R[i]$ in $\mathscr{K}(R)$ with $H(b) = \alpha H(p)$, since F_r is *reflexive. Then we have $H(bq) = H(b)H(q) = \alpha H(pq) = 0$. As we have shown in Theorem 6.1, $\Omega^{r+1}X$ is *torsion-free, we have that $bq = 0$. Then by a property of exact triangles there is an $a : \Omega^r X \to R[i]$ that satisfies $b = ap$. Since $H(p)$ is a surjection, it thus follows that $H(a) = \alpha$. Then one can apply the observation (**) after Definition 2.1 to conclude that $\Omega^r X$ is *reflexive. □

The following lemma is given in [4, Proposition 11.5].

Lemma 6.1. *Let $Y \in \mathscr{K}(R)$, assume the following two conditions:*

*(1) Y is *torsion-free.*
*(2) ΩY is *reflexive.*

Then we have $\mathrm{Ext}^1_R(H(Y), R) = 0$.

Proof. Recall that we have a triangle in $\mathscr{K}(R)$;

$$\Omega Y \longrightarrow F \xrightarrow{\ p\ } Y \longrightarrow \Omega Y[1],$$

where p a right $\mathrm{Add}(R)$-approximation. It follows from this sequence we have the following commutative diagram of cohomology modules where the rows are exact sequences:

$$\begin{array}{ccccccccc}
H((\Omega Y)^*[-1]) & \xrightarrow{\alpha} & H(Y^*) & \xrightarrow{H(p^*)} & H(F^*) & \xrightarrow{\beta} & H((\Omega Y)^*) & \xrightarrow{\alpha[1]} & H(Y^*[1]) \\
 & & \downarrow{\scriptstyle \rho_Y} & & \downarrow{\scriptstyle =} & & \downarrow{\scriptstyle \rho_{\Omega Y}} & & \\
0 & \longrightarrow & H(Y)^* & \xrightarrow{H(p)^*} & H(F)^* & \longrightarrow & H(\Omega Y)^* & \longrightarrow & \mathrm{Ext}^1_R(H(Y), R) \to 0.
\end{array}$$

Since ρ_Y is injective by the assumption, $H(p^*)$ is also injective, hence α (and $\alpha[1]$) is the zero mapping, and thus β is surjective. Since $\rho_{\Omega Y}$ is bijective, we can conclude from the diagram that $\mathrm{Ext}^1_R(H(Y), R) = 0$. □

We finally have the following proposition, which is obtained just as a combination the previous Theorem 6.2 and Lemma 6.1. See [4, Proposition 11.6].

Proposition 6.1. *Let R be a generically Gorenstein ring, and assume that $H(X^*) = 0$ for $X \in \mathscr{K}(R)$. Then we have*

$$\mathrm{Ext}^r_R(H(X), R) = 0,$$

for all $r > 0$.

Now we are going to show an outline of the proof of the main theorem. What we want to prove is the following statement, which is a reproduction of Theorem 1.1:

Let R be a generically Gorenstein ring, and let $X \in \mathscr{K}(R)$. Then, $H(X) = 0$ if and only if $H(X^*) = 0$.

The proof consists of several steps.

(1) It is enough to prove the implication $H(X^*) = 0 \Rightarrow H(X) = 0$, since the other implication follows from this and the duality $X^{**} \cong X$.

(2) We may assume that $(R, \mathfrak{m})$ is a local ring that is generically Gorenstein. This is because of ordinary technique in commutative algebra. One has only to notice that a complex $X \in \mathscr{K}(R)$ is acyclic iff $X_\mathfrak{p} \in \mathscr{K}(R_\mathfrak{p})$ are acyclic for all prime ideal $\mathfrak{p}$ of R.

(3) Now assume $(R, \mathfrak{m})$ is local. We may furthermore assume that $\mathfrak{m}H(X) = 0$.
To show this, let K be a Koszul complex with respect to a generating set $\{x_1, \dots, x_d\}$ of $\mathfrak{m}$, or equivalently K is the tensor product over R of complexes $0 \to R \xrightarrow{x_i} R \to 0$ for $1 \leq i \leq d$. Then by virtue of Nakayama lemma one can show that $H(X) = 0$ iff $H(X \otimes_R K) = 0$. One can also show the isomorphism $(X \otimes_R K)^* \cong X^*[-d] \otimes K$. Therefore we may replace X with $X \otimes_R K$. Since, for any element $x \in \mathfrak{m}$, the multiplication map by x on K is null homotopic, so is on $X \otimes_R K$. Hence $\mathfrak{m}H(X \otimes_R K) = 0$.

(4) Now assume that $H(X^*) = 0$ but $H(X) \neq 0$, and that $(R, \mathfrak{m})$ is a local ring. We may also assume that $H(X)$ is a non-trivial $R/\mathfrak{m}$-module by (3). By Proposition 6.1 above we have $\mathrm{Ext}^r_R(H(X), R) = 0$ for all $r > 0$. Therefore $\mathrm{Ext}^r_R(R/\mathfrak{m}, R) = 0$ for all $r > 0$. This requires that R is a Gorenstein ring of dimension zero. However in this case that $H(X^*) = 0$ implies that $H(X) = 0$, a contradiction. □

Acknowledgement: The author is grateful to Bernhard Keller for advising him to use the symbol Σ for the cosyzygy functor.

References

[1] MAURICE AUSLANDER, MARK BRIDGER, *Stable module theory*, Memoirs of the American Mathematical Society, No. 94 American Mathematical Society, Providence, R.I. (1969), 146 pp.

[2] Luchezar L. Avramov, Ragnar-Olaf Buchweitz, Liana M. Şega, *Extensions of a dualizing complex by its ring: commutative versions of a conjecture of Tachikawa*, J. Pure Appl. Algebra 201 (2005), no. 1-3, 218–239.

[3] David Jorgensen, Liana M. Şega, *Independence of the total reflexivity conditions for modules*, Algebr. Represent. Theory 9 (2006), no. 2, 217–226.

[4] Yuji Yoshino, *Homotopy categories of unbounded projective complexes*, Preprint, arXiv:1805.05705v4.

PART B

General Lectures (Survey Articles)

The Jordan-Hölder property, Grothendieck monoids and Bruhat inversions

H. Enomoto

Graduate School of Mathematics, Nagoya University,
Chikusa-ku, Nagoya, 464-8602, Japan
E-mail: m16009t@math.nagoya-u.ac.jp

In this article, we give a survey on results by the author [3]. We introduce a *Grothendieck monoid* of an exact category, an invariant of an exact category which has more information than the classical Grothendieck group. This monoid has information on simple objects, and can be used to check whether the Jordan-Hölder type theorem holds in a given exact category. As a toy example, we give a criterion for a torsion-free class over the path algebra of type A to satisfy the Jordan-Hölder property, by giving a classification of simple objects in a torsion-free class via combinatorics.

Keywords: Exact categories; Jordan-Hölder property; Bruhat inversions.

1. Introduction

For simplicity, we will use the following notation throughout this article.

- k denotes a fixed field.
- Λ denotes a finite-dimensional k-algebra.
- $\mathsf{mod}\,\Lambda$ denotes the category of finitely generated right Λ-modules.
- All subcategories are assumed to be closed under isomorphisms.

The representation theory of Λ is aimed at investigating the categorical structure of $\mathsf{mod}\,\Lambda$. Recently, not only the whole category $\mathsf{mod}\,\Lambda$, but also subcategories of $\mathsf{mod}\,\Lambda$ have been investigated from the several viewpoint: Cohen-Macaulay representation theory, Gorenstein homological algebra, (τ-)tilting theory, torsion(-free) classes, and so on. Almost all subcategories which naturally arise in the representation theory of algebras are *extension-closed* in $\mathsf{mod}\,\Lambda$ in the following sense:

Definition 1.1. Let $\mathcal{E}$ be a subcategory of $\mathsf{mod}\,\Lambda$. We say that it is *extension-closed* if for every short exact sequence $0 \to X \to Y \to Z \to 0$ of Λ-modules, if X and Z belong to $\mathcal{E}$, then so does Y.

Such a subcategory naturally has a structure of *exact category* in the sense of Quillen. For simplicity, in this article, we say that $\mathcal{E}$ is an *exact category* if it is an extension-closed subcategory of $\operatorname{mod}\Lambda$ for a finite-dimensional k-algebra Λ. This definition does not cover general exact categories, but most of exact categories arising in the representation theory of finite-dimensional algebras arise in this way, so this is harmless to our purpose.

The important thing is that in exact categories, we have the notion of *short exact sequences.*

Definition 1.2. Let $\mathcal{E}$ be an extension-closed subcategory of $\operatorname{mod}\Lambda$. Then a *short exact sequence in* $\mathcal{E}$ is a short exact sequence $0 \to X \to Y \to Z \to 0$ of Λ-modules with $X, Y, Z \in \mathcal{E}$.

By using this short exact sequence inside an exact category $\mathcal{E}$, we can define a *simple* object and a *composition series* inside $\mathcal{E}$.

Definition 1.3. Let $\mathcal{E}$ be an exact category.

- We say that an object $X \in \mathcal{E}$ is a *simple object* if there is no short exact sequence $0 \to L \to X \to N \to 0$ in $\mathcal{E}$ with $L, N \neq 0$. Equivalently, if $\mathcal{E}$ is an extension-closed subcategory of $\operatorname{mod}\Lambda$, then $X \neq 0$ is simple if there is no non-trivial Λ-submodule of L satisfying $L, X/L \in \mathcal{E}$.
- A *composition series* is a chain $0 = X_0 < X_1 < X_2 < \cdots < X_n = X$ of submodules of X such that X_i/X_{i-1} is a simple object in $\mathcal{E}$ for each i.

Since we are in the finite-dimensional situation, by considering dimension, every object in $\mathcal{E}$ has at least one composition series. Nevertheless, the analogue of the classical Jordan-Hölder theorem often fails, which is the main interest in this article. To state it accurately, we introduce the *Jordan-Hölder property* for exact categories.

Definition 1.4. Let $\mathcal{E}$ be an exact category.

- We say that two composition series $0 = X_0 < X_1 < \cdots < X_m = X$ and $0 = X'_0 < X'_1 < \cdots < X'_n = X$ of X in $\mathcal{E}$ are *equivalent* if $m = n$ holds and there is a permutation σ on $\{1, 2, \ldots, n\}$ such that $X_i/X_{i-1} \cong X'_{\sigma(i)}/X'_{\sigma(i)-1}$ holds for each i.
- We say that $\mathcal{E}$ has the *Jordan-Hölder property*, abbreviated by *(JHP)*, if any composition series of X are equivalent for every $X \in \mathcal{E}$.

Example 1.1. Let $\mathcal{E}$ be a subcategory of $\mathsf{mod}\, k$ consisting of vector spaces whose dimensions are not equal to 1. Then $\mathcal{E}$ can be checked to be extension-closed, thus is an exact category. However, we have the following two decompositions of $k^6 \in \mathcal{E}$:

$$k^3 \oplus k^3 = k^6 = k^2 \oplus k^2 \oplus k^2.$$

Obviously, both k^2 and k^3 are simple objects in $\mathcal{E}$, so the above gives a decomposition of k^6 into simple objects. However, since k^2 is not isomorphic to k^3, $\mathcal{E}$ does not satisfy (JHP).

Example 1.2. Let kQ be a path algebra of a quiver $1 \leftarrow 2 \rightarrow 3$. Define a subcategory $\mathcal{F}_1$ of $\mathsf{mod}\,\Lambda$ consisting of direct sums of S_1, S_3, P_2, I_1, I_3, where S_i, P_i and I_i denotes a simple, indecomposable projective and indecomposable injective kQ-module respectively corresponding each vertex $i = 1, 2, 3$. Then $\mathcal{F}_1$ is an extension-closed subcategory of $\mathsf{mod}\, kQ$, and actually is a *torsion-free class* introduced later. It can be checked that S_1, S_3, I_1, I_3 are all the simple objects in $\mathcal{E}$. However, P_2 has two composition series in $\mathcal{E}$:

$$0 < S_1 < P_2 \quad \text{and}$$
$$0 < S_3 < P_2.$$

In the first composition series, composition factors are S_1 and $P_2/S_1 \cong I_3$, and in the second, S_3 and $P_2/S_3 \cong I_1$. Since S_1, I_3, S_3, I_1 are pairwise non-isomorphic, $\mathcal{F}_1$ does not satisfy (JHP).

On the other hand, let $\mathcal{F}_2$ be a subcategory of $\mathsf{mod}\, kQ$ consisting of direct sums of S_1, S_3, P_2, I_3. Then later in Example 4.4 we will see that $\mathcal{F}_2$ satisfies (JHP).

The goal of this article is to answer the following question.

Question 1.1. Let $\mathcal{E}$ be a given exact category. How can one check whether $\mathcal{E}$ satisfies (JHP) or not?

2. Grothendieck monoids of exact categories

We will characterize (JHP) in terms of a certain invariant of an exact category, a *Grothendieck monoid.* This is a natural monoid version of the classical Grothendieck *group*, so first let us recall about the Grothendieck group.

Definition 2.1. Let $\mathcal{E}$ be an exact category. Then the *Grothendieck group* of $\mathcal{E}$ is an abelian group $\mathsf{K}_0(\mathcal{E})$ together with a map $[-]\colon \mathrm{Ob}(\mathcal{E}) \to \mathsf{K}_0(\mathcal{E})$ from a set of objects in $\mathcal{E}$ which satisfies the following universal property.

(1) $[-]$ is *additive*, that is, for every short exact sequence $0 \to X \to Y \to Z \to 0$, we have $[Y] = [X] + [Z]$ in $\mathsf{K}_0(\mathcal{E})$.
(2) $[-]$ is *universal* with respect to the property (1), that is, for every additive map $\varphi\colon \mathrm{Ob}(\mathcal{E}) \to G$ to an abelian group G, there exists a unique group homomorphism $\overline{\varphi}\colon \mathsf{K}_0(\mathcal{E}) \to G$ satisfying $\varphi = \overline{\varphi} \circ [-]$.

It is well-known that $\mathsf{K}_0(\mathsf{mod}\,\Lambda)$ is a free abelian group with basis simple Λ-modules, whose proof uses the classical Jordan-Hölder theorem for Λ-modules. It is natural to wonder whether the converse holds: if $\mathsf{K}_0(\mathcal{E})$ is free, then does $\mathcal{E}$ satisfy (JHP)?

Unfortunately, this is not true. In fact, all the exact categories in Examples 1.1 and 1.2 have the free Grothendieck groups. This shows that the Grothendieck group does not have enough information of the exact category, thus we need a more sophisticated invariant. This leads to the notion of *Grothendieck monoid.* For simplicity, *monoids* are assumed to be commutative with the operation $+$ and the unit 0.

Definition 2.2. Let $\mathcal{E}$ be an exact category. Then the *Grothendieck monoid* of $\mathcal{E}$ is a monoid $\mathsf{M}(\mathcal{E})$ together with a map $[-]\colon \mathrm{Ob}(\mathcal{E}) \to \mathsf{M}(\mathcal{E})$ which satisfies the following universal property.

(1) $[-]$ is *additive*, that is, for every short exact sequence $0 \to X \to Y \to Z \to 0$, we have $[Y] = [X] + [Z]$ and $[0] = 0$ in $\mathsf{M}(\mathcal{E})$.
(2) $[-]$ is *universal* with respect to the property (1), that is, for every additive map $\varphi\colon \mathrm{Ob}(\mathcal{E}) \to M$ to a monoid M, there exists a unique monoid homomorphism $\overline{\varphi}\colon \mathsf{M}(\mathcal{E}) \to M$ satisfying $\varphi = \overline{\varphi} \circ [-]$.

One may agree that this is just a monoid version of the Grothendieck group. Of course, since it is defined by the universal property, one should prove that it actually exists. We omit it, but the usual existence proof applies.

One of the remarkable property of the Grothendieck monoid, compared to the Grothendieck group, is that it remembers simple objects:

Proposition 2.1. *Let $\mathcal{E}$ be an exact category and X an object in $\mathcal{E}$. Then the following are equivalent.*

(1) X is a simple object in $\mathcal{E}$.
(2) $[X]$ is an atom *in $\mathsf{M}(\mathcal{E})$, that is, $[X] = y + z$ implies $y = 0$ or $z = 0$.*

Moreover, the map $[-]$ induces a bijection between isomorphism classes of simple objects in $\mathcal{E}$ and atoms in $\mathsf{M}(\mathcal{E})$.

Note that in a group, there is no non-trivial atom, so we cannot expect to find simples in the Grothendieck group.

3. Main results

Now we can state the main result on (JHP).

Theorem 3.1. *For an exact category $\mathcal{E}$, the following are equivalent:*

(1) $\mathcal{E}$ satisfies (JHP).
(2) $\mathsf{M}(\mathcal{E})$ is a free monoid, that is, isomorphic to a direct sum of the monoid $\mathbb{N}$ of non-negative integers.
(3) Simple objects in $\mathcal{E}$ are linearly independent in $\mathsf{K}_0(\mathcal{E})$.

The second condition completely characterizes (JHP) in terms of the combinatorial property of its Grothendieck monoid. However, it is rather difficult to compute the Grothendieck monoid compared to the Grothendieck group, and the third condition gives a more useful criterion for actual computations. By combining this result to the computation of the Grothendieck group, one can show the following purely numerical criterion.

Theorem 3.2. *If an exact category $\mathcal{E}$ satisfies the condition (*) defined later, the following are equivalent.*

(1) $\mathcal{E}$ satisfies (JHP).
(2) The number of simple objects in $\mathcal{E}$ is equal to the number of indecomposable projective objects in $\mathcal{E}$.

This theorem says that in a nice situation, we only have to count the number of simple objects, and to compare it with that of projective objects (which is often easier to compute than simples). Here we say that an exact category $\mathcal{E}$ *satisfies the condition (*)* if there is a finite-dimensional algebra Λ' and a Λ'-module U with finite injective dimension such that $\mathcal{E}$ is equivalent (as exact categories) to ${}^{\perp}U := \{X \in \mathsf{mod}\,\Lambda' \mid \operatorname{Ext}^i_{\Lambda'}(X, U) = 0 \text{ for } i > 0\}$. For example, every functorially finite torsion(-free) class or the category of Cohen-Macaulay modules over an Iwanaga-Gorenstein algebra satisfies the condition (*) (see Example 5.13 in Ref. 3 for the more details).

4. Torsion-free classes over path algebras of type A

In this section, we apply our general results to investigate particular exact categories which are not difficult to work with but not trivial, *torsion-free classes over path algebras of type A*.

For a finite-dimensional k-algebra Λ, a subcategory $\mathcal{F}$ of $\operatorname{mod} \Lambda$ is a *torsion-free class* if it is extension-closed and closed under submodules, that is, if X belongs to $\mathcal{F}$ then so do all submodules of X. Recently, torsion-free classes (or their dual, torsion classes) have been widely investigated in the representation theory of algebras due to their combinatorial nature and connection with (τ-)tilting theory and mutation theory (see e.g. Ref. 1).

It has been known since Gabriel that the representation theory of quivers (or path algebras) are closely related to algebraic Lie theory and related combinatorics. As for torsion-free classes, they have been classified by using *c-sortable elements in the Coxeter group of* Q [5, 8]. Here in this paper, we restrict our attention to type A case, where everything is explicit and easy to describe.

Let us introduce some terminology. A *quiver of type* A_n is a quiver whose underlying graph is the following.

$$1 - 2 - \cdots - n.$$

For example, $1 \leftarrow 2 \leftarrow 3$ and $1 \leftarrow 2 \rightarrow 3$ are both type A_3 quivers. As usual, we identify right kQ-modules and representations of Q.

The representation theory of type A_n quiver is closely related to its Weyl group, namely, the symmetric group S_{n+1}. We denote by S_{n+1} the symmetric group acting on $\{1, 2, \ldots, n+1\}$ from left. We denote by $s_i = (i\ i+1) \in S_{n+1}$ the adjacent transposition for $1 \leq i \leq n$. In order to equip S_{n+1} with the information on the orientation of the quiver Q, we introduce a *Coxeter element* c_Q of S_{n+1}.

Definition 4.1. A *Coxeter element* of S_{n+1} is an element $c \in S_{n+1}$ which is obtained as the product of all the adjacent transpositions $s_1, \cdots, s_n$ in some order. We say that a Coxeter element c is *associated to* Q if s_i appears before s_j in c whenever there exists an arrow $i \leftarrow j$ in Q. It is easy to construct a Coxeter element associated to a given quiver Q, and we often write it as c_Q.

We can check that a Coxeter element associated Q is uniquely determined from Q, and actually it is known that Coxeter elements are in bijection with orientations of edges in the underlying graph of Q.

Example 4.1. Let Q be a quiver $1 \leftarrow 2 \leftarrow 3$. Then its associated Coxeter element c_Q is obtained by

$$c_Q = s_1 s_2 s_3 = 2341 \in S_4.$$

Here we use the *one-line notation* to indicate elements in the symmetric group S_{n+1}, that is, we write $w = w(1)w(2)\cdots w(n+1)$ for $w \in S_{n+1}$. Similarly, for a quiver $1 \leftarrow 2 \rightarrow 3$, the associated Coxeter element c_Q is

$$c_Q = s_1 s_3 s_2 = 2413.$$

Then Ingalls and Thomas [5] classified torsion-free classes in $\mathsf{mod}\, kQ$ for a type A quiver by using c_Q-*sortable* elements introduced by Reading [7]. We explain this correspondence, following to the article Ref. 8.

Definition 4.2. [7] Let c be a Coxeter element of S_{n+1}. We say that an element w of S_{n+1} is c-*sortable* if there exists a reduced expression of the form $w = c^{(1)}c^{(2)}\cdots c^{(m)}$ such that $c^{(i)}$ is a subword of $c^{(i-1)}$ for $1 \leq i \leq m$, where $c^{(0)} = c$.

Example 4.2. Let Q be a quiver $1 \leftarrow 2 \rightarrow 3$ of type A_3. Its associated Coxeter element is $s_1 s_3 s_2$. Then there are 14 c_Q-sortable elements: $\{e, s_1, s_3, s_2, s_{13}, s_{12}, s_{32}, s_{132}, s_{12|1}, s_{32|3}, s_{132|1}, s_{132|3}, s_{132|13}, s_{132|132}\}$. Here we write as $s_{13} := s_1 s_3$, and write $s_{132|13} = s_{132}s_{13}$ to indicate $c^{(1)} = s_{132}$ and $c^{(2)} = s_{13}$. For later use, put $w_1 := s_{132|13}$ and $w_2 := s_{132|1}$.

To an element of S_{n+1}, we can associate a combinatorially defined set $\mathsf{inv}(w)$, the *inversion set* of w, and its subset $\mathsf{Binv}(w)$, the set of *Bruhat inversions of* w.

Definition 4.3. Let w be an element of S_{n+1}.

(1) An *inversion* of w is a pair (i, j) with $1 \leq i < j \leq n+1$ satisfying $w^{-1}(i) > w^{-1}(j)$. We denote by $\mathsf{inv}(w)$ the set of inversions of w.
(2) A *Bruhat inversion* of w is an inversion (i, j) of w such that there is no m with $i < m < j$ satisfying $w^{-1}(i) > w^{-1}(m) > w^{-1}(j)$. We denote by $\mathsf{Binv}(w)$ the set of Bruhat inversions of w.

Bruhat inversions correspond to covering relations of the Bruhat order of S_{n+1}, see Lemma 2.1.4 of Ref. 2 for the details.

Example 4.3. By definition, (i, j) with $i < j$ is an inversion of w precisely when the one-line notation of w is of the form $\cdots j \cdots i \cdots$, and (i, j) is a *non-Bruhat* inversion precisely when there is m with $i < m < j$ such that w is of the form $\cdots j \cdots m \cdots i \cdots$. For example, one can compute as follows for $w_1 = s_{13213} = 4231$ and $w_2 = s_{1321} = 4213$ in Example 4.2.

$$\begin{aligned}\mathsf{inv}(w_1) &= \{(1,2),(1,3),(1,4),(2,4),(3,4)\},\\ \mathsf{Binv}(w_1) &= \{(1,2),(1,3),(2,4),(3,4)\},\\ \mathsf{inv}(w_2) &= \{(1,2),(1,4),(2,4),(3,4)\},\\ \mathsf{Binv}(w_2) &= \{(1,2),(2,4),(3,4)\}.\end{aligned}$$

Let Q be a quiver of type A_n. For a pair of integers (i,j) with $1 \le i < j \le n+1$, one can define a representation $M_{[i,j)}$ of Q as follows, where $[i,j) = \{m \mid i \le m < j\}$.

- We put k for a vertex m if $m \in [i,j)$, and put 0 otherwise.
- We associate an identity map to each arrow inside $[i,j)$, and 0 for the other arrows.

Then Ingalls and Thomas gives the following classification of torsion-free classes over the path algebra of type A (Theorem 4.3 of Ref. 4).

Theorem 4.1. *Let Q be a quiver of type A_n and w an element of S_{n+1}. Let $\mathcal{F}(w)$ be a subcategory of $\mathsf{mod}\, kQ$ consisting of direct sums of $M_{[i,j)}$ with $(i,j) \in \mathsf{inv}(w)$. Then the following hold.*

(1) w is c_Q-sortable if and only if $\mathcal{F}(w)$ is a torsion-free class.
(2) The assignment $w \mapsto \mathcal{F}(w)$ gives a bijection between c_Q-sortable elements in S_{n+1} and torsion-free classes in $\mathsf{mod}\, kQ$.

Then we can determine simple objects in a torsion-free class $\mathcal{F}(w)$ in $\mathsf{mod}\, kQ$, thereby give a characterization of (JHP) for $\mathcal{F}(w)$.

Theorem 4.2. *Let Q be a quiver of type A_n and w a c_Q-sortable element of S_{n+1}. Then the following hold for a torsion-free class $\mathcal{F} := \mathcal{F}(w)$.*

(1) We have a bijection between Bruhat inversions of w and simple objects in $\mathcal{F}$ by sending (i,j) to $M_{[i,j)}$.
(2) $\mathcal{F}$ satisfies (JHP) if and only if the number of Bruhat inversions of w is equal to that of supports of w.

Here a *support* of an element w of S_{n+1} is the set of i with $1 \le i \le n$ such that s_i appears in a reduced expression of w.

We give a sketch of the proof. Theorem 4.2 (1) is proved directly by the detailed analysis of the relation between (non-)Bruhat inversions of w and the submodule structure of indecomposable representations. Theorem 4.2

(2) then easily follows from Theorem 3.2, since the number of indecomposable projectives is known to be equal to that of supports of w.

Example 4.4. Let Q be a quiver $1 \leftarrow 2 \rightarrow 3$. Then $w_1 = s_{13213} = 4231$ and $w_2 = s_{1321} = 4213$ are c_Q-sortable by Example 4.2. By Example 4.3, the torsion-free class $\mathcal{F}(w_1)$ consists of direct sums of $M_{[1,2)} = S_1, M_{[1,3)} = I_1, M_{[1,4)} = P_2, M_{[2,4)} = I_3, M_{[3,4)} = S_3$. Thus $\mathcal{F}(w_1)$ is equal to $\mathcal{F}_1$ in Example 1.2. According to Theorem 4.2 and Example 4.3, there are four simples in $\mathcal{F}(w_1)$, but the support of w_1 (and w_2) consists of $1, 2, 3$. Thus $\mathcal{F}(w_1)$ does not satisfy (JHP).

On the other hand, $\mathcal{F}(w_2)$ consists of direct sums of $M_{[1,2)} = S_1, M_{[1,4)} = P_2, M_{[2,4)} = I_3, M_{[3,4)} = S_3$, which is equal to $\mathcal{F}_2$ in Example 1.2. Since the number of Bruhat inversions of w_2 is three by Example 4.3, Theorem 4.2 implies that $\mathcal{F}(w_2)$ satisfies (JHP).

Finally, let us mention a generalization of Theorem 4.2 given in Ref. 4. Theorem 4.1 holds for any acyclic quiver Q, so it is natural to ask whether the analogue of Theorem 4.2 holds for a quiver which is not of type A. In Ref. 4, the author proved that this is the case for any quiver of Dynkin type, i.e., of type A_n, D_n, E_6, E_7, E_8. Actually, the author proved the analogous result for *preprojective algebras of Dynkin type* (whose torsion-free classes were classified by Mizuno [6]) by using some tools in root systems, and we deduced the path algebra case from the preprojective algebra case.

References

[1] T. Adachi, O. Iyama, I. Reiten, *τ-tilting theory*, Compos. Math. 150 (2014), no. 3, 415–452.

[2] A. Björner, F. Brenti, *Combinatorics of Coxeter groups*, Graduate Texts in Mathematics, 231. Springer, New York, 2005.

[3] H. Enomoto, *The Jordan-Hölder property and Grothendieck monoids of exact categories*, arXiv:1908.05446.

[4] H. Enomoto, *Bruhat inversions in Weyl groups and torsion-free classes over preprojective algebras*, arXiv:2002.09205.

[5] C. Ingalls, H. Thomas, *Noncrossing partitions and representations of quivers*, Compos. Math. 145 (2009), no. 6, 1533–1562.

[6] Y. Mizuno, *Classifying τ-tilting modules over preprojective algebras of Dynkin type*, Math. Z. 277 (2014), no. 3-4, 665–690.

[7] N. Reading, *Clusters, Coxeter-sortable elements and noncrossing partitions*, Trans. Amer. Math. Soc. 359 (2007), no. 12, 5931–5958.

[8] H. Thomas, *Coxeter groups and quiver representations*, In Surveys in representation theory of algebras, volume 716 of Contemp. Math., pages 173–186. Amer. Math. Soc., Providence, RI, 2018.

Action functor formalism

K. Shimizu

Department of Mathematical Sciences, Shibaura Institute of Technology,
307 Fukasaku, Minuma-ku, Saitama-shi, Saitama 337-8570, Japan
E-mail: kshimizu@shibaura-it.ac.jp

For a finite tensor category $\mathcal{C}$, the action functor $\rho : \mathcal{C} \to \mathrm{Rex}(\mathcal{C})$ is defined by $\rho(X) = X \otimes (-)$ for $X \in \mathcal{C}$, where $\mathrm{Rex}(\mathcal{C})$ is the category of linear right exact endofunctors on $\mathcal{C}$. We show that ρ has a left and a right adjoint and demonstrate that adjoints of ρ are useful for dealing with certain (co)ends in $\mathcal{C}$. As an application, we give relations between some ring-theoretic notions and particular (co)ends.

Keywords: Hopf algebra, tensor category, (co)end, Hochschild (co)homology.

1. Introduction

Tensor categories are widely used as a basic framework in several areas of mathematics and mathematical physics. A modular tensor category [1, 2] is one of important classes of semisimple tensor categories being actively investigated in connection with representation theory, rational conformal field theory (CFT), topological quantum field theory (TQFT), and quantum computing.

Lyubashenko proposed a definition of a 'non-semisimple' modular tensor category and established its basic theory [3–5]. Recently non-semisimple modular tensor categories are studied with motivation coming from 'non-rational' (or logarithmic) CFT and 'non-semisimple' TQFT. To define a non-semisimple modular tensor category, a special type of colimit, called a *coend* [6, 7], is required. The use of coends and its dual notion, *ends* [6, 7], seems to be essential in the theory of non-semisimple modular tensor categories and related study of finite tensor categories. This article reviews the author's framework [8] for dealing with particular (co)ends in a finite tensor category.

This article is organized as follows: In Section 2, we recall basic definitions of a monoidal category and related notions. For more on the notions recalled in this section, see [2, 6, 7, 9, 10]. Section 3 reviews the background of this study. In this section, we also explain why some particular (co)ends are important in the study of 'non-semisimple' modular tensor categories and related categories.

In Section 4, we introduce the action functor and give its basic properties. For a finite tensor category $\mathcal{C}$, the action functor $\rho : \mathcal{C} \to \mathrm{Rex}(\mathcal{C})$ is defined by $\rho(X) = X \otimes (-)$, where $\mathrm{Rex}(\mathcal{C})$ is the category of right exact linear endofunctors on $\mathcal{C}$. An important observation is that ρ has a left adjoint ρ^{la} and a right adjoint ρ^{ra}. Furthermore, these adjoints can be expressed by certain (co)ends. In particular, $\mathbb{L}_{\mathcal{C}} = \rho^{\mathrm{la}}(\mathrm{id}_{\mathcal{C}})$ is the coend used by Lyubashenko to define a non-semisimple modular tensor category. The end $\mathbb{A}_{\mathcal{C}} = \rho^{\mathrm{ra}}(\mathrm{id}_{\mathcal{C}})$ is used in [11–13] to extend some fundamental results in the Hopf algebra theory, such as the character theory and the integral theory, to the setting of finite tensor categories.

Adjoints of the action functor can be used to study relations between some ring-theoretic notions and particular (co)ends. Our basic strategy is explained at the end of Section 4 and summarized as follows: Let $\mathcal{C}$ be a finite tensor category, and let A be a finite-dimensional algebra such that $\mathcal{C} \approx A$-mod. Then $\mathrm{Rex}(\mathcal{C})$ is equivalent to the category A-mod-A of finite-dimensional A-bimodules. Some ring-theoretic notions can be formulated in terms of the category of bimodules. If a ring-theoretic notion, say X, which we aim to investigate has such a description, then one can transport it to the category $\mathcal{C}$ through the equivalence A-mod-$A \approx \mathrm{Rex}(\mathcal{C})$ and the functor ρ^{la} or ρ^{ra}. This allows us to discuss a relation between X and some (co)ends in $\mathcal{C}$.

In Section 5, we review some applications of adjoints of ρ obtained by following this strategy. We show that $\mathrm{Ext}^{\bullet}_{\mathcal{C}}(\mathbb{1}, \mathbb{A}_{\mathcal{C}})$ is isomorphic to the Hochschild cohomology $\mathrm{HH}^{\bullet}(A)$. Noteworthy, this result extends the projective $\mathrm{SL}_2(\mathbb{Z})$-action on the Hochschild cohomology of a ribbon factorisable Hopf algebra [14] to the setting of non-semisimple modular tensor categories. Furthermore, under the assumption that the double dual functor on $\mathcal{C}$ is isomorphic to the identity functor, $\mathrm{Ext}^{\bullet}_{\mathcal{C}}(\mathbb{A}_{\mathcal{C}}, 1)$ is shown to be the dual of the Hochschild homology $\mathrm{HH}_{\bullet}(A)$ by a similar argument and an abstract treatment of the Nakayama functor established in [15]. More applications are also mentioned in this article.

Notation 1.1. Throughout this article, we work over an algebraically closed field k. By an algebra, we mean an associative unital algebra over the field k. Given an algebra A, we denote by A-mod and mod-A the categories of finite-dimensional left and right A-modules, respectively. Given two algebras A and B, we denote by A-mod-B the category of finite-dimensional A-B-bimodules.

2. Basics of monoidal categories

2.1. *Monoidal categories*

A *monoidal category* [6, 10] is a category $\mathcal{C}$ equipped with a functor $\otimes : \mathcal{C} \times \mathcal{C} \to \mathcal{C}$ (called the tensor product), an object $\mathbb{1} \in \mathcal{C}$ (called the unit object) and natural isomorphisms $a_{X,Y,Z} : (X \otimes Y) \otimes Z \to X \otimes (Y \otimes Z)$, $l_X : \mathbb{1} \otimes X \to X$ and $r_X : X \otimes \mathbb{1} \to X$ $(X, Y, Z \in \mathcal{C})$ satisfying the pentagon and the triangle axioms. We say that a monoidal category $\mathcal{C}$ is *strict* if the natural isomorphism a, l and r are identities. Mac Lane's coherence theorem implies that every monoidal category is equivalent to a strict one. In view of this theorem, we may (and do) assume that every monoidal category is strict.

Let L and R be objects of $\mathcal{C}$. We say that L is a *left dual object* of R and R is a *right dual object* of L if there are morphisms $\varepsilon : L \otimes R \to \mathbb{1}$ and $\eta : \mathbb{1} \to R \otimes L$ in $\mathcal{C}$ such that the equations $(\varepsilon \otimes \mathrm{id}_L)(\mathrm{id}_L \otimes \eta) = \mathrm{id}_L$ and $(\mathrm{id}_R \otimes \varepsilon)(\eta \otimes \mathrm{id}_R) = \mathrm{id}_R$ hold. A monoidal category $\mathcal{C}$ is said to be *left rigid* if every object of $\mathcal{C}$ has a left dual object. Suppose that $\mathcal{C}$ is left rigid. If we denote a left dual object of $X \in \mathcal{C}$ by X^*, then the assignment $X \mapsto X^*$ extends to a contravariant endofunctor on $\mathcal{C}$, which we call the *left duality*. We note that there are natural isomorphisms

$$\mathrm{Hom}_{\mathcal{C}}(V \otimes X, W) \cong \mathrm{Hom}_{\mathcal{C}}(V, W \otimes X^*),$$
$$\mathrm{Hom}_{\mathcal{C}}(V, X \otimes W) \cong \mathrm{Hom}_{\mathcal{C}}(X^* \otimes V, W)$$

for $V, W, X \in \mathcal{C}$.

A monoidal category is said to be *right rigid* if every object has a right dual object. For a right rigid monoidal category, the right duality is defined. A monoidal category is said to be *rigid* if it is left rigid and right rigid. In a rigid monoidal category, the left duality and the right duality are mutually inverse to each other.

Example 2.1. If H is a Hopf algebra [16] with bijective antipode, then H-mod has a structure of a rigid monoidal category. We note that the antipode of a finite-dimensional Hopf algebra is bijective. Thus H-mod for a finite-dimensional Hopf algebra H is a rigid monoidal category.

2.2. *Braided and ribbon categories*

A *braided monoidal category* is a monoidal category $\mathcal{C}$ equipped with a *braiding*, that is, a natural isomorphism $\sigma_{X,Y} : X \otimes Y \to Y \otimes X$ $(X, Y \in \mathcal{C})$ satisfying the hexagon axiom [9, 10].

Example 2.2. For a monoidal category $\mathcal{C}$, there is a canonical braided monoidal category $\mathcal{Z}(\mathcal{C})$ called the *Drinfeld center* of $\mathcal{C}$. An object of this category is a pair (V, ξ) consisting of an object $V \in \mathcal{C}$ and a natural isomorphism $\xi_X : V \otimes X \to X \otimes V$ $(X \in \mathcal{C})$ such that the equation

$$\xi_{X \otimes Y} = (\mathrm{id}_X \otimes \xi_Y)(\xi_X \otimes \mathrm{id}_Y)$$

holds for all objects $X, Y \in \mathcal{C}$. See, *e.g.*, [9, 10] for details.

A *ribbon category* is a braided rigid monoidal category equipped with a *twist*, that is, a natural isomorphism $\theta_X : X \to X$ $(X \in \mathcal{C})$ such that the equations $\theta_{X \otimes Y} = \sigma_{Y,X}\sigma_{X,Y}(\theta_X \otimes \theta_Y)$ and $\theta_{X^*} = (\theta_X)^*$ hold for all objects $X, Y \in \mathcal{C}$, where σ is the braiding of $\mathcal{C}$. Ribbon categories give a systematic approach to knot invariants. Roughly speaking, a k-valued knot invariant can be constructed from a simple object of a k-linear ribbon category [2, 9].

Example 2.3. For each semisimple Lie algebra $\mathfrak{g}$, there is a Hopf algebra $U_q(\mathfrak{g})$ defined as a 'q-deformation' of $U(\mathfrak{g})$. It is known that $U_q(\mathfrak{g})$-mod has a structure of a ribbon category. By applying the above-mentioned construction to a simple $U_q(\mathfrak{g})$-module, we can obtain so many knot invariants including the celebrated Jones polynomial.

2.3. *Finite tensor categories and fusion categories*

Although its precise meaning varies, the term 'tensor category' is usually used in recent literature to mean a monoidal category with somewhat representation-theoretic flavor. In this article, we mainly deal with *finite tensor categories* in the sense of Etingof and Ostrik [17].

Definition 2.1. A *finite abelian category* is a k-linear category being equivalent to A-mod for some finite-dimensional algebra A. A *finite tensor category* is a rigid monoidal category $\mathcal{C}$ that is a finite abelian category at the same time such that the tensor product $\otimes : \mathcal{C} \times \mathcal{C} \to \mathcal{C}$ is k-linear in each variable and the unit object $\mathbb{1} \in \mathcal{C}$ is a simple object.

The category H-mod for a finite-dimensional Hopf algebra H is a typical example of a finite tensor category. Some results in the representation theory of finite-dimensional Hopf algebras have been generalized to finite tensor categories.

Although the semisimplicity is a strong condition from the viewpoint of representation theory, semisimple finite tensor categories are extensively studied with motivation coming from various areas of mathematics and mathematical physics. The decomposition rule of the tensor product is often called the *fusion rule* in such a context. Thus,

Definition 2.2. A semisimple finite tensor category is called a *fusion category* [18].

3. Background: Non-semisimple modular tensor categories

3.1. *Modular tensor categories*

A modular tensor category [1, 10] is a fusion category equipped with a 'non-degenerate' ribbon structure. The precise definition is as follows:

Definition 3.1. Let $\mathcal{C}$ be a ribbon fusion category with braiding σ with twist θ, and let $\{V_i\}_{i\in I}$ be representatives of the isomorphism classes of simple objects of $\mathcal{C}$. For $i, j \in I$, we define $s_{ij} \in k$ and $t_i \in k$ by

$$s_{ij} = \mathrm{tr}(\sigma_{V_j^*,V_i}\sigma_{V_i,V_j^*}), \quad \text{and} \quad t_i \,\mathrm{id}_{V_i} = \theta_{V_i},$$

where tr means the quantum trace [2, 9]. The matrix $S_{\mathcal{C}} = (s_{ij})_{i,j\in I}$ and the diagonal matrix $T_{\mathcal{C}} = (\delta_{ij}t_i)_{i,j\in I}$ are called the S-matrix and the T-matrix of $\mathcal{C}$, respectively. A *modular tensor category* is a ribbon fusion category with invertible S-matrix.

The adjective 'modular' comes from the fact that a modular tensor category $\mathcal{C}$ yields a projective representation of the modular group $SL_2(\mathbb{Z})$ in the following way: It is known that $SL_2(\mathbb{Z})$ is the group generated by $\mathfrak{s}$ and $\mathfrak{t}$ subject to the relations $(\mathfrak{s}\mathfrak{t})^3 = \mathfrak{s}$ and $\mathfrak{s}^4 = 1$. If $\mathcal{C}$ is a modular tensor category, then the assignment $\mathfrak{s} \mapsto S_{\mathcal{C}}$, $\mathfrak{t} \mapsto T_{\mathcal{C}}$ gives rise to a projective representation of $SL_2(\mathbb{Z})$.

From the viewpoint of low-dimensional topology, it should be noted that a modular tensor category yields a 3-dimensional TQFT. In particular, a modular tensor category gives an invariant of closed 3-manifolds and a projective representation of the mapping class group of a closed surface. If we take the torus $S^1 \times S^1$, then we obtain the projective representation of $SL_2(\mathbb{Z})$ given in the above.

3.2. *Ends and coends*

Lyubashenko [3–5] introduced a 'non-semisimple' modular tensor category and showed that one can obtain an invariant of closed 3-manifolds and a projective representation of the mapping class group of a closed surface from such a 'non-semisimple' modular tensor category.

To explain Lyubashenko's definition of a 'non-semisimple' modular tensor category, we need the notion of coends [6, 7]. We first recall the definitions of a coend and related notions: Let $\mathcal{A}$ and $\mathcal{V}$ be categories, and let G and H be functors from $\mathcal{A}^{\mathrm{op}} \times \mathcal{A}$ to $\mathcal{V}$. Then a *dinatural transformation* from G to H is a family $\alpha = \{\alpha_X : G(X,X) \to H(X,X)\}_{X \in \mathrm{Obj}(\mathcal{A})}$ of morphisms in $\mathcal{V}$ such that the equation

$$H(\mathrm{id}_X, f) \circ \alpha_X \circ G(f, \mathrm{id}_Y) = H(f, \mathrm{id}_Y) \circ \alpha_Y \circ G(\mathrm{id}_X, f)$$

holds for all morphisms $f : X \to Y$ in $\mathcal{A}$. We write $\alpha : G \stackrel{..}{\longrightarrow} H$ to mean that α is a dinatural transformation from G to H.

We regard an object of $\mathcal{V}$ as a constant functor from $\mathcal{A}^{\mathrm{op}} \times \mathcal{A}$ to $\mathcal{V}$. Then a *coend* of H is a pair (C, i) consisting of an object $C \in \mathcal{V}$ and a dinatural transformation $i : H \stackrel{..}{\longrightarrow} C$ with the following universal property: If C' is an object of $\mathcal{V}$ and $i' : H \stackrel{..}{\longrightarrow} C$ is a dinatural transformation, then there exists a unique morphism $\phi : C \to C'$ in $\mathcal{V}$ such that $\phi \circ i_X = i'_X$ for all $X \in \mathcal{A}$. A coend of H is unique up to isomorphism if it exists. According to [6], we denote the coend of H by $\int^{X \in \mathcal{A}} H(X,X)$.

The dual notion of a coend is also important: An *end* of G is a pair (E, π) consisting of an object $E \in \mathcal{V}$ and a dinatural transformation $\pi : E \stackrel{..}{\longrightarrow} G$ that is universal in a certain sense. If it exists, an end of G is unique up to isomorphisms. We usually denote it by $\int_{X \in \mathcal{A}} G(X,X)$.

Before proceeding further, we recall some useful formulas for ends and coends from [6, 7]. First, since the Hom functor preserves limits, we have

$$\mathrm{Hom}_{\mathcal{V}}\left(V, \int_{X \in \mathcal{A}} G(X,X)\right) \cong \int_{X \in \mathcal{A}} \mathrm{Hom}_{\mathcal{V}}(V, G(X,X))$$

whenever the end on the left-hand side exists. Similarly, we have

$$\mathrm{Hom}_{\mathcal{V}}\left(\int^{X \in \mathcal{A}} H(X,X), V\right) \cong \int_{X \in \mathcal{A}} \mathrm{Hom}_{\mathcal{V}}(H(X,X), V)$$

whenever the coend on the left-hand side exists.

Now let $F : \mathcal{A} \to \mathcal{B}$ and $F' : \mathcal{A} \to \mathcal{B}$ be functors sharing the domain and the codomain. We denote by $\mathrm{Nat}(F, F')$ the class of natural transformations from F to F'. If $\mathcal{A}$ is essentially small, then we have

$$\int_{X \in \mathcal{A}} \operatorname{Hom}_{\mathcal{B}}(F(X), F'(X)) \cong \operatorname{Nat}(F, F').$$

3.3. *Non-semisimple modular tensor categories*

Lyubashenko showed that the coend $\mathbb{L}_{\mathcal{C}} = \int^{X \in \mathcal{C}} X^* \otimes X$ exists for every finite tensor category $\mathcal{C}$. Now we suppose that $\mathcal{C}$ is braided. Then, by the universal property, we can endow $\mathbb{L}_{\mathcal{C}}$ with a structure of a Hopf algebra in $\mathcal{C}$ in a similar manner as we do in Tannaka theory. Furthermore, $\mathbb{L}_{\mathcal{C}}$ has a pairing $\omega_{\mathcal{C}} : \mathbb{L}_{\mathcal{C}} \otimes \mathbb{L}_{\mathcal{C}} \to \mathbb{1}$ compatible with its Hopf algebra structure.

Definition 3.2. We say that $\mathcal{C}$ is *non-degenerate* if the pairing $\omega_{\mathcal{C}}$ is non-degenerate. A *modular tensor category* [5] is a non-degenerate braided finite tensor category equipped with a twist.

Lyubashenko constructed an invariant of closed 3-manifolds and a projective representation of the mapping class group of a closed surface from a modular tensor category in the above sense. One can show that a ribbon fusion category is modular in the above sense if and only if its S-matrix is invertible. Because of these facts, Definition 3.2 seems to be a 'correct' non-semisimple generalization of Definition 3.1.

The following theorem may be useful to give examples of modular tensor categories:

Theorem 3.1 ([19]). *For a braided finite tensor category $\mathcal{C}$, the following conditions are equivalent:*

(1) $\mathcal{C}$ is non-degenerate in the sense of Definition 3.2.
(2) $\mathcal{C}$ is factorisable in the sense of Etingof, Nikshych and Ostrik [20].
(3) Every transparent object of $\mathcal{C}$ is isomorphic to the direct sum of finitely many copies of $\mathbb{1}$. Here, an object T of $\mathcal{C}$ is said to be transparent if the equation $\sigma_{X,T}\sigma_{T,X} = \mathrm{id}_{T \otimes X}$ holds for every object X of $\mathcal{C}$.
(4) The following linear map is injective:

$$\Omega_{\mathcal{C}} : \operatorname{Hom}_{\mathcal{C}}(\mathbb{1}, \mathbb{L}_{\mathcal{C}}) \to \operatorname{Hom}_{\mathcal{C}}(\mathbb{L}_{\mathcal{C}}, \mathbb{1}), \quad a \mapsto \omega_{\mathcal{C}} \circ (a \otimes \mathrm{id}).$$

If $\mathcal{C}$ is a ribbon fusion category, then the source and the target of the linear map $\Omega_{\mathcal{C}}$ have natural bases and $\Omega_{\mathcal{C}}$ is represented by the S-matrix with respect to those bases [19]. In the general case, we do not have nice bases of the source and the target of $\Omega_{\mathcal{C}}$. Hence we are led to the study of the vector spaces $\operatorname{Hom}_{\mathcal{C}}(\mathbb{L}_{\mathcal{C}}, \mathbb{1})$ and $\operatorname{Hom}_{\mathcal{C}}(\mathbb{1}, \mathbb{L}_{\mathcal{C}})$.

3.4. *The internal character theory*

Let $\mathcal{C}$ be a finite tensor category (which is not necessarily braided). Then the space $\mathrm{Hom}_{\mathcal{C}}(\mathbb{L}_{\mathcal{C}}, \mathbb{1})$ can be thought of as the 'center' since there are isomorphisms

$$\begin{aligned}\mathrm{Hom}_{\mathcal{C}}(\mathbb{L}_{\mathcal{C}}, \mathbb{1}) = \mathrm{Hom}_{\mathcal{C}}(\textstyle\int^{X \in \mathcal{C}} X^* \otimes X, \mathbb{1}) &\cong \textstyle\int_{X \in \mathcal{C}} \mathrm{Hom}_{\mathcal{C}}(X^* \otimes X, \mathbb{1}) \\ &\cong \textstyle\int_{X \in \mathcal{C}} \mathrm{Hom}_{\mathcal{C}}(X, X) \cong \mathrm{Nat}(\mathrm{id}_{\mathcal{C}}, \mathrm{id}_{\mathcal{C}})\end{aligned}$$

[5, Proposition 5.2.5]. It is difficult to answer what $\mathrm{Hom}_{\mathcal{C}}(\mathbb{1}, \mathbb{L}_{\mathcal{C}})$ is. A vector space canonically isomorphic to it has been studied in an attempt to extend the Hopf algebra theory to tensor categories [11–13]. To explain this, we consider the forgetful functor $U : \mathcal{Z}(\mathcal{C}) \to \mathcal{C}$, $(V, \xi) \mapsto V$. The functor U has a right adjoint [17]. We fix a right adjoint R of U and then define $\mathbb{A}_{\mathcal{C}} := UR(\mathbb{1})$. The functor UR is expressed by $UR(V) = \int_{X \in \mathcal{C}} X \otimes V \otimes X^*$ and, in particular, we have $\mathbb{A}_{\mathcal{C}} = \int_{X \in \mathcal{C}} X \otimes X^*$. By this expression, $\mathbb{A}_{\mathcal{C}}$ is shown to be a left dual object of $\mathbb{L}_{\mathcal{C}}$. Hence, in particular, we have

$$\mathrm{Hom}_{\mathcal{C}}(\mathbb{1}, \mathbb{L}_{\mathcal{C}}) \cong \mathrm{Hom}_{\mathcal{C}}(\mathbb{A}_{\mathcal{C}}, \mathbb{1}).$$

A basic result on monoidal functors equips $\mathbb{A}_{\mathcal{C}}$ with a canonical structure of an algebra in $\mathcal{C}$. If $\mathcal{C} = H$-mod for some finite-dimensional Hopf algebra H, then $\mathbb{A}_{\mathcal{C}}$ is the adjoint representation of H. Hence we have

$$\mathrm{Hom}_{\mathcal{C}}(\mathbb{A}_{\mathcal{C}}, \mathbb{1}) \cong \{f \in H^* \mid f(ab) = f(bS^2(a)) \text{ for all } a, b \in H\},$$

where S is the antipode of H. In particular, if $H = kG$ for some finite group G, then the vector space $\mathrm{Hom}_{\mathcal{C}}(\mathbb{A}_{\mathcal{C}}, \mathbb{1})$ can be identified with the space of class functions on G. In view of these facts,

Definition 3.3. We call $\mathbb{A}_{\mathcal{C}}$ and $\mathrm{CF}(\mathcal{C}) := \mathrm{Hom}_{\mathcal{C}}(\mathbb{A}_{\mathcal{C}}, \mathbb{1})$ the *adjoint algebra* and the space of *class functions* of $\mathcal{C}$, respectively.

We now assume that $\mathcal{C}$ is pivotal, that is, it is equipped with an isomorphism $X \cong X^{**}$ ($X \in \mathcal{C}$) of tensor functors. Then the internal character $\mathrm{ch}(X) \in \mathrm{CF}(\mathcal{C})$ of an object $X \in \mathcal{C}$ is defined. A basic theory on such a generalized character was established in [11] by using the theory of Hopf monads [21–23]. The theory of internal characters was used in a part of the proof of Theorem 3.1. It was also used to establish the 'non-semisimple' categorical Verlinde formula in [24, 25].

We note that we call $\mathrm{CF}(\mathcal{C})$ the space of class functions just by analogy with group theory. Unlike the space $\mathrm{Hom}_{\mathcal{C}}(\mathbb{L}_{\mathcal{C}}, \mathbb{1})$, a ring-theoretical meaning of $\mathrm{CF}(\mathcal{C})$ is unclear. This question is one of important motivations of the techniques introduced in the next section.

4. The action functor and its properties

4.1. *The action functor*

We have reviewed recent results on 'non-semisimple' modular tensor categories. In this theory, we encounter with several (co)ends. To overcome technical difficulties in dealing with such (co)ends and related vector spaces (such as the space of class functions), the author proposed an approach of the 'action functor'.

Throughout this section, we fix a finite tensor category $\mathcal{C}$. Given finite abelian categories $\mathcal{M}$ and $\mathcal{N}$, we denote by $\mathrm{Rex}(\mathcal{M},\mathcal{N})$ the category of k-linear right exact functors from $\mathcal{M}$ to $\mathcal{N}$. If $\mathcal{M}=\mathcal{N}$, then $\mathrm{Rex}(\mathcal{M},\mathcal{N})$ is abbreviated as $\mathrm{Rex}(\mathcal{M})$. Since the tensor product of $\mathcal{C}$ is k-linear and exact in each variable [17], the following definition makes sense:

Definition 4.1. The *action functor* of $\mathcal{C}$ is defined by

$$\rho : \mathcal{C} \to \mathrm{Rex}(\mathcal{C}), \quad \rho(X)(V) = X \otimes V \quad (X, V \in \mathcal{C}).$$

A similar functor has been considered in [26]; see Remark 4.1. We aim to explain that some (co)ends can be investigated by considering adjoints of ρ. First of all, we shall prove that ρ has a left adjoint and a right adjoint. The following variant of the Eilenberg-Watts theorem is essential:

Lemma 4.1. *The following functor is an equivalence of k-linear categories:*

$$B\text{-mod-}A \to \mathrm{Rex}(A\text{-mod}, B\text{-mod}), \quad M \mapsto M \otimes_A (-).$$

Let $\mathcal{M}$ and $\mathcal{N}$ be finite abelian categories, and let A and B be finite-dimensional algebras such that $\mathcal{M} \approx A$-mod and $\mathcal{N} \approx B$-mod. The above lemma yields several useful consequences:

(1) $\mathrm{Rex}(\mathcal{M},\mathcal{N})$ is a finite abelian category. Indeed, by the above lemma, there are equivalences of k-linear categories

$$\mathrm{Rex}(\mathcal{M},\mathcal{N}) \approx B\text{-mod-}A \approx (A^{\mathrm{op}} \otimes_k B)\text{-mod}.$$

(2) A sequence $0 \to F \to G \to H \to 0$ in $\mathrm{Rex}(\mathcal{M})$ is exact if and only if the induced sequence $0 \to F(P) \to G(P) \to H(P) \to 0$ is exact in $\mathcal{N}$ for every projective object $P \in \mathcal{M}$. This claim is easily verified if we view $0 \to F \to G \to H \to 0$ as a sequence of B-A-bimodules.
(3) A k-linear functor $F : \mathcal{M} \to \mathcal{N}$ has a right adjoint if and only if F is right exact. Indeed, if F is right exact, then $F \cong M \otimes_A (-)$ for some $M \in B$-mod-A (under the identification $\mathcal{M} = A$-mod and

$\mathcal{N} = B$-mod). Hence F has a right adjoint $\mathrm{Hom}_B(M, -)$. The converse is standard.

(4) A k-linear functor $F : \mathcal{M} \to \mathcal{N}$ has a left adjoint if and only if F is left exact. This proposition can be proved by applying the above argument to the functor $F^{\mathrm{op}} : \mathcal{M}^{\mathrm{op}} \to \mathcal{N}^{\mathrm{op}}$. Notice that $\mathcal{M}^{\mathrm{op}}$ is equivalent to A^{op}-mod and, in particular, it is finite abelian.

Now we show that $\rho : \mathcal{C} \to \mathrm{Rex}(\mathcal{C})$ is exact: Let $0 \to X \to Y \to Z \to 0$ be a short exact sequence in $\mathcal{C}$. Since the tensor product of $\mathcal{C}$ is k-linear in each variable, we obtain a short exact sequence $0 \to X \otimes V \to Y \otimes V \to Z \otimes V \to 0$ for every object $V \in \mathcal{C}$. This implies that $0 \to \rho(X) \to \rho(Y) \to \rho(Z) \to 0$ is an exact sequence in $\mathrm{Rex}(\mathcal{C})$.

Since the source and the target of the functor ρ are finite abelian categories, it has a left adjoint and a right adjoint. We denote a left adjoint and a right adjoint of ρ by ρ^{la} and ρ^{ra}, respectively.

4.2. *The action functor and (co)ends*

Next, we observe that the functors ρ^{la} and ρ^{ra} can be expressed by certain (co)ends. We note the following existence theorem of (co)ends: Let **Vec** denote the category of all vector spaces over k. Given a functor $G : \mathcal{C}^{\mathrm{op}} \times \mathcal{C} \to \mathcal{C}$, we can define the contravariant functor

$$G^{\sharp} : \mathcal{C} \to \mathbf{Vec}, \quad V \mapsto \int_{X \in \mathcal{C}} \mathrm{Hom}_{\mathcal{C}}(V, G(X, X)) \quad (V \in \mathcal{C})$$

since **Vec** is complete. Since the Yoneda functor preserves and reflects limits, we see that the functor $G^{\sharp}$ is representable if and only if the end of G exists. Furthermore, if an object $E \in \mathcal{C}$ represents $G^{\sharp}$, then E has a natural structure of an end of G. Similarly, the functor

$$G^{\flat} : \mathcal{C} \to \mathbf{Vec}, \quad V \mapsto \int_{X \in \mathcal{C}} \mathrm{Hom}_{\mathcal{C}}(G(X, X), V) \quad (V \in \mathcal{C})$$

is representable if and only if a coend of G exists. An object representing $G^{\flat}$, if it exists, has a natural structure of a coend of G.

By using this result and some basic properties of (co)ends, we can express ρ^{la} and ρ^{ra} as follows: For $V \in \mathcal{C}$ and $F \in \mathrm{Rex}(\mathcal{C})$, we have

$$\begin{aligned} \mathrm{Hom}_{\mathcal{C}}(V, \rho^{\mathrm{ra}}(F)) \cong \mathrm{Nat}(\rho(V), F) &\cong \textstyle\int_{X \in \mathcal{C}} \mathrm{Hom}_{\mathcal{C}}(V \otimes X, F(X)) \\ &\cong \textstyle\int_{X \in \mathcal{C}} \mathrm{Hom}_{\mathcal{C}}(V, F(X) \otimes X^*). \end{aligned}$$

This means $\rho^{\mathrm{ra}}(F) = \int_{X \in \mathcal{C}} F(X) \otimes X^*$. Similarly, we have

$$\begin{aligned}\mathrm{Hom}_{\mathcal{C}}(\rho^{\mathrm{la}}(F), V) \cong \mathrm{Nat}(F, \rho(V)) &\cong \textstyle\int_{X \in \mathcal{C}} \mathrm{Hom}_{\mathcal{C}}(F(X), V \otimes X) \\ &\cong \textstyle\int_{X \in \mathcal{C}} \mathrm{Hom}_{\mathcal{C}}(F(X^*), V \otimes X^*) \\ &\cong \textstyle\int_{X \in \mathcal{C}} \mathrm{Hom}_{\mathcal{C}}(F(X^*) \otimes X, V),\end{aligned}$$

where the third isomorphism follows from the fact that the duality functor is an anti-autoequivalence on $\mathcal{C}$. Hence, $\rho^{\mathrm{la}}(F) = \int^{X \in \mathcal{C}} F(X^*) \otimes X$. We note that the above argument does not only express adjoints of ρ by (co)ends but also show that particular (co)ends exist in $\mathcal{C}$.

Remark 4.1. The above formulas for ρ^{ra} and ρ^{la} are found in [26]. Unlike [26], the codomain of ρ is not $[\mathcal{C}, \mathcal{C}]$ but its full subcategory $\mathrm{Rex}(\mathcal{C})$, which is a finite abelian category. We also note that the centralizer of a Hopf monad [22] is given by a similar formula to $\rho^{\mathrm{la}}(F)$.

4.3. *Abstract Eilenberg-Watts equivalence*

We introduce several tools to investigate properties of adjoints of the action functor. Given k-linear abelian categories $\mathcal{M}$ and $\mathcal{N}$, we denote by $\mathcal{M} \boxtimes \mathcal{N}$ the Deligne tensor product of $\mathcal{M}$ and $\mathcal{N}$ [10, 27]. Like the tensor product of vector spaces, $\mathcal{M} \boxtimes \mathcal{N}$ is defined to be a k-linear category equipped with a k-bilinear functor $\boxtimes : \mathcal{M} \times \mathcal{N} \to \mathcal{M} \boxtimes \mathcal{N}$ that is right exact in each variable and universal among such functors. In general, the Deligne tensor product of $\mathcal{M}$ and $\mathcal{N}$ does not exist. If $\mathcal{M} = A$-mod and $\mathcal{N} = B$-mod for some finite-dimensional algebras A and B, then their Deligne tensor product exists and is given by $\mathcal{M} \boxtimes \mathcal{N} = (A \otimes_k B)$-mod with $\boxtimes = \otimes_k$.

Let $\mathcal{M}$ and $\mathcal{N}$ be finite abelian categories, and let A and B be finite-dimensional algebras such that $\mathcal{M} \approx A$-mod and $\mathcal{N} \approx B$-mod. Then there is an equivalence of k-linear categories

$$\Phi_{\mathcal{M},\mathcal{N}} := \Big(\mathcal{M}^{\mathrm{op}} \boxtimes \mathcal{N} \xrightarrow{\ \approx\ } B\text{-mod-}A \xrightarrow[\text{Lemma 4.1}]{\approx} \mathrm{Rex}(\mathcal{M}, \mathcal{N})\Big).$$

The equivalence $\Phi_{\mathcal{M},\mathcal{N}}$ does not depend on the choice of A and B and is called the *abstract Eilenberg-Watts equivalence* [11, 15]. This equivalence has been used to prove:

Lemma 4.2 ([13, Lemma 2.5]). *Let F be an object of* $\mathrm{Rex}(\mathcal{M}, \mathcal{N})$, *and let G be a right adjoint of F. The object F is projective if and only if the following two conditions (1) and (2) are satisfied:*

(1) $F(M)$ is projective for every object $M \in \mathcal{M}$.

(2) $G(N)$ is injective for every object $N \in \mathcal{N}$.

Proof. Since $\Phi_{\mathcal{M},\mathcal{N}}$ is an equivalence, F is a projective object if and only if the following functor $\mathcal{Y}_F$ is exact:

$$\mathcal{Y}_F := \left(\mathcal{M}^{\mathrm{op}} \boxtimes \mathcal{N} \xrightarrow{\Phi_{\mathcal{M},\mathcal{N}}} \mathrm{Rex}(\mathcal{M},\mathcal{N}) \xrightarrow{\mathrm{Nat}(F,-)} \mathbf{Vec}\right).$$

By a property of the Deligne tensor product [27, Proposition 5.13], the functor $\mathcal{Y}_F$ is exact if and only if the k-bilinear functor

$$\mathcal{Y}'_F := \left(\mathcal{M}^{\mathrm{op}} \times \mathcal{N} \xrightarrow{\boxtimes} \mathcal{M}^{\mathrm{op}} \boxtimes \mathcal{N} \xrightarrow{\mathcal{Y}_F} \mathbf{Vec}\right)$$

is exact in each variable. By some technical computation on the space of natural transformations, we obtain natural isomorphisms

$$\mathcal{Y}'_F(M,N) \cong \mathrm{Hom}_{\mathcal{N}}(F(M),N) \cong \mathrm{Hom}_{\mathcal{M}}(M,G(N))$$

for $M \in \mathcal{M}$ and $N \in \mathcal{N}$. Hence the claim follows. □

4.4. *The Nakayama functor*

Let $\mathcal{M}$ and $\mathcal{N}$ be finite abelian categories. We denote by $\mathrm{Lex}(\mathcal{M},\mathcal{N})$ the category of k-linear left exact functors from $\mathcal{M}$ to $\mathcal{N}$. In a similar way as above, we have an equivalence

$$\Psi_{\mathcal{M},\mathcal{N}} : \mathcal{M}^{\mathrm{op}} \boxtimes \mathcal{N} \to \mathrm{Lex}(\mathcal{M},\mathcal{N}), \quad M \boxtimes N \mapsto \mathrm{Hom}_{\mathcal{M}}(M,-) \otimes_k N$$

of k-linear categories [12, 15], where $\otimes_k$ is characterized by the natural isomorphism $\mathrm{Hom}_{\mathcal{M}}(V \otimes_k X, X') \cong \mathrm{Hom}_k(V, \mathrm{Hom}_{\mathcal{M}}(X,X'))$ for $X, X' \in \mathcal{M}$ and $V \in k\text{-mod}$.

Definition 4.2. The Nakayama functor [15] of $\mathcal{M}$ is given by

$$\mathbb{N}_{\mathcal{M}} = \Phi_{\mathcal{M},\mathcal{M}} \Psi^{-1}_{\mathcal{M},\mathcal{M}}(\mathrm{id}_{\mathcal{M}}) \in \mathrm{Rex}(\mathcal{M}).$$

There is a formula that expresses $\Psi^{-1}_{\mathcal{M},\mathcal{M}}$ by certain coends. By using that formula, we may also define the Nakayama functor by

$$\mathbb{N}_{\mathcal{M}}(M) = \int^{X \in \mathcal{M}} \mathrm{Hom}_{\mathcal{M}}(M,X)^* \otimes_k X \quad (M \in \mathcal{M}).$$

Let A be a finite-dimensional algebra. Then $A^* := \mathrm{Hom}_k(A,k)$ is an A-bimodule in a natural way. The Nakayama functor of A-mod is (isomorphic to) the functor $A^* \otimes_A (-)$. This justifies the terminology.

Below we list useful properties of the Nakayama functor [15]. Most of them are obtained just by rephrasing known results on the bimodule A^* in the language of categories and functors:

(1) The Nakayama functor $\mathbb{N}_{\mathcal{M}}$ is an equivalence if and only if $\mathcal{M}$ is Frobenius. There is an isomorphism $\mathbb{N}_{\mathcal{M}} \cong \mathrm{id}_{\mathcal{M}}$ if and only if $\mathcal{M}$ is symmetric Frobenius. Here, $\mathcal{M}$ is said to be (symmetric) *Frobenius* if $\mathcal{M} \approx A$-mod for some (symmetric) Frobenius algebra A.
(2) If $F : \mathcal{M} \to \mathcal{N}$ is a k-linear exact functor between finite abelian categories, then there is an isomorphism $\mathbb{N}_{\mathcal{M}} \circ F^{\mathrm{la}} \cong F^{\mathrm{ra}} \circ \mathbb{N}_{\mathcal{N}}$. Here F^{la} and F^{ra} are a left and a right adjoint of F, respectively.
(3) The Nakayama functor of $\mathrm{Rex}(\mathcal{M}, \mathcal{N})$ is given by

$$\mathbb{N}_{\mathrm{Rex}(\mathcal{M},\mathcal{N})}(F) = \mathbb{N}_{\mathcal{N}} \circ F \circ \mathbb{N}_{\mathcal{M}} \quad (F \in \mathrm{Rex}(\mathcal{M}, \mathcal{N})).$$

4.5. *Properties of adjoints of the action functor*

Now we are ready to give the proof of:

Theorem 4.1. *The functors ρ^{la} and ρ^{ra} are k-linear faithful exact functors and preserve projective objects.*

Proof. We first note that $\mathcal{C}$ is Frobenius [17]. Thus the Nakayama functor of $\mathcal{C}$ and that of $\mathrm{Rex}(\mathcal{C})$ are equivalences. By the results on the Nakayama functors reviewed in the above, we have $\rho^{\mathrm{la}} \cong \mathbb{N}_{\mathrm{Rex}(\mathcal{C})}^{-1} \circ \rho^{\mathrm{ra}} \circ \mathbb{N}_{\mathcal{C}}$ and hence it suffices to show that ρ^{ra} has the desired properties.

Exactness. To show that ρ^{ra} is exact, we remark that ρ preserves projective objects. Indeed, let P be a projective object of $\mathcal{C}$. Then $P \otimes X$ is projective for every $P \in \mathcal{C}$. Let Q be a right dual object of P. Then Q is injective and $Q \otimes X$ is injective for every $X \in \mathcal{C}$. We note that $\rho(Q)$ is right adjoint to $\rho(P)$. By Lemma 4.2, $\rho(P) \in \mathrm{Rex}(\mathcal{C})$ is projective.

Now the exactness of ρ^{ra} is proved as follows: Let G be a progenerator of $\mathcal{C}$. Since $\mathrm{Hom}_{\mathcal{C}}(G, -) \circ \rho^{\mathrm{ra}} \cong \mathrm{Nat}(\rho(G), -)$, and since $\rho(G)$ is projective, the functor $\mathrm{Hom}_{\mathcal{C}}(G, -) \circ \rho^{\mathrm{ra}}$ is exact. Since G is a progenerator, we conclude that the functor ρ^{ra} is exact.

Faithfulness. This part essentially relies on [28, Proposition 2.6 (ii)]. Let $F \in \mathrm{Rex}(\mathcal{C})$ be a non-zero object. By that proposition, there is an object $X \in \mathcal{C}$ such that F is a quotient of $\rho(X)$. Thus we have

$$\mathrm{Hom}_{\mathcal{C}}(X, \rho^{\mathrm{ra}}(F)) \cong \mathrm{Nat}(\rho(X), F) \neq 0.$$

This implies that $\rho^{\mathrm{ra}}(F) \neq 0$. By this result and the exactness of ρ^{ra}, we conclude that ρ^{ra} is faithful.

Preservation of projectivity. Since $\mathcal{C}$ and $\mathrm{Rex}(\mathcal{C})$ are Frobenius, it suffices to show that ρ^{ra} preserves injective objects. This is fairly easy: If E

is an injective object of $\mathrm{Rex}(\mathcal{C})$, then $\rho^{\mathrm{ra}}(E)$ is injective since

$$\mathrm{Hom}_{\mathcal{C}}(\rho^{\mathrm{ra}}(E), -) \cong \mathrm{Nat}(E, -) \circ \rho : \mathcal{C} \to \mathbf{Vec}$$

is exact. The proof is done. □

4.6. *General strategy*

The content of this section is summarized as follows: We first defined the action functor $\rho : \mathcal{C} \to \mathrm{Rex}(\mathcal{C})$. We then show that it has a left adjoint ρ^{la} and a right adjoint ρ^{ra}. It turns out that they are expressed by

$$\rho^{\mathrm{la}}(F) = \int^{X \in \mathcal{C}} F(X^*) \otimes X \quad \text{and} \quad \rho^{\mathrm{ra}}(F) = \int_{X \in \mathcal{C}} F(X) \otimes X^*$$

for $F \in \mathrm{Rex}(\mathcal{C})$ and have nice properties as exhibited in Theorem 4.1.

We are interested in particular (co)ends in $\mathcal{C}$. The coend $\mathbb{L}_{\mathcal{C}}$ appeared in Definition 3.2 and the adjoint algebra $\mathbb{A}_{\mathcal{C}}$, which is dual to $\mathbb{L}_{\mathcal{C}}$, are given by $\mathbb{L}_{\mathcal{C}} = \rho^{\mathrm{la}}(\mathrm{id}_{\mathcal{C}})$ and $\mathbb{A}_{\mathcal{C}} = \rho^{\mathrm{ra}}(\mathrm{id}_{\mathcal{C}})$, respectively. These formulas allow us to investigate $\mathbb{L}_{\mathcal{C}}$ and $\mathbb{A}_{\mathcal{C}}$ systematically.

Our strategy is as follows: There is a finite-dimensional algebra A such that $\mathcal{C} \approx A$-mod. Some ring-theoretic notions for A can be formulated in terms of A-bimodules. If a ring-theoretic notion for A which we aim to investigate has such a description, then we can transport it to the category $\mathcal{C}$ through the equivalence A-mod-$A \approx \mathrm{Rex}(\mathcal{C})$ and adjoints of ρ. In the next section, we demonstrate how this works.

5. Applications

Let $\mathcal{C}$ be a finite tensor category, and define $\mathbb{L}_{\mathcal{C}}$ and $\mathbb{A}_{\mathcal{C}}$ as above. Following the strategy explained in the last section, we have established several results on the structure and properties of $\mathbb{L}_{\mathcal{C}}$ and $\mathbb{A}_{\mathcal{C}}$ in [8]. In this section, we review and give remarks on some results in [8].

5.1. *A description of the space of class functions*

Let A be a finite-dimensional algebra such that $\mathcal{C} \approx A$-mod. Unlike the case of Hopf algebras, there is no obvious relation between the monoidal structure of $\mathcal{C}$ and the algebra structure of A. Nevertheless, under the assumption that the double dual functor on $\mathcal{C}$ is isomorphic to the identity, one can show that $\mathrm{CF}(\mathcal{C})$ is isomorphic to the space

$$\mathrm{SLF}(A) = \{ f \in A^* \mid f(ab) = f(ba) \text{ for all } a, b \in A \}$$

of symmetric linear functions on A. To see this, the following intrinsic definition of the space of symmetric linear functions is important: For a finite abelian category $\mathcal{M}$, we set $\mathrm{SLF}(\mathcal{M}) = \mathrm{Nat}(\mathrm{id}_{\mathcal{M}}, \mathbb{N}_{\mathcal{M}})$, where $\mathbb{N}_{\mathcal{M}}$ is the Nakayama functor on $\mathcal{M}$. If $\mathcal{M} \approx A$-mod for some finite-dimensional algebra A, then we have isomorphisms

$$\mathrm{SLF}(\mathcal{M}) \cong \mathrm{SLF}(A\text{-mod}) \cong \mathrm{Hom}_{A\text{-mod-}A}(A, A^*) \cong \mathrm{SLF}(A).$$

Now the problem is whether $\mathrm{CF}(\mathcal{C})$ and $\mathrm{SLF}(\mathcal{C})$ are isomorphic or not. By utilizing basic properties of the Nakayama functor, we have proved:

Theorem 5.1 ([8, Lemma 6.8]). *There is a natural isomorphism*

$$\mathrm{Hom}_{\mathcal{C}}(\rho^{\mathrm{ra}}(\mathbb{S}F), \mathbb{S}(X)) \cong \mathrm{Nat}(F, \rho(X) \circ \mathbb{N}_{\mathcal{M}}) \quad (X \in \mathcal{C}, F \in \mathrm{Rex}(\mathcal{C})),$$

where $\mathbb{S}(X) = X^{**}$ *is the double left dual functor on* $\mathcal{C}$.

In general, $\mathrm{CF}(\mathcal{C})$ and $\mathrm{SLF}(\mathcal{C})$ are not isomorphic. If $\mathbb{S}$ is isomorphic to the identity functor, then we have

$$\mathrm{Hom}_{\mathcal{C}}(\rho^{\mathrm{ra}}(F), X) \cong \mathrm{Nat}(F, \rho(X) \circ \mathbb{N}_{\mathcal{C}}) \quad (X \in \mathcal{C}, F \in \mathrm{Rex}(\mathcal{C}))$$

by the above theorem. By letting $F = \mathrm{id}_{\mathcal{C}}$ and $X = \mathbb{1}$, we have

Theorem 5.2. $\mathrm{CF}(\mathcal{C}) \cong \mathrm{SLF}(\mathcal{C})$ *if* $\mathbb{S} \cong \mathrm{id}_{\mathcal{C}}$.

5.2. *Ideals and topologizing full subcategories*

There is no doubt that an ideal is an important notion in the ring theory. Let A be a finite-dimensional algebra such that $\mathcal{C} \approx A$-mod. In terms of bimodules, an ideal of A is nothing but a subbimodule of ${}_A A_A$. Since ${}_A A_A$ corresponds to the identity functor through the equivalence $\mathrm{Rex}(\mathcal{C}) \approx A\text{-mod-}A$, we refer to a subobject of $\mathrm{id}_{\mathcal{C}}$ in $\mathrm{Rex}(\mathcal{C})$ as an *ideal* of $\mathcal{C}$, by abuse of terminology.

A full subcategory $\mathcal{S}$ of $\mathcal{C}$ is said to be *topologizing* if it is closed under finite direct sums and subquotients. We denote by $\mathfrak{I}(\mathcal{C})$ and $\mathfrak{T}(\mathcal{C})$ the set of ideals of $\mathcal{C}$ and the set of topologizing full subcategories of $\mathcal{C}$. There is a bijective map $\psi_{\mathcal{C}} : \mathfrak{I}(\mathcal{C}) \to \mathfrak{Top}(\mathcal{C})$. If we identify $\mathcal{C}$ with A-mod, then $\psi_{\mathcal{C}}$ sends an ideal I of A (in the usual sense) to $A/I\text{-mod} \in \mathfrak{T}(\mathcal{C})$.

Now, for a topologizing full subcategory $\mathcal{S} \in \mathfrak{T}(\mathcal{C})$, we consider

$$\mathbb{A}_{\mathcal{S}} := \int_{X \in \mathcal{S}} X \otimes X^* \quad \text{and} \quad \mathbb{L}_{\mathcal{S}} := \int^{X \in \mathcal{S}} X^* \otimes X.$$

Strictly speaking, $\mathbb{A}_{\mathcal{S}}$ indicates an end of the functor $\mathcal{S}^{\mathrm{op}} \times \mathcal{S} \to \mathcal{C}$ given by $(X, Y) \mapsto Y \otimes X^*$. $\mathbb{L}_{\mathcal{S}}$ is a coend of a similar functor. We shall discuss whether they exist and, if they do, relations between $\mathbb{A}_{\mathcal{S}}$ and $\mathbb{L}_{\mathcal{S}}$ with $\mathbb{A}_{\mathcal{C}}$ and $\mathbb{L}_{\mathcal{C}}$. The use of adjoints of the action functor gives the following result:

Theorem 5.3 ([8, Theorems 4.6, 4.7]). *Let $\mathcal{S}$ be a topologizing full subcategory of $\mathcal{C}$, and let $t_{\mathcal{S}}$ be the ideal of $\mathcal{C}$ corresponding to $\mathcal{S}$. Then the coend $\mathbb{L}_{\mathcal{S}}$ and the end $\mathbb{A}_{\mathcal{S}}$ exist and there are canonical isomorphisms*

$$\mathbb{L}_{\mathcal{S}} \cong \rho^{\mathrm{la}}(t_{\mathcal{S}}) \quad \text{and} \quad \mathbb{A}_{\mathcal{S}} \cong \rho^{\mathrm{ra}}(\mathrm{id}_{\mathcal{C}}/t_{\mathcal{S}}).$$

Furthermore, the following map preserves and reflects the order:

$$\mathfrak{T}(\mathcal{C}) \to \{\text{subobjects of } \mathbb{L}_{\mathcal{C}}\}, \quad \mathcal{S} \mapsto \mathbb{L}_{\mathcal{S}}.$$

The following map also preserves and reflects the order:

$$\mathfrak{T}(\mathcal{C}) \to \{\text{quotient objects of } \mathbb{A}_{\mathcal{C}}\}, \quad \mathcal{S} \mapsto \mathbb{A}_{\mathcal{S}}.$$

5.3. *A filtration of the space of class functions*

Theorem 5.3 is motivated by study of class functions of a finite tensor category. To prove the linear independence of internal characters of simple objects in [11], we have considered the full subcategory $\mathcal{C}_1$ of semisimple objects of $\mathcal{C}$ and the vector space $\mathrm{CF}_1(\mathcal{C}) := \mathrm{Hom}_{\mathcal{C}}(\mathbb{A}_{\mathcal{C}_1}, \mathbb{1})$. A key observation is that $\mathbb{A}_{\mathcal{C}_1}$ is a quotient of $\mathbb{A}_{\mathcal{C}}$ and thus $\mathrm{CF}_1(\mathcal{C})$ can be regarded as a subspace of the space $\mathrm{CF}(\mathcal{C})$ of class functions. Theorem 5.3 generalizes this observation: If $\mathcal{S}$ is a topologizing full subcategory, then $\mathrm{Hom}_{\mathcal{C}}(\mathbb{A}_{\mathcal{S}}, \mathbb{1})$ can be regarded as a subspace of $\mathrm{CF}(\mathcal{C})$.

We desire to construct a 'nice' basis of $\mathrm{CF}(\mathcal{C})$. The vector space $\mathrm{CF}_1(\mathcal{C})$ has a basis consisting of internal characters of simple objects. Hence it is natural to find a basis of $\mathrm{CF}(\mathcal{C})$ extending that of $\mathrm{CF}_1(\mathcal{C})$. To attack this problem, in [8], the author considered the subspace

$$\mathrm{CF}_\ell(\mathcal{C}) := \mathrm{Hom}_{\mathcal{C}}(\mathbb{A}_{\mathcal{C}_\ell}, \mathbb{1}) \subset \mathrm{CF}(\mathcal{C}),$$

where $\mathcal{C}_\ell$ is the full subcategory of $\mathcal{C}$ consisting of all objects with Loewy length $\leq \ell$. By Theorem 5.3, we have a filtration

$$\mathrm{CF}_1(\mathcal{C}) \subset \mathrm{CF}_2(\mathcal{C}) \subset \cdots \subset \mathrm{CF}(\mathcal{C}).$$

Now we suppose $\mathbb{S} \cong \mathrm{id}_{\mathcal{C}}$. Then, by Theorem 5.2, there is an isomorphism $\mathrm{CF}(\mathcal{C}) \cong \mathrm{SLF}(\mathcal{C})$. The point is that this isomorphism is obtained as a part of a natural isomorphism. The naturality actually implies:

Theorem 5.4 ([8, Theorem 6.9]). *Suppose $\mathbb{S} \cong \mathrm{id}_{\mathcal{C}}$. Let A be a finite-dimensional algebra such that $\mathcal{C} \approx A$-mod, and let J be the Jacobson radical of A. Then there are isomorphisms $\mathrm{CF}_\ell(\mathcal{C}) \cong \mathrm{SLF}(A/J^\ell)$ for all $\ell \geq 1$.*

Proof. We identify $\mathcal{C}$ and $\mathrm{Rex}(\mathcal{C})$ with A-mod and A-mod-A, respectively. Then $\mathcal{C}_\ell$ is identified with A/J^ℓ-mod. Hence we have isomorphisms

$$\mathrm{CF}_\ell(\mathcal{C}) \cong \mathrm{Hom}_{\mathcal{C}}(\rho^{\mathrm{ra}}(A/J^\ell), \mathbb{1}) \cong \mathrm{Hom}_{A\text{-mod-}A}(A/J^\ell, A^*) \cong \mathrm{SLF}(A/J^\ell)$$

by Theorem 5.1. □

How can we give a basis of $\mathrm{CF}(\mathcal{C})$ respecting the filtration? At this time, there is no general answer to this question. Anyway, the above theorem enables us to apply the ring theory for the study of $\mathrm{CF}_\ell(\mathcal{C})$. For $\ell = 2$, this approach has yielded the following result:

Theorem 5.5 ([8, Theorem 6.11]). *Let $\{V_i\}_{i \in I}$ be representatives of the isomorphism classes of simple objects of $\mathcal{C}$. Suppose that the double dual functor on $\mathcal{C}$ is isomorphic to $\mathrm{id}_{\mathcal{C}}$. Then there is an isomorphism*

$$\mathrm{CF}_2(\mathcal{C}) \cong \mathrm{CF}_1(\mathcal{C}) \oplus \bigoplus_{i \in I} \mathrm{Ext}^1_{\mathcal{C}}(V_i, V_i).$$

The proof is outlined as follows: Let A be a finite-dimensional basic algebra such that $\mathcal{C} \approx A$-mod and, for simplicity, identify $\mathcal{C}$ with A-mod. By applying the Taft-Wilson theorem [16] to the pointed coalgebra A^*, we obtain an isomorphism $\mathrm{SLF}(A/J^2) \cong \mathrm{SLF}(A/J) \oplus \bigoplus_{i \in I} \mathrm{Ext}^1_A(V_i, V_i)$. Now the result follows from Theorem 5.4.

5.4. *Hochschild (co)homology*

Finally, we give relations between the adjoint algebra $\mathbb{A}_{\mathcal{C}}$ and the Hochschild (co)homology. We first introduce the following definition: Given a finite abelian category $\mathcal{M}$, we set

$$\mathrm{HH}^\bullet(\mathcal{M}) = \mathrm{Ext}^\bullet_{\mathcal{R}}(\mathrm{id}_{\mathcal{M}}, \mathrm{id}_{\mathcal{M}}) \quad \text{and} \quad \mathrm{HH}_\bullet(\mathcal{M}) = (\mathrm{Ext}^\bullet_{\mathcal{R}}(\mathrm{id}_{\mathcal{M}}, \mathbb{N}_{\mathcal{M}}))^*,$$

where $\mathcal{R} = \mathrm{Rex}(\mathcal{M})$ and $(-)^*$ denotes the dual vector space. These notations are justified as follows: If A is a finite-dimensional algebra such that $\mathcal{M} \approx A$-mod, then we have

$$\mathrm{HH}^\bullet(\mathcal{M}) \cong \mathrm{Ext}^\bullet_{A\text{-mod-}A}(A, A) \cong \mathrm{HH}^\bullet(A),$$

the Hochschild cohomology of A. $\mathrm{HH}_\bullet(\mathcal{M})$ is isomorphic to the Hochschild homology $\mathrm{HH}_\bullet(A)$. To see this, we fix an object $X \in A$-mod-A and its projective resolution $X \leftarrow P^\bullet$. By definition, $\mathrm{Tor}^{A^e}_\bullet(A, M)$ is the homology of the sequence $0 \leftarrow A \otimes_{A^e} P^0 \leftarrow A \otimes_{A^e} P^1 \leftarrow \cdots$. By the tensor-Hom adjunction, we have an isomorphism

$$(A \otimes_{A^e} P^\bullet)^* = \mathrm{Hom}_k(A \otimes_{A^e} P^\bullet, k) \cong \mathrm{Hom}_{A^e}(P^\bullet, A^*)$$

and therefore $(\mathrm{Tor}_\bullet^{A^e}(A,X))^* \cong \mathrm{Ext}_{A^e}^\bullet(X,A^*)$. Hence,

$$\mathrm{HH}_\bullet(\mathcal{M}) \cong (\mathrm{Ext}^\bullet_{A\text{-mod-}A}(A,A^*))^* \cong \mathrm{Tor}_\bullet^{A^e}(A,A)^* \cong \mathrm{HH}_\bullet(A).$$

Theorem 5.6 ([8, Corollary 7.5]). *There is an isomorphism*

$$\mathrm{Ext}^\bullet_{\mathcal{C}}(\mathbb{1},\mathbb{A}_{\mathcal{C}}) \cong \mathrm{HH}^\bullet(\mathcal{C}).$$

If $\mathbb{S} \cong \mathrm{id}_{\mathcal{C}}$, *then there is also an isomorphism* $(\mathrm{Ext}^\bullet_{\mathcal{C}}(\mathbb{A}_{\mathcal{C}},\mathbb{1}))^* \cong \mathrm{HH}_\bullet(\mathcal{C})$.

Proof. Let X be an object of $\mathcal{C}$ and let $X \leftarrow P^\bullet$ be a projective resolution of X. We fix an object F of $\mathrm{Rex}(\mathcal{C})$. By applying $\mathrm{Hom}_{\mathcal{C}}(-,\rho^{\mathrm{ra}}(F))$ to the projective resolution of X, we obtain the following commutative diagram:

$$\begin{array}{ccccccc}
0 & \longrightarrow & \mathrm{Hom}_{\mathcal{C}}(P^0,\rho^{\mathrm{ra}}(F)) & \longrightarrow & \mathrm{Hom}_{\mathcal{C}}(P^1,\rho^{\mathrm{ra}}(F)) & \longrightarrow & \cdots \\
 & & \downarrow\cong & & \downarrow\cong & & \\
0 & \longrightarrow & \mathrm{Hom}_{\mathcal{C}}(\rho(P^0),F) & \longrightarrow & \mathrm{Hom}_{\mathcal{C}}(\rho(P^1),F) & \longrightarrow & \cdots
\end{array}$$

The cohomology of the first row is $\mathrm{Ext}^\bullet_{\mathcal{C}}(X,\rho^{\mathrm{ra}}(F))$. On the other hand, since ρ is exact and preserves projective objects (see the proof of Theorem 4.1), the sequence $\rho(X) \leftarrow \rho(P^\bullet)$ is a projective resolution of $\rho(X)$. Hence the cohomology of the second row is $\mathrm{Ext}^\bullet_{\mathcal{C}}(\rho(X),F)$. In conclusion, we have $\mathrm{Ext}^\bullet_{\mathcal{C}}(X,\rho^{\mathrm{ra}}(F)) \cong \mathrm{Ext}^\bullet_{\mathcal{C}}(\rho(X),F)$. By letting $X = \mathbb{1}$ and $F = \mathrm{id}_{\mathcal{C}}$, we obtain the first isomorphism of this theorem.

Now we suppose $\mathbb{S} \cong \mathrm{id}_{\mathcal{C}}$. To obtain the second isomorphism, we fix an object F of $\mathrm{Rex}(\mathcal{C})$ and consider its projective resolution $F \leftarrow P^\bullet$. Let X be an object of $\mathcal{C}$. By applying $\mathrm{Nat}(-,\rho(X)\circ\mathbb{N}_{\mathcal{C}})$ to the projective resolution of F, we obtain the following commutative diagram:

$$\begin{array}{ccccccc}
0 & \longrightarrow & \mathrm{Nat}(P^0,\rho(X)\circ\mathbb{N}_{\mathcal{C}}) & \longrightarrow & \mathrm{Hom}_{\mathcal{C}}(P^1,\rho(X)\circ\mathbb{N}_{\mathcal{C}}) & \longrightarrow & \cdots \\
 & & \text{Theorem 5.1}\downarrow\cong & & \downarrow\cong & & \\
0 & \longrightarrow & \mathrm{Hom}_{\mathcal{C}}(\rho^{\mathrm{ra}}(P^0),X) & \longrightarrow & \mathrm{Hom}_{\mathcal{C}}(\rho^{\mathrm{ra}}(P^1),X) & \longrightarrow & \cdots.
\end{array}$$

We recall that ρ^{ra} is exact and preserves projective object. By the similar argument as above, we have an isomorphism

$$\mathrm{Ext}^\bullet_{\mathrm{Rex}(\mathcal{C})}(F,\rho(X)\circ\mathbb{N}_{\mathcal{C}}) \cong \mathrm{Ext}^\bullet_{\mathcal{C}}(\rho^{\mathrm{ra}}(F),X),$$

which yields the second isomorphism of this theorem. □

Let H be a finite-dimensional ribbon factorisable Hopf algebra. Then H-mod is a modular tensor category and hence $SL_2(\mathbb{Z})$ acts projectively on the center of H by a result of Lyubashenko. We note that the center of H is the 0-th Hochschild cohomology of H. Lentner, Mierach, Schweigert and Sommerhäuser [14] showed that $SL_2(\mathbb{Z})$ also acts projectively on the higher Hochschild cohomology of H. In [8], the above theorem is used to extend their result to modular tensor categories.

Acknowledgments

The author is grateful to the referee for careful reading of the manuscript. The author is supported by JSPS KAKENHI Grant Number JP16K17568.

References

[1] B. Bakalov and A. Kirillov, Jr., *Lectures on tensor categories and modular functors*, University Lecture Series, Vol. 21 (American Mathematical Society, Providence, RI, 2001).

[2] V. G. Turaev, *Quantum invariants of knots and 3-manifolds*, de Gruyter Studies in Mathematics, Vol. 18 (Walter de Gruyter & Co., Berlin, 1994).

[3] V. Lyubashenko, Modular transformations for tensor categories, *J. Pure Appl. Algebra* **98**, 279 (1995).

[4] V. V. Lyubashenko, Invariants of 3-manifolds and projective representations of mapping class groups via quantum groups at roots of unity, *Comm. Math. Phys.* **172**, 467 (1995).

[5] T. Kerler and V. V. Lyubashenko, *Non-semisimple topological quantum field theories for 3-manifolds with corners*, Lecture Notes in Mathematics, Vol. 1765 (Springer-Verlag, Berlin, 2001).

[6] S. Mac Lane, *Categories for the working mathematician*, Graduate Texts in Mathematics, Vol. 5, second edn. (Springer-Verlag, New York, 1998).

[7] F. Loregian, Coend calculus, preprint 2019, `arXiv:1501.02503v5`.

[8] K. Shimizu, Further results on the structure of (co)ends in finite tensor categories, *Appl. Categ. Structures* **28**, 237 (2020).

[9] C. Kassel, *Quantum groups*, Graduate Texts in Mathematics, Vol. 155 (Springer-Verlag, New York, 1995).

[10] P. Etingof, S. Gelaki, D. Nikshych and V. Ostrik, *Tensor categories*, Mathematical Surveys and Monographs, Vol. 205 (American Mathematical Society, Providence, RI, 2015).

[11] K. Shimizu, The monoidal center and the character algebra, *J. Pure Appl. Algebra* **221**, 2338 (2017).

[12] K. Shimizu, On unimodular finite tensor categories, *Int. Math. Res. Not. IMRN*, 277 (2017).

[13] K. Shimizu, Integrals for finite tensor categories, *Algebr. Represent. Theory* **22**, 459 (2019).
[14] S. Lentner, S. N. Mierach, C. Schweigert and Y. Sommerhäuser, Hochschild cohomology and the modular group, *J. Algebra* **507**, 400 (2018).
[15] J. Fuchs, G. Schaumann and C. Schweigert, Eilenberg-Watts calculus for finite categories and a bimodule Radford S^4 theorem, *Trans. Amer. Math. Soc.* **373**, 1 (2020).
[16] S. Montgomery, *Hopf algebras and their actions on rings*, CBMS Regional Conference Series in Mathematics, Vol. 82 (Published for the Conference Board of the Mathematical Sciences, Washington, DC, 1993).
[17] P. Etingof and V. Ostrik, Finite tensor categories, *Mosc. Math. J.* **4**, 627 (2004).
[18] P. Etingof, D. Nikshych and V. Ostrik, On fusion categories, *Ann. of Math. (2)* **162**, 581 (2005).
[19] K. Shimizu, Non-degeneracy conditions for braided finite tensor categories, *Adv. Math.* **355**, 106778, 36 (2019).
[20] P. Etingof, D. Nikshych and V. Ostrik, An analogue of Radford's S^4 formula for finite tensor categories, *Int. Math. Res. Not.*, 2915 (2004).
[21] A. Bruguières and A. Virelizier, Hopf monads, *Adv. Math.* **215**, 679 (2007).
[22] A. Bruguières and A. Virelizier, Quantum double of Hopf monads and categorical centers, *Trans. Amer. Math. Soc.* **364**, 1225 (2012).
[23] A. Bruguières, S. Lack and A. Virelizier, Hopf monads on monoidal categories, *Adv. Math.* **227**, 745 (2011).
[24] A. M. Gainutdinov and I. Runkel, The non-semisimple Verlinde formula and pseudo-trace functions, *J. Pure Appl. Algebra* **223**, 660 (2019).
[25] A. M. Gainutdinov and I. Runkel, Projective objects and the modified trace in factorisable finite tensor categories, *Compos. Math.* **156**, 770 (2020).
[26] G. Janelidze and G. M. Kelly, A note on actions of a monoidal category 2001/02 pp. 61–91. CT2000 Conference (Como).
[27] P. Deligne, Catégories tannakiennes, in *The Grothendieck Festschrift, Vol. II*, Progr. Math. Vol. 87 (Birkhäuser Boston, Boston, MA, 1990) pp. 111–195.
[28] P. Etingof and S. Gelaki, Exact sequences of tensor categories with respect to a module category, *Adv. Math.* **308**, 1187 (2017).

Boolean graphs - A survey

Tongsuo Wu
School of Mathematical Sciences, Shanghai Jiao Tong University, Shanghai, 200240, China
E-mail: tswu@sjtu.edu.cn
http://math.sjtu.edu.cn/faculty/wuts/

A Boolean graph is the zero divisor graph of a Boolean ring. For a positive integer n, let $[n] = \{1, 2, \ldots, n\}$, and $2^{[n]}$ the power set of $[n]$. A finite Boolean graph B_n is isomorphic to a graph defined on the vertex set $2^{[n]} \setminus \{[n], \emptyset\}$, where two vertices are adjacent if and only if their meet is empty. In this paper, we give a survey of some works done in the area of research related to Boolean graphs, in both graph theoretic and algebraic aspects. We also introduce some most recent works by the author and others.

Keywords: Boolean graphs; uniqueness; characterizations; spectra; Cohen-Macaulayness.

1. Introduction and preliminaries

Unless otherwise specified, all rings in this paper are commutative with identity and all graphs are simple, undirected and connected. We use [1] and [2] as basic references for theory of commutative rings and graphs, respectively. For a commutative semigroup S, recall from [3] that the zero-divisor graph $\Gamma(S)$ is a graph with all nonzero zero divisors as its vertices and, two vertices u and v are adjacent if and only if $uv = 0$. The concept was first introduced and studied by Beck in [4] and later, modified and developed by Anderson and Livingston in [5]; for the detailed history and bibliography, see the recent comprehensive survey works of [6] for commutative rings and, [7] for commutative semigroups.

Recall that a ring R is called *Boolean* if and only if $r^2 = r$ holds for every element r in R. Clearly, a Boolean ring is always commutative with characteristic 2. A graph is called *Boolean* if it is isomorphic to the zero-divisor graph of a Boolean ring. It seems that the definition of a Boolean graph is formally given in [8], and systematic studies began in [9–12] among others. For a finite Boolean graph G, it was noted that in [12] that the complement $\overline{G}$ is isomorphic to the intersection graph on $2^{[n]} \setminus \{[n], \emptyset\}$. Now for any finite or infinite nonempty set S, the complement of the intersection graph defined on $2^S \setminus \{\emptyset, S\}$ is called a *strongly Boolean graph* and is denoted by B_S. It was proved in [12] that each strongly Boolean graph has a unique corresponding semigroup; later in [8], this result was generalized to any Boolean graph, thus each Boolean graph also has a unique Boolean ring. The result is essentially independently proved in [10].

Clearly, a finite Boolean graph B_n is easy to get, but as n grows big, the graph becomes complicated with many vertices and edges, though it is still highly symmetric. We draw in Figure 1 the Boolean graphs B_3 and B_4, for an immediate impression:

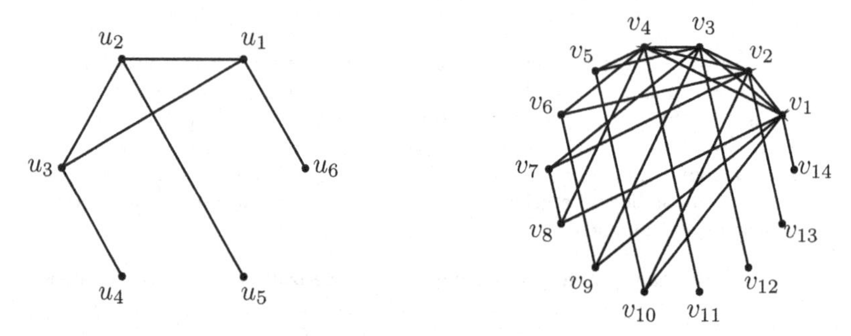

Figure 1. The graphs B_3 and B_4

In this paper, we are going to give a brief survey on the works of D.F. Anderson, L. Chen, D.S. Dillery, J. Guo, J.D. LaGrange, R. Levy, A. Liu, D.C. Lu, A. Mohammadian, J. Shapiro, and the author. We will also introduce some most recent works on the further algebraic combinatorial property of Boolean graphs. This paper is organized as follows. In section 2, we introduce theorems that concern Boolean algebras uniquely determined by their graphs; in section 3, we introduce various characterizations of Boolean graphs, as well as blow-ups of (strongly) Boolean graphs; in section 4, we mainly introduce series works of J.D. LaGrange (and sometimes, together with other coauthors) on the eigenvalue properties of finite Boolean graphs; in section 5, we introduce some most recent works on some properties (e.g., unmixedness and shellability) of finite Boolean graphs B_n, as well as their complement graphs $\overline{B_n}$.

2. Uniqueness of corresponding algebras

We begin this section with the following example:

Example 2.1. *For any finite or infinite set A with α elements, there is an associated commutative semigroup $P_\alpha = \{x_C \mid C \subseteq A\}$. The multiplication of P_α is defined by $x_C x_D = x_{C\cap D}$. It is straightforward to verify that P_α is a commutative semigroup with the identity element x_A. Also $x_\emptyset$ is the zero element of P_α, i.e., $x_\emptyset x_B = x_\emptyset$, for each element $x_B \in P_\alpha$. Then we have semigroup isomorphisms*

$$P_\alpha \cong (2^A, \cap) \cong (2^A, \cup),$$

where 2^A is the power set of A.

Let $\mathbb{Z}_2$ be the ring of integers modulo 2. For any finite number n, let $\mathbb{Z}_2^{(n)}$ be the ring direct sum of n copies of $\mathbb{Z}_2$ and consider its multiplicative semigroup $(Z_2^{(n)}, \cdot)$. When $|A| = n$, it is straightforward to verify that the map

$$\sigma : (2^A, \cap) \to (Z_2^{(n)}, \cdot), B \mapsto (y_1, y_2, \cdots, y_n), \textit{where } y_i = \begin{cases} 0 \textit{ if } i \notin B \\ 1 \textit{ if } i \in B \end{cases}$$

is a semigroup isomorphism.

Denote by B_α the zero-divisor graph of P_α, and call it *a strongly Boolean graph* as in [13]. B_α is a symmetric graph with a moderate number of edges. Below we list some properties of $G = B_\alpha$ for any finite $\alpha = n$:

1. $V(G) = P_n - \{0, 1\}$ and hence, it contains $2^n - 2$ vertices.

2. For any $x_D \in G$ with $|D| = i$, let $N(x_D)$ be the neighborhood of x_D, i.e.,

$$N(x_D) = \{y \in V(G) \mid x_D \textit{ is adjacent to } y \textit{ in } G\}.$$

Then

$$|N(x_D)| = \binom{n-i}{1} + \binom{n-i}{2} + \cdots + \binom{n-i}{n-i} = 2^{n-i} - 1.$$

3. The edge number of B_n is

$$2^{n-2}\binom{n}{1} + 2^{n-3}\binom{n}{2} + \cdots + 2^0\binom{n}{n-1} - 2^{n-1} + 1.$$

4. The clique number of B_n is n. When $n \geq 3$, the diameter of B_n is 3 and B_n has n end vertices. Furthermore, B_n has a unique maximum clique, which is a clique with clique number n.

5. The automorphism group of B_n is the symmetric group S_n on $[n]$, thus this graph is highly symmetric.

A commutative semigroup S is called a *zero-divisor* semigroup, if S contains a zero element 0, and each element of S is a zero-divisor. Notice that each commutative semigroup T with a zero-divisor element has a unique maximal sub-semigroup S which is a zero-divisor semigroup, and $\Gamma(T) = \Gamma(S)$. In the following, α is assumed to be a finite or an infinite cardinal number.

Proposition 2.2. *([12, Theorem 2.2]) Assume $|A| = \alpha \geq 3$ and let $P_\alpha = \{x_B \mid B \subseteq A\}$ be the commutative semigroup defined on 2^A with graph B_α. If S is a commutative zero-divisor semigroup whose graph $\Gamma(S)$ is isomorphic to B_α, then S is isomorphic to the zero-divisor semigroup $P_\alpha - \{1\}$.*

At about the same time, the above uniqueness result was extended in [8] to commutative rings.

Theorem 2.3. *([8, Theorem 4.2]) Let G be a Boolean graph containing more than two vertices. Then G has a unique corresponding zero-divisor semigroup.*

The key for proving the above theorem is the following famous *Stone Representation Theorem* (see [8] for an argument and references):

Theorem 2.4. *For a Boolean algebra B, there exists a compact Hausdorff zero-dimensional topological space X such that B is isomorphic to the Boolean algebra of clopen subsets of X.*

Corollary 2.5. *([8, Corollary 4.3]) Let G be a Boolean graph containing more than two vertices. Then G has a unique corresponding ring.*

Given a commutative ring R, the set of idempotents of R forms a Boolean algebra, denoted $B(R)$, under the operations $a \wedge b = ab$ and $a \vee b = a + b - ab$, with largest element 1, smallest element 0, and complement given by $1 - a$. For two nonzero elements c, d of a von Neumann regular ring R, recall from [9, Lemma 3.1] that $cR = dR$ holds if and only if $\text{ann}(c) = \text{ann}(d)$. Set $c \equiv d$ if $cR = dR$, and let $[c]$ be the equivalent class determined by the equivalence relation $\equiv$, where $c \in [c]$. It is easy to see that each equivalent class $[c]$ has a unique idempotent element e such that $e^2 = e$ and $[c] = [e]$ hold.

In order to derive Corollary 2.5 from Theorem 2.3, the following fundamental result is needed:

Theorem 2.6. *([9, Theorem 4.1]) Let R and S be commutative von Neumann regular rings. Then $\Gamma(R)$ and $\Gamma(S)$ are isomorphic as graphs if and only if there is a Boolean algebra isomorphism $\varphi : B(R) \to B(S)$ such that $||[e]|| = ||[\varphi(e)]||$ for each $e \in B(R)$ with $1 \neq e$.*

Surely, Corollary 2.5 will also follow either from [14, Theorem 2.1], or from [15, Theorem 3.8], together with Theorem 2.3.

The following theorem is essentially identical with Corollary 2.5, but with a quite different approach. It was discovered by LaGrange at about the same time independently.

Theorem 2.7. *([10, Theorem 4.1]) Let R be a ring with nonzero zero-divisors, not isomorphic to $\mathbb{Z}_9$ or $\mathbb{Z}_3[x]/(x^2)$. If S is a Boolean ring such that $\Gamma(R) \cong \Gamma(S)$, then $R \cong S$. In particular, if R and S are Boolean rings, then $\Gamma(R) \cong \Gamma(S)$ if and only if $R \cong S$.*

We remark that the above result is generalized by Ali Mohammadian in [16, Theorem 7] by omitting the commutative with $1 \neq 0$ hypotheses. The following is another related result:

Theorem 2.8. *([16, Theorem 11]) For Boolean rings R and S, $\Gamma(R[x]) \cong \Gamma(S[x])$ if and only if $R \cong S$.*

Recall that a von Neumann regular unitary ring R (may be noncommutative) is called *abelian*, if every idempotent is in the center of R. We end this section by recording two additional uniqueness results on abelian regular rings, which are related to Boolean rings and graphs:

Theorem 2.9. *([14, Theorem 2.5]) Let R and S be abelian regular rings. Then the following conditions are equivalent:*

(1) $\Gamma(B(R)) \cong \Gamma(B(S))$.

(2) $(K_0(R), [R]) \cong (K_0(S), [S])$ *as partially-ordered abelian groups with order units, where $K_0(R)$ is the Grothendieck group of R.*

(3) $B(R) \cong B(S)$ *as Boolean rings.*

Proposition 2.10. *([14, Proposition 2.10]) Let R be an abelian regular ring, which is not a division ring. Then the following conditions are equivalent:*

(1) *The graph $\Gamma(R)$ is Boolean.*

(2) *The natural group homomorphism $\varphi : Aut(R) \to Aut(\Gamma(R))$ is isomorphic.*

3. (Blow-ups of) strongly Boolean graphs: Characterizations

Recall from [9] that a graph G is called *complemented*, if for each vertex u of G, there is a vertex $v \neq u$ such that $\{u, v\} \in E(G)$ and the edge is not in a triangle in G. In this case, v is called a *complement of* u and, the relation is denoted as $u \perp v$. Further, a complemented graph G is called *uniquely complemented*, if for each vertex u of G, $u \perp v$ and $u \perp w$ implies $N(v) = N(w)$. We begin with the following characterization of a Boolean graph:

Theorem 3.1. *([15, Theorem 3.8]) A simple connected graph G is a Boolean graph if and only if the following conditions hold:*

(1) *The graph G is N-determined; i.e. no distinct vertices v, w have a same neighborhood.*

(2) *The graph G is (uniquely) complemented.*

(3) *For any $x, y \in V(G)$ with $N(x) \cap N(y) \neq \emptyset$, there exists some $z \in V(G)$ such that $N(x) \cap N(y) = N(z)$.*

(4) *There exists a semilattice S with zero element 0 such that $G = \Gamma(S)$.*

The following characterizes Boolean rings by the property of the graph $\Gamma(R)$:

Theorem 3.2. *([10, Theorem 2.5]) Let R be a ring such that the graph $\Gamma(R)$ has at least three vertices. Then R is Boolean if and only if each vertex of $\Gamma(R)$ has a unique complement.*

Actually, in [9, 10] the authors have had systematic studies on several subclasses of Boolean rings R, including *rationally complete Boolean rings.* Recall that a Boolean algebra is complete if every subset has an infimum. A Boolean ring R is said to be rationally complete, if the Boolean algebra $B(R)$ is complete. In particular, the subclass can be characterized via the graph property of $\Gamma(R)$. For this, we introduce a new concept. For a nonempty subset S of $R \setminus \{1, 0\}$, if a nonzero element e in R is such that $e \cdot S = 0$, then e is called a *central* vertex of S in $\Gamma(R)$. For example, it follows from $e(1-e) = 0$ that e is a central vertex of the set $(1-e)R \setminus \{0\}$.

Theorem 3.3. *([10, Theorem 3.4]) For a Boolean ring R, let $G = \Gamma(R)$. Then the following conditions are equivalent:*

(1) *R is rationally complete.*

(2) *For any nonempty vertex subset S of G, if S has a central vertex in G, then S has a central vertex such that its complement in G is adjacent to every central vertex of S.*

Theorem 3.4. *([17, Corollary 2.5]) Let $G = \Gamma(R)$, where R is a commutative ring neither isomorphic to $\mathbb{Z}_2$ or $\mathbb{Z}_2 \oplus \mathbb{Z}_2$. Then the following conditions are equivalent:*

(1) *R is a Boolean ring.*

(2) $\Gamma(R)$ *is* N*-determined and,* $N(A) = N(x)$ *for some distinct* $\{x\} \subseteq V(G), A \subseteq V(G)$.

A semigroup is called *reduced* if it contains no nonzero nilpotent elements. For a graph G, let

$$\mathcal{H}(G) = \{N(v) \mid v \in V(G)\} \cup \{\emptyset, V(G)\},$$

and let $\mathcal{L}(G)$ be the complete lattice consisting of all possible intersection of elements from $\mathcal{H}(G)$. The following is contained in [18, Corollary 1.2]

Theorem 3.5. *([18, Corollary 1.2]) Let* G *be a graph with* $|V(G)| \geq 2$*. Then the following conditions are equivalent:*

(1) $G \cong \Gamma(S)$ *for a reduced semigroup (Boolean semigroup or meet semilattice, respectively)* S.

(2) $\mathcal{H}(G)$ *is a join-semilattice and,* $\mathcal{L}(G)$ *is a Boolean algebra.*

(3) $\mathcal{H}(G)$ *is a join-semilattice and,* $G \cong \Gamma(P)$ *for some poset* P.

Now we return to the characterizations of strongly Boolean graphs introduced in the beginning of section 2. First, we have

Theorem 3.6. *([19, Theorem 2.2]) Let* G *be a graph with a maximum clique* S*. Then* G *is isomorphic to the strongly Boolean graph* B_S *if and only if the following properties are satisfied:*

(1) G *is connected and for each nontrivial subset* A *of* $V(S)$*, there exists a vertex* $v \in V(G)$ *such that* $A = N(v) \cap V(S)$.

(2) G *is* $S \cap N$*-determined, i.e.,* $V(S) \cap N(x) = V(S) \cap N(y)$ *implies* $x = y$ *for vertices* $x, y \in V(G)$.

(3) *For vertices* $x, y \in V(G)$*,* $V(S) \subseteq N(x) \cup N(y)$ *holds if and only if* $x \in N(y)$.

We call a graph G *satisfying the neighbourhood condition* (abbreviated as the N-condition), if for each pair of nonadjacent vertices $u, v \in V(G)$, there exists a vertex w such that $N(u) \cup N(v) \subseteq N(w)$. Recall that this is one of the four necessary conditions given for a graph G in [20, Theorem 1], such that $G \cong \Gamma(T)$ holds for a semigroup T. We also need the following condition introduced in [19]:

Definition 3.7. *Let* G *be a graph with a maximum clique.* G *is called satisfying the* M*-condition, if for a maximum clique* S *of* G *and each induced discrete subgraph* D *of* G *with* $V(S) \nsubseteq \cup_{x \in V(D)} N(x)$*, there exists a vertex* $z \in V(G)$*, such that the following are satisfied:*

(1) $\cup_{x\in V(D)} N(x) \subseteq N(z)$;
(2) $V(S) \cap (\cup_{x\in V(D)} N(x)) = V(S) \cap N(z)$.

Theorem 3.8. *([19, Theorem 5.3]) Let G be a graph with a maximum clique S. Then G is a Boolean graph if and only if G is uniquely complemented and satisfies conditions M and N.*

Recall that to *blow-up a graph* G is to replace every vertex v of G by a set T_v to get a possibly new and larger graph G_T, where $v \in T_v$ holds for each vertex v. The induced subgraph of G_T on T_v is a discrete graph, i.e., a graph without any edge, while for distinct vertices x, y of G, each vertex of T_x is adjacent to all vertices of T_y in G_T if and only if x is adjacent to y in G, see [19] for the listed references. Recall also another concept, *graph expanding*, and note that G is a blow-up of a graph H if and only if $\overline{G}$ is an expanding of $\overline{H}$. We illustrate these concepts and relation via the following figure:

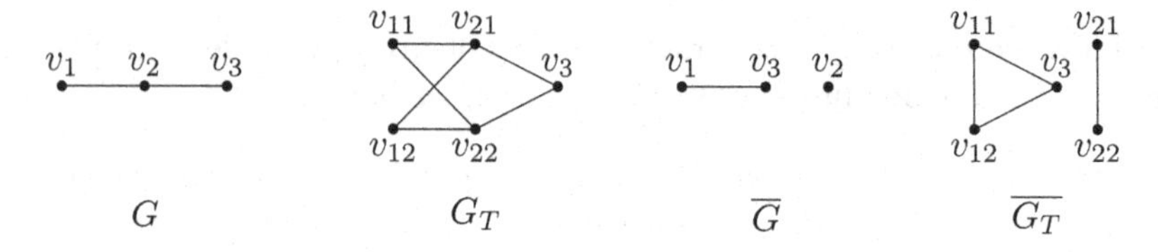

Figure 2. Graph blow-up and expanding

Now we are ready to introduce the following:

Theorem 3.9. *([19, Theorem 5.1]) Let G be a graph with a maximum clique S. Then G is a graph blow-up of the strongly Boolean graph B_S if and only if the following properties are satisfied:*

(1) For each vertex $v \in V(G)$, $N(v) \cap V(S)$ is a nontrivial subset of $V(S)$.

(2) For each nontrivial subset $A \subseteq V(S)$, there exists a vertex $v \in V(G)$ such that $N(v) \cap V(S) = A$.

(3) For vertices $x, y \in V(G)$, $V(S) \subseteq N(x) \cup N(y)$ holds if and only if $x \in N(y)$.

Theorem 3.10. *([19, Theorem 2.6]) Let G be a graph with a maximum clique S. Then G is a blow-up of a Boolean graph if and only if G is a complemented graph and satisfies conditions M and N.*

Recall that the comaximal ideal graph $\mathcal{C}(R)$ of a ring R is a simple connected graph, with vertex set $\{I \mid I \text{ is a proper ideal of } R, \text{ and } I \nsubseteq J(R)\}$, where I is adjacent to J if and only if $I + J = R$. Clearly, S is a maximum clique of $\mathcal{C}(R)$, which is induced by $Max(R)$ in $\mathcal{C}(R)$; see [21, 22] for further discussions. It is observed in [19] that the graph $\mathcal{C}(R)$ satisfies the conditions (N) and (M); furthermore, if $\mathcal{C}(R)$ has a finite clique number, then $\mathcal{C}(R)$ is complemented and thus it is a blow-up of a Boolean graph by Theorem 3.8, a fact discovered first in [23]. Note that $\mathcal{C}(R)$ may not be a blow-up of B_∞ when $\omega(\mathcal{C}(R)) = \infty$, see [21].

For any commutative ring R, there is a compressed graph $\Gamma_E(R)$ assigned to $\Gamma(R)$, see [24] for more details and further studies. In fact, having a same annihilating ideal in R defines an equivalence relation in R, and the set of equivalence classes forms a commutative monoid, and this monoid is Boolean if the ring R is reduced. The zero-divisor graph of this monoid is denoted as $\Gamma_E(R)$. Clearly, for a reduced ring R, the graph $\Gamma(R)$ is a blow-up of the compressed graph $\Gamma_E(R)$. Among many interesting results in [24], we record the following:

Theorem 3.11. *([24, Theorem 2.11]) Let R be a reduced commutative ring, which is not a domain. Then the following statements are equivalent:*

(1) $\Gamma_E(R) \cong \Gamma(R)$.

(2) *The natural map* $\Gamma(R) \to \Gamma_E(R)$ *is a graph isomorphism.*

(3) R *is a Boolean ring.*

If we consider this theorem and the characterizations, it is natural to pose the following question: For what kind of reduced rings R are the graphs $\Gamma_E(R)$ Boolean?

Finally, let us turn to the *annihilating ideal graph* of a ring, which was first introduced and studied in [25]. This graph provides another excellent setting for studying some aspects of algebraic properties of a commutative ring, especially, the ideal structure of a ring. For a ring R, let $\mathbb{I}(R)$ be the set of ideals of R, $\mathbb{A}(R)$ the set of annihilating-ideals of R, where a nonzero ideal I of R is called an *annihilating-ideal* if there exists a nonzero ideal J of R such that $IJ = \{0\}$. Recall that the annihilating-ideal graph $\mathbb{AG}(R)$ of R is a simple graph with vertex set $\mathbb{A}(R)$, such that distinct vertices I and J are adjacent if and only if $IJ = \{0\}$.

Theorem 3.12. *([13, Theorem 3.5]) For a commutative ring R, let $G = \mathbb{AG}(R)$ be its annihilating-ideal graph. If G has a maximum clique S with $3 \le |V(S)| \le \infty$, then the following statements are equivalent:*

(1) R *is a reduced ring.*

(2) G *is a blow-up of a strong Boolean graph.*

(3) *G is a complemented graph.*

We remark that this theorem is analogous to Theorem 3.5 of [9]. Also, it is shown by Theorem 3.5 of [9] that the zero-divisor graph $\Gamma(R)$ of a ring R is complemented if and only if the total quotient ring $T(R)$ of R is von Neumann regular. However, it is observed that this is not true for annihilating-ideal graphs, as shown in [13, Example 3.6].

The following provides an analogue to Theorem 2.5 of [10]:

Theorem 3.13. *([13, Theorem 3.8]) Let R be a commutative ring, and let $G = \mathbb{AG}(R)$ be its annihilating-ideal graph. If G has a maximum clique S with $3 \leq |V(S)| < \infty$, then the following statements are equivalent:*

(1) *R is a finite direct product of fields.*

(2) *G is a strongly Boolean graph.*

(3) *Every element of G has a unique complement.*

4. Spectra of Boolean graphs

For any finite graph G with vertex set $\{v_i \mid 1 \leq i \leq n\}$, there is an $n \times n$ adjacency matrix $A =: A(G)$ related to G. For distinct i, j, the (i, j)-entry of A is 1 if v_i is adjacent to v_j, and 0 otherwise; while each diagonal entry is 0. Note that $A(G)$ is a symmetric $(0, 1)$-matrix, thus all eigenvalues of $A(G)$ are real numbers. The family of eigenvalues of A are also called the eigenvalues of the graph G, and they form the spectra of G. It is known that there exists a lot of literature on the spectrum of a graph, and LaGrange's works show that a finite Boolean graph is fully characterized by properties of its spectra, as the following typical result shows:

Theorem 4.1. *([26, Corollary 3.5]) Let R be a finite commutative ring that is not isomorphic to $\mathbb{Z}_2, \mathbb{Z}_9$ or $\mathbb{Z}_3[x]/(x^2)$ and let A be the adjacency matrix of the graph $\Gamma(R)$. Then the following are equivalent:*

(1) *R is a Boolean ring.*

(2) *Every vertex of $\Gamma(R)$ has a unique complement.*

(3) *The spectra of $\Gamma(R)$ can be partitioned into 2-element subsets of the form $\{\lambda, \pm 1/\lambda\}$.*

(4) *The determinant of A equals to -1.*

(5) *The eigen polynomial $f(x)$ of A is absolutely palindromic, i.e., if $f(x) = x^m + a_{m-1}x^{m-1} + \cdots + a_0$, then $|a_r| = |a_{m-r}|$ for each integer $0 \leq r \leq m$.*

More reciprocal eigenvalue properties are investigated in [26]. Furthermore, two $(k-1) \times (k-1)$ matrices, P_k and Q_k, are introduced as follows:

$\forall i, j \in [1, k-1]$,

$$P_k(i,j) = \begin{cases} \begin{pmatrix} i \\ k-j \end{pmatrix}, & if\, i+j \geq k \\ 0, & if\, i+j < k \end{cases} \quad (1)$$

$$Q_k(i,j) = \begin{cases} \begin{pmatrix} i-1 \\ k-j-1 \end{pmatrix}, & if\, i+j \geq k \\ 0, & if\, i+j < k \end{cases} \quad (2)$$

and, connections of their spectras with that of the Boolean graph B_k are investigated, see [27] and [29] for the technical details.

In [28], it is proved that the adjacency matrix of a finite Boolean graph can always be obtained by reducing the entries of Pascal's triangle (mod 2), and also where it is proved that a finite commutative ring with identity and more than four elements is a Boolean ring if and only if the eigenvalues over the algebraic closure of $\mathbb{F}_2$ are precisely the nonzero elements of $\mathbb{F}_4$ (neglecting multiplicities).

5. Some further algebraic combinatorial properties of finite graphs B_n

Recall that a graph G is said to be *unmixed*, if all minimal vertex covers of G have the same cardinality; a graph is unmixed if and only if the clique complex Δ of the complement $\overline{G}$ is unmixed, i.e., all facets of Δ have a same cardinality, see e.g., the book [30]. Recall that a graph G is *vertex decomposable* if either it has no edges, or else has some vertex x such that we have as follows:

(1) Both $G \setminus N_G[x]$ and $G \setminus x$ are vertex decomposable, where $N_G[x] = N_G(x) \cup \{x\}$.

(2) For every independent set S in $G \setminus N_G[x]$, there exists a vertex $y \in N_G(x)$ such that $S \cup \{y\}$ is independent in $G \setminus x$.

Recall that a simplicial complex Δ is called *(nonpure) shellable*, if there is a shelling order $F_1, \ldots, F_r$ of all facets, such that for each $1 \leq i \leq r-1$, the simplicial complex $\langle F_1, \ldots, F_i \rangle \cap \langle F_{i+1} \rangle$ is pure of dimension $dim\, F_{i+1} - 1$. Note that $F_1, \ldots, F_r$ of all facets of Δ is a shelling order if and only if for each pair (i,j) with $1 \leq i < j \leq r$, there exists an integer k with $1 \leq k < j$, such that both $|F_j \setminus F_k| = 1$ and $F_j \setminus F_k \subseteq F_j \setminus F_i$ holds; if it is assumed further that $F_k \setminus F_j \subseteq F_i$, then $F_1, \ldots, F_r$ is called a *strong shelling order*. By

[31], a nonpure simplicial complex Δ is called strongly shellable, provided that there exists a strong shelling order in $\mathcal{F}(\Delta)$. Recall from [31] that there is no implicit relationship between those two concepts; refer to [32] for further discussions on strongly shellable clutters.

Note that an unmixed vertex decomposable graph G is Cohen-Macaulay, thus its edge ideal $I(G)$ is Cohen-Macaulay, i.e., the ring $k[x_1, \ldots, x_n]/I(G)$ is Cohen-Macaulay, see [30] or [33] for details; while a strongly shellable simplicial complex Δ has the property that both the Stanley-Reisner ideal $I_{\Delta^\vee}$ (of the Alexander dual complex of Δ) and the facet ideal $I(\Delta)$ have linear quotients, thus also have linear minimal free resolution.

Theorem 5.1. *Let $G = B_n$ be the finite Boolean graph. Then*
(1) *([34, Theorem 2.1]) G is unmixed, and* $\text{height}(I(G)) = \frac{|V(G)|}{2}$.
(2) *([34, Theorem 2.4]) G is vertex decomposable, thus also Cohen-Macaulay.*
(3) *([35, Proposition 3.3]) The clique complex of $\overline{G}$ is strongly shellable.*

We remark that the proof to [34, Theorem 2.4] is an algorithm for finding the series of decomposable vertices such that the second condition in the definition is automatically fulfilled in each step. In spite of this, we can check for $2 \leq n \leq 7$ in a common PC via the algorithm.

Note that the complement $\overline{B_n}$ of the graph B_n is the well-known *intersection graph*, thus we are able to have an induction proof to the following, in addition to another algorithm:

Theorem 5.2. *([34, Theorem 3.1]) Let $G = \overline{B_n}$ with $n \geq 4$. Then the graph G is nonpure shellable, thus the edge ideal $I(G)$ is sequentially Cohen-Macaulay.*

Let G be either the Boolean graph B_n or its complement $\overline{B_n}$, with $n \geq 4$, and let Δ be the clique complex of the graph G. Then the edge ideal $I(G)$ does not have a linear minimal free resolution, and it implies that $I(G)$ does not have linear quotients, or equivalently, the Alexander dual complex $\Delta^\vee$ is not shellable. In contrast, the graph B_3 has very good properties, e.g., $I(B_3)$ has 2-linear resolution, while $I_{\Delta^\vee}$ has 3-linear resolution, see [34] for details.

ACKNOWLEDGMENTS. The author expresses his sincere gratitude to the referees for helpful suggestions. The author was supported by NSF of Shanghai (No. 19ZR1424100) and partly by NNSF No. 11971338.

References

[1] D. Eisenbud. *Commutative Algebra with a View Toward Algebraic Geometry*. (Springer Science Business Media, Inc 2004.)
[2] R. Diestel. *Graph Theory*, 2nd edition, GTM 173 (Springer, Berlin, Heidelberg, New York, 2000.)
[3] F.R. DeMeyer, T. McKenzie and K. Schneider, The zero-divisor graph of a commutative semigroup, Semigroup Forum **206** 65(2002).
[4] I. Beck. Coloring of commutative rings. J. Algebra **208** 116(1988).
[5] D.F. Anderson and P.S. Livingston. The zero-divisor graph of a commutative ring. J. Algebra **434** 217(1999).
[6] D.F. Anderson, M.C. Axtell and J.A. Stickles, Zero-divisor graphs in commutative rings, Commutative Algebra, Noetherian and Non-Noetherian Perspectives (M. Fontana, S.E. Kabbaj, B. Olberding, I. Swanson eds.), Springer Science+Business Media **434-447** (2011).
[7] D.F. Anderson and A. Badawi. The zero-divisor graph of a commutative semigroup: a survey. Groups, modules, and model theory-surveys and recent developments, 23 – 39, Springer, Cham (2017).
[8] D.C. Lu, T.S. Wu. The zero-divisor graphs which are uniquely determined by neighborhoods. Comm. Algebra **3855** 35 : 12(2007).
[9] D.F. Anderson, R. Levy and J. Shapiro. Zero-divisor graphs, von Neumann regular rings, and Boolean algebras. J. Pure Appl. Algebra **221** 180(2003).
[10] J.D. LaGrange. Complemented zero-divisor graphs and Boolean rings. J. Algebra **600** 315(2007).
[11] R. Levy, J. Shapiro. The zero-divisor graph of von Neumann regular rings. Comm. Algebra **745** 30 : 2(2002).
[12] T.S. Wu, L. Chen. Simple graphs and commutative zero-divisor semigroups. Algebra Colloq. **211** 16 : 2(2009).
[13] J. Guo, T.S. Wu and H.Y. Yu. On rings whose annihilating-ideal graphs are blow-ups of a Boolean graph. J. Korean Math. Soc. **847** 54 : 3(2017).
[14] D.C. Lu, W.T. Tong. The zero-divisor graphs of Abelian regular rings. Northeast Math. J. **339** 20 : 3(2004).
[15] T.S. Wu, D.C. Lu. Sub-semigroups determined by the zero-divisor graph. Discrete Math. **122** 308 : 22(2008).
[16] A. Mohammadian. On zero-divisor graphs of Boolean rings. Pacific J. Math. **375** 251 : 2(2011).
[17] J.D. LaGrange. Characterizations of three classes of zero-divisor

graphs. Canad. Math. Bull. **127** 55 : 1(2012).
[18] J.D. LaGrange. Annihilators in zero-divisor graphs of semilattices and reduced commutative semigroups. J. Pure Appl. Algebra **2955** 220 : 8((2016).
[19] J. Guo, T.S. Wu and M. Ye. Complemented graphs and blow-ups of Boolean graphs, with applications to comaximal ideal graphs. Filomat **897** 29 : 4(2015).
[20] F.R. DeMeyer, L. Demeyer. Zero-divisor graphs of semigroups. J. Algebra **190** 283 : 1 (2005).
[21] M. Ye, T.S. Wu, Q. Liu and J. Guo. Graph blow-up and its applications in comaximal ideal graphs. J. Algebra Appl. 14 : 3(2015), 1550027 (13 pages).
[22] M. Ye, T.S. Wu, Q. Liu and H.Y. Yu. Implements of graph blow-up in comaximal ideal graphs. Comm. Algebra **2476** 42 : 6(2014).
[23] S.M. Moconja, Z.Z. Petrovic. On the structure of comaximal graphs of commutative rings with identity. Bull. Aust. Math. Soc. **11** 83(2011).
[24] D.F. Anderson, J.D. LaGrange. Commutative Boolean monoids, reduced rings, and the compressed zero-divisor graph. J. Pure Appl. Algebra **1626** 216 : 7(2012).
[25] M. Behboodi, Z. Rakeei. The annihilating-ideal graph of commutative rings I. J. Algebra Appl. **727** 10 : 4(2011).
[26] J.D. LaGrange. Boolean rings and reciprocal eigenvalue properties. Linear Algebra Appl. **1863** 436 : 7(2012).
[27] J.D. LaGrange. Spectra of Boolean graphs and certain matrices of binomial coefficients. Int. Electron. J. Algebra **78** 9(2011).
[28] D.S. Dillery and J.D. LaGrange. Spectra of Boolean graphs over finite fields of characteristic two. Canad. Math. Bull. **58** 63 : 1(2020).
[29] J.D. LaGrange. Eigenvalues of Boolean graphs and Pascal-type matrices. Int. Electron. J. Algebra **109** 13(2013).
[30] R.H. Villarreal. *Monomial Algebra.* Second Edition (Taylor & Francis Group, LLC 2015; First Edition: Marcel Dekker Inc, New York, 2001.)
[31] J. Guo, Y.H. Shen and T.S. Wu. Strongly shellable simplicial complexes. J. Korean Math. Soc. **1613** 56 : 6(2019).
[32] J. Guo, Y.H. Shen and T.S. Wu. Edgewise strongly shellable clutters. J. Algebra Appl. 17 : 1(2018) 1850018 (19 pages).
[33] T.S. Wu and J. Guo. Complexes and Cohen-Macaulay property. 220pp. (Preprint 2019).
[34] A.-M. Liu, T.S. Wu. Boolean graphs are Cohen-Macaulay. Comm. Algebra **4498** 46 : 10 (2018).

[35] A.-M. Liu, T.S. Wu. A construction of sequentially Cohen-Macaulay graphs. Algbera Colloq. (To appear)

PART C

General Lectures (Original Articles)

On two-sided Harada rings constructed from QF rings

Yoshitomo Baba

Department of Mathematics Education, Osaka Kyoiku University, Osaka 582-8582, Japan

E-mail: ybaba@cc.osaka-kyoiku.ac.jp

In [8] M. Harada studied a left artinian ring R such that every non-small left R-module contains a non-zero injective submodule. (We can see the results also in his lecture note [9, §10.2].) In [10] K. Oshiro called the ring a left H-ring and later in [11] he called it a left Harada ring. Since then many significant results are invented. We can see many results on left Harada rings in [4] and many equivalent conditions in [3, Theorem B]. In [5] we introduce "H-epimorphism" and "co-H-sequence" and, in a two-sided Harada ring, we characterize the structure of a right Harada ring using a well-indexed set of a left Harada ring. Further in [6] we introduce another new concept "weak co-H-sequence" and study two-sided Harada ring. In this paper, from a given QF ring, we construct two-sided Harada rings.

Keywords: Harada ring, Artinian ring, quasi-Frobenius ring.

1. Introduction

Throughout this section, let R be a basic artinian ring with $J = J(R)$.

Let $\{e_i\}_{i=1}^n$ be a complete set of orthogonal primitive idempotents of R and let $\{f_i\}_{i=1}^k \subseteq \{e_i\}_{i=1}^n$. A sequence f_1R, f_2R, $\ldots$, f_kR is called a *right co-H-sequence* of R if the following (CHS1), (CHS2), (CHS3) hold:

(CHS1) For each $i = 1, 2, \ldots, k-1$, there exists an R-isomorphism ξ_i : $f_iR_R \to f_{i+1}J_R$.

(CHS2) The last term f_kR_R is injective.

(CHS3) f_1R, f_2R, $\ldots$, f_kR is the longest sequence among the sequences which satisfy (CHS1), (CHS2), i.e., there does not exist an R-isomorphism: $fR_R \to f_1J_R$, where $f \in \{e_i\}_{i=1}^n$.

Similarly, we define a *left co-H-sequence* $Rf_1, Rf_2, \ldots, Rf_k$ of R.

R is called a *left Harada ring* if there exists a basic set $\{e_{i,j}\}_{i=1,j=1}^{m\ \ n(i)}$ of orthogonal primitive idempotents of R such that $e_{i,n(i)}R$, $e_{i,n(i)-1}R$, $\ldots$, $e_{i,1}R$ is a right co-H-sequence of R for all $i = 1, 2, \ldots, m$.

And then we call the set $\{e_{i,j}\}_{i=1,j=1}^{m\ n(i)}$ a *well-indexed set of left Harada ring* or a *left well-indexed set.*

We call $\varphi : f_1R_R \to f_2J_R$ (resp. ${}_RRf_1 \to {}_RJf_2$) for primitive idempotents f_1, f_2 of R a *right* (resp. *left*) *H-epimorphism* if φ is a non-zero R-epimorphism with $J \cdot \mathrm{Ker}\,\varphi = 0$ (resp. $\mathrm{Ker}\,\varphi \cdot J = 0$).

Let $f_1, f_2, \ldots, f_k$ be distinct elements in $\{e_{i,j}\}_{i=1,j=1}^{m\ n(i)}$. A sequence Rf_1, Rf_2, ..., Rf_k is called a *left weak co-H-sequence* (or simply *left w-co-H-sequence*) if the following (WCHS1), (WCHS2) hold.

(WCHS1) For any $i = 1, 2, \ldots, k-1$, there exists a left H-epimorphism $\zeta_i : {}_RRf_i \to {}_RJf_{i+1}$.

(WCHS2) There exists neither a left H-epimorphism $\zeta : {}_RRf \to {}_RJf_1$ nor a left H-epimorphism $\zeta' : {}_RRf_k \to {}_RJf'$ for any $f,\ f' \in \{e_{i,j}\}_{i=1,j=1}^{m\ n(i)} - \{f_1, f_2, \ldots, f_k\}$, i.e., Rf_1, Rf_2, ..., Rf_k is the longest sequence in the set of all sequences which consist of distinct terms and satisfy (WCHS 1).

Further a left w-co-H-sequence Rf_1, Rf_2, ..., Rf_k is called a *left cyclic weak co-H-sequence* if there exists a left H-epimorphism $\zeta_k : {}_RRf_k \to {}_RJf_1$. Similarly, we define a right (*cyclic*) *weak co-H-sequence* f_1R, $f_2R, \ldots, f_kR$.

According to [1, Theorem 3.1], for primitive idempotents e, f of R, we call

$$(eR,\ Rf)$$

is an *i-pair* if $S(eR_R) \cong T(fR_R)$ and $S({}_RRf) \cong T({}_RRe)$, where $S(eR_R)$ and $T(fR_R)$ mean the socle of eR_R and the top fR/fJ of fR_R, respectively.

Throughout this paper, for primitive idempotents e, f and g, we use the following terminologies.

- If $S({}_{eRe}eRf) = S(eRf_{fRf})$ and it is simple both as a left eRe-module and as a right fRf-module, then we abbreviate it to
$$S(eRf)\,.$$
- For $a \in R$, we write the left (resp. right) multiplication map by a
$$(a)_L \ \ (\text{resp. } (a)_R\,)\,.$$

2. Left QF-well-indexed set of QF rings

Definition 2.1. Let Q be an indecomposable basic QF ring. Then we call $\{f'_{i,s}\}_{i=1,s=1}^{m'\ \delta'_i}$ a *left* QF-*well-indexed set* of Q if $\{f'_{i,s}\}_{i=1,s=1}^{m'\ \delta'_i}$ is a complete set of orthogonal primitive idempotents of Q which satisfies the following two conditions:

(QFWI 1) $Qf'_{i,1}, Qf'_{i,2}, \ldots, Qf'_{i,\delta'_i}$ is a left w-co-H-sequence for any $i = 1, 2, \ldots, m'$.

(QFWI 2) If $\delta'_i \geq 2$, then $(f'_{i,s}Q, Qf'_{i,s})$ is an i-pair for any $s = 1, 2, \ldots, \delta'_i$.

Left QF-well-indexed sets have the following equivalent conditions.

Lemma 2.1. *Let Q be an indecomposable basic QF ring and let $\{ f'_{i,s} \}_{i=1,s=1}^{m' \ \delta'_i}$ be a complete set of orthogonal primitive idempotents of Q which satisfies (QFWI 2). The following are equivalent.*

(a) *$\{ f'_{i,s} \}_{i=1,s=1}^{m' \ \delta'_i}$ satisfies (QFWI 1), i.e., $\{ f'_{i,s} \}_{i=1,s=1}^{m' \ \delta'_i}$ is a left* QF*-well-indexed set of Q.*

(b) (i) *If $\delta'_i \geq 2$, then ${}_QQf'_{i,s}/S({}_QQf'_{i,s}) \cong {}_QJ(Q)f'_{i,s+1}$ for any $s = 1, 2, \ldots, \delta'_i - 1$.*

(ii) *For any $i = 1, 2, \ldots, m'$ and $f \in \{ f'_{j,t} \}_{j=1,t=1}^{m' \ \delta'_j} - \{ f'_{i,s} \}_{s=1}^{\delta'_i}$ with (fQ, Qf) an i-pair, both ${}_QQf/S({}_QQf) \not\cong {}_QJ(Q)f'_{i,1}$ and ${}_QQf'_{i,\delta'_i}/S({}_QQf'_{i,\delta'_i}) \not\cong {}_QJ(Q)f$ hold.*

(a′) *$f'_{i,\delta'_i}Q, f'_{i,\delta'_i-1}Q, \ldots, f'_{i,1}Q$ is a right w-co-H-sequence for any $i = 1, 2, \ldots, m'$.*

(b′) (i) *If $\delta'_i \geq 2$, then $f'_{i,s+1}Q_Q/S(f'_{i,s+1}Q_Q) \cong f'_{i,s}J(Q)_Q$ for any $s = 1, 2, \ldots, \delta'_i - 1$.*

(ii) *For any $i = 1, 2, \ldots, m'$ and $f \in \{ f'_{j,t} \}_{j=1,t=1}^{m' \ \delta'_j} - \{ f'_{i,s} \}_{s=1}^{\delta'_i}$ with (fQ, Qf) an i-pair, both $fQ_Q/S(fQ_Q) \not\cong f'_{i,\delta'_i}J(Q)_Q$ and $f'_{i,1}Q/S(f'_{i,1}Q_Q) \not\cong fJ(Q)_Q$ hold.*

Proof. $(a) \Leftrightarrow (b)$ We claim that there exists a left H-epimorphism $\zeta : {}_QQf \to {}_QJ(Q)g$ if and only if $\mathrm{Ker}\,\zeta = S({}_QQf)$, i.e., ${}_QQf/S({}_QQf) \cong {}_QJ(Q)g$. In fact, $(\Rightarrow)$ follows from [5, Theorem 2.2 (I)(2)(i)] and $(\Leftarrow)$ is clear. So the statement holds since $\{ f'_{i,s} \}_{i=1,s=1}^{m' \ \delta'_i}$ satisfies (QFWI 2).
$(a) \Leftrightarrow (a')$ It follows from [6, Theorem 2.2 (1)].
$(a') \Leftrightarrow (b')$ We see by the same way as in $(a) \Leftrightarrow (b)$. □

Further left QF-well-indexed sets have the following properties.

Lemma 2.2. *Let Q be an indecomposable basic QF ring with a left* QF*-well-indexed set $\{ f'_{i,s} \}_{i=1,s=1}^{m' \ \delta'_i}$. Then, since QF rings are two-sided Harada rings, bijection $\psi : \{ 1, 2, \ldots, m' \} \to \{ 1, 2, \ldots, m' \}$ given in [6, §3] is defined. With respect to ψ, the following hold:*

(1) *For any* $i = 1, 2, \ldots, m'$ *and any* $s = 1, 2, \ldots, \delta'_i$, $(f'_{\psi(i),s}Q,\ Qf'_{i,s})$ *is an* i*-pair. So* $S(\, f'_{\psi(i),s}Qf'_{i,s}\,)$ *is defined.*

(2) *In particular, if* $\delta'_i \geq 2$, *then* $\psi(i) = i$.

(3) *If* $\delta'_i = 1$, *then* $\delta'_{\psi(i)} = 1$.

Proof. (1) For any $i \in \{1, 2, \ldots, m'\}$, as we point out just after the definition of ψ in [6, §3], $(f'_{\psi(i),1}Q,\ Qf'_{i,p_i(1)})$ is an i-pair. So, since Q is a QF ring, $(f'_{\psi(i),1}Q,\ Qf'_{i,1})$ is an i-pair. Therefore $(f'_{\psi(i),2}Q,\ Qf'_{i,2})$ is an i-pair by [6, Theorem 2.2 and Lemma 2.4 (I)]. And inductively we see that $(f'_{\psi(i),s}Q,\ Qf'_{i,s})$ is an i-pair for any $s = 1, 2, \ldots, \delta'_i$. Hence we can define $S(\, f'_{\psi(i),s}Qf'_{i,s}\,)$ by, for instance, [2, Lemma 1 (2),(3)].

(2) It is clear from (1) and the definition of a left QF-well-indexed set.

(3) We put $X \overset{put}{:=} \{\, i \in \{1, 2, \ldots, m'\} \mid \delta'_i \geq 2\,\}$, then $\{\, \psi(i) \mid i \in X\,\} = X$ from (2). So the statement holds. □

Definition 2.2. Let Q be an indecomposable basic QF ring with a left QF-well-indexed set $\{\, f'_{i,s}\,\}_{i=1,s=1}^{m'\ \delta'_i}$. For each $i \in \{1, 2, \ldots, m'\}$, we put

$$r'_i(1) = 1, \qquad x_{i,1} = 1$$

and we take positive integers

$$\delta_i\ , \qquad \gamma_i$$

to satisfy

$$\delta'_i \leq \delta_i \leq \gamma_i\,.$$

Moreover, we take

$$r_i(u),\ p_i(u)\ \in \{1, 2, \ldots, \gamma_i\} \qquad (u = 1, 2, \ldots, \delta_i\,)$$

to satisfy the following (1), (2), (3):

(1) The following (†-1) holds.

(†-1)(*i*) $1 \leq p_i(1) < p_i(2) < \cdots < p_i(\,\delta_i\,) = \gamma_i$

(*ii*) $1 = r_i(1) < r_i(2) < \cdots < r_i(\,\delta_i\,) \leq \gamma_i$ (So $r_i(\,x_{i,1}\,) = r'_i(\,1\,) = 1$.)

(2) If $\delta'_i = 1$ and $i = \psi(i)$, then the following (†-2) holds.

(†-2) $r_i(\,u\,) \leq p_i(\,u-1\,)$ for all $u = 2, 3, \ldots, \gamma_i$.

(3) If $\delta'_i \geq 2$ (we note that, then $i = \psi(i)$ from Lemma 2.2 (2)), then the following (†-3) holds, where we let

$$\begin{cases} r'_i(\,s\,) \in \{1, 2, \ldots, \gamma_i\} & (\,s = 2, 3, \ldots, \delta'_i\,) \\ p'_i(\,t\,) \in \{1, 2, \ldots, \gamma_i\} & (\,t = 1, 2, \ldots, \delta'_i - 1\,) \\ x_{i,s} \in \{2, 3, \ldots, \delta_i\} & (\,s = 1, 2, \ldots, \delta'_i\,) \\ y_{i,t} \in \{1, 2, \ldots, \delta_i - 1\} & (\,t = 1, 2, \ldots, \delta'_i - 1\,)\,. \end{cases}$$

$$
\begin{array}{lll}
(\dagger\text{-}3)(i) & 1 = x_{i,1} \le y_{i,1} < x_{i,2} \le y_{i,2} < \cdots < x_{i,\delta'_i-1} \le y_{i,\delta'_i-1} < x_{i,\delta'_i} & \\
(ii) & r_i(x_{i,s}) = r'_i(s) & (s = 2,3,\ldots,\delta'_i) \\
(iii) & p_i(y_{i,t}) = p'_i(t) & (t = 1,2,\ldots,\delta'_i - 1) \\
(iv) & p_i(x_{i,s}-1) < r_i(x_{i,s}) \le p_i(x_{i,s}) & (s = 2,3,\ldots,\delta'_i) \\
(v) & r_i(y_{i,t}) \le p_i(y_{i,t}) < r_i(y_{i,t}+1) & (t = 1,2,\ldots,\delta'_i - 1) \\
(vi) & r_i(u+1) \le p_i(u) & \begin{pmatrix} x_{i,t} \le u < y_{i,t}, \\ \text{where } t = 1,2,\ldots,\delta'_i - 1 \end{pmatrix} \\
(vii) & p_i(u) < r_i(u) & \begin{pmatrix} y_{i,t} < u < x_{i,t+1}, \\ \text{where } t = 1,2,\ldots,\delta'_i - 1 \end{pmatrix}
\end{array}
$$

Then the following holds.

Lemma 2.3. *Let Q be an indecomposable basic QF ring with a left* QF*-well-indexed set* $\{f'_{i,s}\}_{i=1,s=1}^{m'\ \delta'_i}$. *Suppose that* $\delta'_i \ge 2$. *Then*

$$1 = r'_i(1) \le p'_i(1) < r'_i(2) \le p'_i(2) < \cdots < r'_i(\delta'_i - 1) \le p'_i(\delta'_i - 1) < r'_i(\delta'_i) \le \gamma_i .$$

Proof. It follows from $(\dagger\text{-}1)(ii)$ and $(\dagger\text{-}3)(i), (ii), (iii), (iv), (v)$. □

3. Two-sided Harada rings constructed from QF rings

Definition 3.1. For each $i = 1,2,\ldots,m'$ and $s = 1,2,\ldots,\gamma_i$, we put

$$\tau_i^r(s) \overset{put}{:=} \max\{\, u \in \{1,2,\ldots,\delta_i\} \mid r_i(u) \le s \,\}.$$

That is, $\tau_i^r(s) \in \{1,2,\ldots,\delta_i\}$ such that

$$r_i(\tau_i^r(s)) \le s < r_i(\tau_i^r(s)+1),$$

where we let $r_i(\delta_i + 1) = \gamma_i + 1$.

Now we construct two-sided Harada rings. Let Q be an indecomposable basic QF ring with a left QF-well-indexed set $\{f'_{i,s}\}_{i=1,s=1}^{m'\ \delta'_i}$ and we use the terminologies that now we define. For each $i, j = 1,2,\ldots,m'$, $k = 1,2,\ldots,\delta'_i$ and $l = 1,2,\ldots,\delta'_j$, we put

$$Q_{i,k;j,l} \overset{put}{:=} f'_{i,k} Q f'_{j,l} .$$

And we put

$$Q_{i,k} \overset{put}{:=} Q_{i,k;i,k}, \quad J_{i,k} \overset{put}{:=} J(Q_{i,k}), \quad S_{\psi(j),l;j,l} \overset{put}{:=} S(f'_{\psi(j),l} Q f'_{j,l}).$$

(We note that $S_{\psi(j),l;j,l}$ is defined by Lemma 2.2 (1).) Moreover, we put

$$m_{i,k} \overset{put}{:=} r'_i(k+1) - r'_i(k),$$

where we let $r'_i(\delta'_i + 1) = \gamma_i + 1$,

$$\mathrm{Q}_{i,k;j,l} \overset{put}{:=} \begin{cases} \begin{pmatrix} Q_{i,k} & \cdots & \cdots & Q_{i,k} \\ J_{i,k} & \ddots & & \vdots \\ \vdots & \ddots & \ddots & \vdots \\ J_{i,k} & \cdots & J_{i,k} & Q_{i,k} \end{pmatrix} & \text{if } (i,k)=(j,l) \\ \begin{pmatrix} Q_{i,k;j,l} & \cdots & Q_{i,k;j,l} \\ \vdots & & \vdots \\ Q_{i,k;j,l} & \cdots & Q_{i,k;j,l} \end{pmatrix} & \text{if } (i,k)\neq(j,l) \end{cases} \quad : (m_{i,k},\, m_{j,l})\text{-matrix},$$

$$\mathrm{M}_{i,j} \overset{put}{:=} \begin{pmatrix} \mathrm{Q}_{i,1;j,1} & \mathrm{Q}_{i,1;j,2} & \cdots & \mathrm{Q}_{i,1;j,\delta'_j} \\ \mathrm{Q}_{i,2;j,1} & \mathrm{Q}_{i,2;j,2} & \cdots & \mathrm{Q}_{i,2;j,\delta'_j} \\ \vdots & \vdots & \ddots & \vdots \\ \mathrm{Q}_{i,\delta'_i;j,1} & \mathrm{Q}_{i,\delta'_i;j,2} & \cdots & \mathrm{Q}_{i,\delta'_i;j,\delta'_j} \end{pmatrix} : \ (\gamma_i,\, \gamma_j)\text{-matrix},$$

(then we note that the (p,q)-component of $\mathrm{Q}_{i,k;j,l}$ is the $\big(r'_i(k)+p-1,\, r'_j(l)+q-1\big)$-component of $\mathrm{M}_{i,j}$) and

$$\tilde{R} \overset{put}{:=} \begin{pmatrix} \mathrm{M}_{1,1} & \mathrm{M}_{1,2} & \cdots & \mathrm{M}_{1,m'} \\ \mathrm{M}_{2,1} & \mathrm{M}_{2,2} & \cdots & \mathrm{M}_{2,m'} \\ \vdots & \vdots & \ddots & \vdots \\ \mathrm{M}_{m',1} & \mathrm{M}_{m',2} & \cdots & \mathrm{M}_{m',m'} \end{pmatrix}.$$

Further, for each $p=1,2,\ldots,m_{i,k}$ and $q=1,2,\ldots,m_{j,l}$, we put

$$A_{i,k;j,l}^{\,p,q} \overset{put}{:=} \begin{cases} S_{i,k;j,l} & \text{if } i=\psi(j),\ k=l \text{ and } r'_j(l) \leq \\ & \quad p_j\big(\tau^r_{\psi(j)}\big(r'_{\psi(j)}(k)+p-1\big)\big) < r'_j(l)+q-1 \\ 0 & \text{otherwise} \end{cases}$$

and

$$\mathrm{A}_{i,k;j,l} \overset{put}{:=} \begin{pmatrix} A_{i,k;j,l}^{\,1,1} & A_{i,k;j,l}^{\,1,2} & \cdots & A_{i,k;j,l}^{\,1,m_{j,l}} \\ A_{i,k;j,l}^{\,2,1} & A_{i,k;j,l}^{\,2,2} & \cdots & A_{i,k;j,l}^{\,2,m_{j,l}} \\ \vdots & \vdots & \ddots & \vdots \\ A_{i,k;j,l}^{\,m_{i,k},1} & A_{i,k;j,l}^{\,m_{i,k},2} & \cdots & A_{i,k;j,l}^{\,m_{i,k},m_{j,l}} \end{pmatrix} \quad (\ :\ \text{subset of } \mathrm{Q}_{i,k;j,l}.)$$

For example, when $\delta'_i = \delta'_j = 1$ and $i = \psi(j)$, we put $S \overset{put}{:=} S_{i,1;j,1}$, and

$$
\mathrm{A}_{i,1;j,1} = \begin{array}{r} r_i(1) \\ \\ \\ \\ r_i(2) \\ \\ \\ \\ r_i(3) \\ \\ \\ \\ \vdots \\ r_i(\delta_i-1) \\ \\ \\ r_i(\delta_i) \\ \\ \\ \end{array}
\left(\begin{array}{cccccccc}
 & p_j(1) & p_j(2) & p_j(3) & \cdots & p_j(\delta_j-1) & p_j(\delta_j) \\
0 \cdots 0 & S \cdots & & & & & \cdots S \\
0 \cdots 0 & S \cdots & & & & & \cdots S \\
\vdots & \vdots & & & & & \vdots \\
0 \cdots 0 & S \cdots & & & & & \cdots S \\
0 \cdots\cdots & & \cdots 0 \; S \cdots & & & & \cdots S \\
0 \cdots\cdots & & \cdots 0 \; S \cdots & & & & \cdots S \\
\vdots & & \vdots & & & & \vdots \\
0 \cdots\cdots & & \cdots 0 \; S \cdots & & & & \cdots S \\
0 \cdots\cdots & & & \cdots 0 \; S \cdots & & & \cdots S \\
0 \cdots\cdots & & & \cdots 0 \; S \cdots & & & \cdots S \\
\vdots & & & \vdots & & & \vdots \\
0 \cdots\cdots & & & \cdots 0 \; S \cdots & & & \cdots S \\
\vdots & & & & \ddots & & \vdots \\
0 \cdots\cdots & & & & & \cdots 0 \; S \cdots & S \\
0 \cdots\cdots & & & & & \cdots 0 \; S \cdots & S \\
\vdots & & & & & \vdots & \vdots \\
0 \cdots\cdots & & & & & \cdots 0 \; S \cdots & S \\
0 \cdots\cdots & & & & & & \cdots 0 \\
0 \cdots\cdots & & & & & & \cdots 0 \\
\vdots & & & & & & \vdots \\
0 \cdots\cdots & & & & & & \cdots 0
\end{array}\right).
$$

When $\delta'_i \geq 2$ and $i = \psi(i)$, we put $S' \overset{put}{:=} S_{i,k;i,k}$, and for $k = 1, 2, \ldots, \delta'_i - 1$

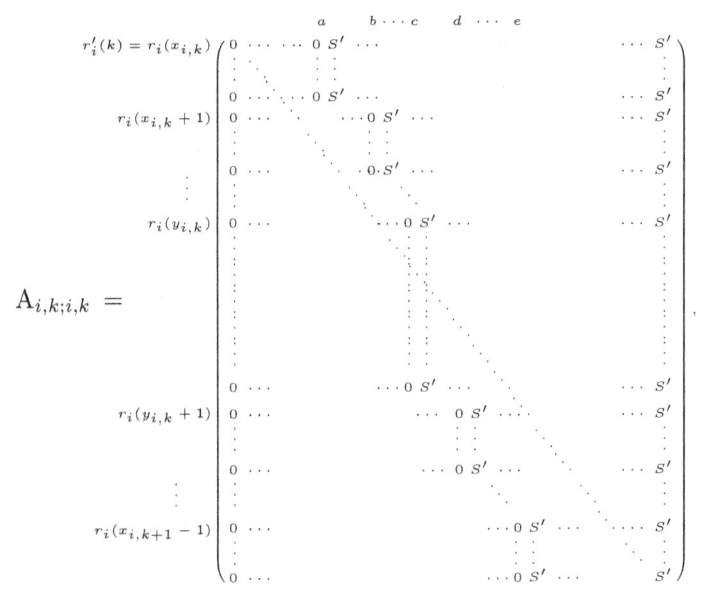

where we put $a \overset{put}{:=} p_i(x_{i,k})$, $b \overset{put}{:=} p_j(x_{i,k} + 1)$, $c \overset{put}{:=} p_j(y_{i,k})$, $d \overset{put}{:=} p_j(y_{i,k} + 1)$ and $e \overset{put}{:=} p_j(x_{i,k+1} - 1)$, and

$$
\mathrm{A}_{i,\delta'_i;i,\delta'_i} = \begin{array}{r} \\ r'_i(\delta'_i) = r_i(x_{i,\delta'_i}) \\ \\ \\ r_i(x_{i,\delta'_i}+1) \\ \\ \\ r_i(x_{i,\delta'_i}+2) \\ \vdots \\ \\ r_i(\delta_i - 1) \\ \\ \\ r_i(\delta_i) \\ \\ \\ \end{array}
\begin{array}{c} \qquad\qquad a' \qquad b' \qquad\qquad c' \cdots d' \qquad\quad e' \\
\left(\begin{array}{l}
0 \cdots\cdots 0\ S' \cdots \hfill \cdots S' \\
\vdots \ \ddots \quad \vdots\ \vdots \hfill \vdots \\
0 \cdots \cdots 0\ S' \cdots \hfill \cdots S' \\
0 \cdots \quad \ddots \quad \cdots 0\ S' \cdots \hfill \cdots S' \\
\vdots \qquad\qquad\quad \vdots\ \vdots \hfill \vdots \\
0 \cdots \qquad\quad \ddots 0\ S' \cdots \hfill \cdots S' \\
0 \cdots \qquad\qquad \ddots \qquad \cdots 0\ S' \cdots \hfill \cdots S' \\
\vdots \qquad\qquad\qquad\qquad\quad \vdots\ \vdots \hfill \vdots \\
0 \cdots \qquad\qquad\qquad \ddots \quad \cdots 0\ S' \cdots \hfill \cdots S' \\
\vdots \qquad\qquad\qquad\qquad \ddots \qquad \ddots \hfill \vdots \\
0 \cdots \qquad\qquad\qquad\qquad\quad \ddots \quad \cdots 0\ S' \cdots S' \\
\vdots \qquad\qquad\qquad\qquad\qquad\qquad\quad \vdots\ \vdots \hfill \vdots \\
0 \cdots \qquad\qquad\qquad\qquad\qquad\quad \ddots 0\ S' \cdots S' \\
0 \cdots \qquad\qquad\qquad\qquad\qquad\qquad \ddots \hfill \cdots 0 \\
\vdots \qquad\qquad\qquad\qquad\qquad\qquad\qquad \ddots \hfill \vdots \\
\vdots \qquad\qquad\qquad\qquad\qquad\qquad\qquad\quad \ddots \hfill \vdots \\
0 \cdots \hfill \cdots 0
\end{array}\right)
\end{array},
$$

where we put $a' \overset{put}{:=} p_i(x_{i,\delta'_i})$, $b' \overset{put}{:=} p_i(x_{i,\delta'_i} + 1)$, $c' \overset{put}{:=} p_i(x_{i,\delta'_i+2})$, $d' \overset{put}{:=} p_i(\delta_i - 1)$ and $e' \overset{put}{:=} p_i(\delta_i) = \gamma_i$.

For each $i, j = 1, 2, \ldots, m'$, we put

$$
\mathrm{N}_{i,j} \overset{put}{:=} \begin{pmatrix} \mathrm{A}_{i,1;j,1} & \mathrm{A}_{i,1;j,2} & \cdots & \mathrm{A}_{i,1;j,\delta'_j} \\ \mathrm{A}_{i,2;j,1} & \mathrm{A}_{i,2;j,2} & \cdots & \mathrm{A}_{i,2;j,\delta'_j} \\ \vdots & \vdots & \ddots & \vdots \\ \mathrm{A}_{i,\delta'_i;j,1} & \mathrm{A}_{i,\delta'_i;j,2} & \cdots & \mathrm{A}_{i,\delta'_i;j,\delta'_j} \end{pmatrix} \quad (:\ \text{subset of } \mathrm{M}_{i,j}\ )
$$

and

$$
\tilde{I} \overset{put}{:=} \begin{pmatrix} \mathrm{N}_{1,1} & \mathrm{N}_{1,2} & \cdots & \mathrm{N}_{1,m'} \\ \mathrm{N}_{2,1} & \mathrm{N}_{2,2} & \cdots & \mathrm{N}_{2,m'} \\ \vdots & \vdots & \ddots & \vdots \\ \mathrm{N}_{m',1} & \mathrm{N}_{m',2} & \cdots & \mathrm{N}_{m',m'} \end{pmatrix} \quad (:\ \text{subset of } \tilde{R}\ ).
$$

And $\tilde{R}$ is an artinian ring by usual addition and multiplication of matrix and $\tilde{I}$ is its ideal since $S_{\psi(j),l;j,l}$ is simple both as a left $Q_{\psi(j),l}$-module and as a right $Q_{j,l}$-module and, for any $p' \leq p$, $p_j\big(\tau^r_{\psi(j)}\big(r'_{\psi(j)}(k) + p' - 1\big)\big) \leq p_j\big(\tau^r_{\psi(j)}\big(r'_{\psi(j)}(k) + p - 1\big)\big)$ by (†-1)(i). Hence we consider a factor ring

$$
R \overset{put}{:=} \tilde{R}/\tilde{I}\,.
$$

From the definition of $\tilde{R}$, an element $\tilde{r}$ of $\tilde{R}$ is

$$
\tilde{r} = \big(\tilde{a}^{\,p,q}_{i,k;j,l}\big)^{m'\ \ \delta'_i\ \ \delta'_j\ \ m_{i,k}\ \ m_{j,l}}_{i,j=1,\,k=1,\,l=1,\,p=1,\,q=1}\,,
$$

where $\tilde{a}_{i,k;j,l}^{\;p,q}$ $(p = 1, 2, \ldots, m_{i,k},\ \ q = 1, 2, \ldots, m_{j,l})$ is a (p, q)-component of $Q_{i,k;j,l}$ $(k = 1, 2, \ldots, \delta'_i,\ \ l = 1, 2, \ldots, \delta'_j)$ which is a part of $M_{i,j}$. Further we put

$$\begin{cases} s \overset{put}{:=} r'_i(k) + p - 1 \\ t \overset{put}{:=} r'_j(l) + q - 1 \end{cases} \text{and } \tilde{a}_{i,s;j,t} \overset{put}{:=} \tilde{a}_{i,k;j,l}^{\;p,q}.$$

Then

$$\tilde{r} = \left(\tilde{a}_{i,s;j,t}\right)_{i,j=1,\, s=1,\, t=1}^{m' \quad \gamma_i \quad \gamma_j}.$$

So an element r of R is

$$r = \left(a_{i,k;j,l}^{\;p,q}\right)_{i,j=1,\, k=1,\, l=1,\, p=1,\, q=1}^{m' \quad \delta'_i \quad \delta'_j \quad m_{i,k} \quad m_{j,l}} = \left(a_{i,s;j,t}\right)_{i,j=1,\, s=1,\, t=1}^{m' \quad \gamma_i \quad \gamma_j},$$

where we put

$$a_{i,k;j,l}^{\;p,q} = a_{i,s;j,t} \overset{put}{:=} \begin{cases} \tilde{a}_{i,s;j,t} + S_{i,k;j,l} & \text{if } i = \psi(j),\ \ k = l,\ \ p_j\left(\tau_i^r(s)\right) < t \\ \tilde{a}_{i,s;j,t} & \text{otherwise.} \end{cases}$$

Furthermore we put

$$A_{i,s;j,t} \overset{put}{:=} A_{i,k;j,l}^{\;p,q}.$$

On the other hand, for any $i, j = 1, 2, \ldots, m'$, $s = 1, 2, \ldots, \gamma_i$ and $t = 1, 2, \ldots, \gamma_j$, we take

$$\begin{cases} k_s \in \{1, 2, \ldots, \delta'_i\}, & p_s \in \{1, 2, \ldots, m_{i,k_s}\} \\ l_t \in \{1, 2, \ldots, \delta'_j\}, & q_t \in \{1, 2, \ldots, m_{j,l_t}\} \end{cases}$$

to satisfy

$$\begin{cases} s = r'_i(k_s) + p_s - 1 \\ t = r'_j(l_t) + q_t - 1\,. \end{cases}$$

And, for each $i = 1, 2, \ldots, m'$ and $s = 1, 2, \ldots, \gamma_i$, we define an element

$$\tilde{f}_{i,s} = \left(\tilde{a}_{i',s';j',t'}\right)_{i',j'=1,\, s'=1,\, t'=1}^{m' \quad \gamma_i \quad \gamma_j}$$

of $\tilde{R}$ by

$$\tilde{a}_{i',s';j',t'} = \begin{cases} 1_{Q_{i,k_s}} & \text{if } i' = j' = i \text{ and } s' = t' = s \\ 0_{Q_{i',k_{s'};j',l_{t'}}} & \text{otherwise,} \end{cases}$$

and an element $f_{s,t}$ of R by

$$f_{s,t} \overset{put}{:=} \tilde{f}_{s,t} + \tilde{I}\,.$$

Then

$$A_{i,s;j,t} = \begin{cases} S_{i,k_s;j,l_t} & \text{if } i = \psi(j),\ \ k_s = l_t \text{ and } p_j\left(\tau_i^r(s)\right) < t \\ 0 & \text{otherwise.} \end{cases}$$

Hence, from the definition of R, $f_{i,s}Rf_{j,t}$ is as follows:

(1) We assume that $i = j$.

(i) Suppose that $\delta'_i \geq 2$. (Then $i = \psi(i)$ by Lemma 2.2 (2).)

In the case $k_s = l_t$,

$$f_{i,s}Rf_{i,t} = \begin{cases} Q_{i,k_s} & \text{if } s \leq t \text{ and } p_i(\tau_i^r(s)) \geq t \\ Q_{i,k_s}/S_{i,k_s} & \text{if } s \leq t \text{ and } p_i(\tau_i^r(s)) < t \\ J_{i,k_s} & \text{if } s > t \text{ and } p_i(\tau_i^r(s)) \geq t \\ J_{i,k_s}/S_{i,k_s} & \text{if } s > t \text{ and } p_i(\tau_i^r(s)) < t. \end{cases}$$

In the case $k_s \neq l_t$, $f_{i,s}Rf_{i,t} = Q_{i,k_s;i,l_t}$.

(ii) Suppose that $\delta'_i = 1$. (Then $k_s = l_t = 1$ for any $s, t = 1, 2, \ldots, \gamma_i$.)

In the case $i = \psi(i)$, $f_{i,s}Rf_{i,t}$ coincides with one in the case (i) $k_s = l_t$.

In the case $i \neq \psi(i)$,

$$f_{i,s}Rf_{i,t} = \begin{cases} Q_{i,1} & \text{if } s \leq t \\ J_{i,1} & \text{if } s > t. \end{cases}$$

(2) Next we assume that $i \neq j$.

(i) Suppose that $\delta'_j \geq 2$. Then $f_{i,s}Rf_{j,t} = Q_{i,k_s;j,l_t}$.

(ii) Suppose that $\delta'_j = 1$.

In the case $i = \psi(j)$,

$$f_{i,s}Rf_{j,t} = \begin{cases} Q_{i,k_s;j,l_t} & \text{if } p_j(\tau_i^r(s)) \geq t \\ Q_{i,k_s;j,l_t}/S_{i,k_s;j,l_t} & \text{if } p_j(\tau_i^r(s)) < t. \end{cases}$$

In the case $i \neq \psi(j)$, (we note that, if $\delta'_i \geq 2$, then $i \neq \psi(j)$ by Lemma 2.2 (2))

$$f_{i,s}Rf_{j,t} = Q_{i,k_s;j,l_t}.$$

Throughout this paper, we use these terminologies.

Lemma 3.1. *Let $\delta'_j \geq 2$ and $t \in \{1, 2, \ldots, \gamma_j - 1\}$. Suppose that $t + 1 = r'_j(l_{t+1})$. Then the following hold:*

(1) (i) $p_j(x_{j,l_{t+1}} - 1) \leq t$

(ii) *If $p_j(x_{j,l_{t+1}} - 1) = t$, then*

(I) $y_{j,l_t} = x_{j,l_{t+1}} - 1$ *and*

(II) $p'_j(l_t) = t$.

(iii) The following are equivalent:

(a) $p'_j(l_t) < t$

(b) $p_j(x_{j,l_{t+1}} - 1) < t$

(c) $t \notin \{p_j(u)\}_{u=1}^{\delta_j}$

(2) *We consider the following two cases.*

(A) $p'_j(l_t) < t$ *and* $r'_j(l_t) \leq s < r'_j(l_{t+1})$.

(B) $p'_j(l_t) = t$ *and* $r'_j(l_t) \leq s < r_j(y_{j,l_{t+1}-1})$.

Then

$$A_{i,s;j,t} = \begin{cases} S_{j,l_t} & \text{if } i = j \text{ and } (A), (B) \\ 0 & \text{otherwise.} \end{cases}$$

Proof. We put $l \overset{put}{:=} l_{t+1}$, i.e., $t+1 = r'_j(l)$, to read easy.

(1)(i) From (†-3)(ii), (iv) and the assumption, $p_j(x_{j,l} - 1) < r_j(x_{j,l}) = r'_j(l) = t+1$. So $p_j(x_{j,l} - 1) \leq t$.

(ii)(I) We assume that $y_{j,l-1} \neq x_{j,l} - 1$. Then $y_{j,l-1} < x_{j,l} - 1 < x_{j,l}$ by (†-3)(i). So $t = p_j(x_{j,l} - 1) < r_j(x_{j,l} - 1) < r_j(x_{j,l}) = r'_j(l) = t+1$ by the assumption and (†-3)(ii), (vii), a contradiction.

(II) $p'_j(l_t) = p'_j(l-1) = p_j(y_{j,l-1}) = p_j(x_{j,l} - 1) = t$ by the assumption, (†-3)(iii) and (I).

(iii) $(a) \Rightarrow (b)$ Assume that $p_j(x_{j,l} - 1) \geq t$. Then $p_j(x_{j,l} - 1) = t$ since $t+1 = r'_j(l) = r_j(y_{j,l}) \geq r_j(x_{j,l}) > p_j(x_{j,l} - 1)$ by (†-3)(i), (ii), (iv). So $p'_j(l_t) = t$ by (1)(ii)(II), a contradiction.

$(b) \Rightarrow (a)$ Obvious.

$(b) \Leftrightarrow (c)$ $t+1 = r'_j(l) = r_j(y_{j,l})$ by (†-3)(ii) and $p_j(x_{j,l-1}) < t+1 \leq p_j(x_{j,l})$ by (†-3)(iv). So the statement follows.

(2) In the case $i \neq j$ or $i = j$ but $1 \leq s < r'_j(l-1)$ or $r'_j(l) \leq s \leq \delta_j$, $A_{i,s;j,t} = 0$ holds by the definition of $A_{i,s;j,t}$.

We assume that $i = j$ and $r'_j(l-1) \leq s < r'_j(l) = t+1$. Then $\tau^r_j(s) \leq x_{j,l} - 1$ since $r_j(x_{j,l}) = r'_j(l) = t+1$. So, in the case (A), $A_{j,s;j,t} = S_{j,l_t}$ by (1)(iii) and the definition of $A_{j,s;j,t}$.

Last we consider the case that $p'_j(l-1) = t$. Then $r_j(x_{j,l}) \leq r_j(y_{j,l-1} + 1)$ since $p_j(y_{j,l-1}) < r_j(y_{j,l-1} + 1)$ by (†-3)(v) and $p_j(y_{j,l-1}) = p'_j(l-1) = t < t+1 = r'_j(l) = r_j(x_{j,l})$. So $x_{j,l} \leq y_{j,l-1} + 1$. Therefore $x_{j,l} = y_{j,l-1} + 1$ because $x_{j,l} \geq y_{j,l-1} + 1$ by (†-3)(i). Hence there does not exist l' in $\{1, 2, \ldots, \delta_j\}$ which satisfy $r_j(y_{j,l-1}) < r_j(l') < r_j(x_{j,l})$. Hence the following hold:

(*i*) For $r_j(y_{j,l-1}) \leq s < r_j(x_{j,l})$, i.e., (B) does not hold, $\tau_j^r(s) = y_{j,l-1}$. So $p_j(\tau_j^r(s)) = p_j(y_{j,l-1}) = p'_j(l-1) = t$. Therefore $A_{j,s;j,t} = 0$ by the definition of $A_{j,s;j,t}$.

(*ii*) For $r_j(x_{j,l-1}) \leq s' < r_j(y_{j,l-1})$, i.e., (B) holds, $\tau_j^r(s) < y_{j,l-1}$. So $p_j(\tau_j^r(s)) < p_j(y_{j,l-1}) = p'_j(l-1) = t$. Therefore $A_{j,s;j,t} = S_{j.l_t}$ by the definition of $A_{j,s;j,t}$. □

Lemma 3.2. *For any $j = 1, 2, \ldots, m'$ and $t = 1, 2, \ldots, \gamma_j - 1$, we define*

$$\tilde{a}_{j,t} = \left(\tilde{a}_{i',s';j',t'}\right)_{i',j'=1,\, s'=1,\, t'=1}^{m' \quad \gamma_i \quad \gamma_j} \in \tilde{R}$$

by the following (1) *and* (2)*:*

(1) *In the case $t+1 \neq r'_j(l_{t+1})$, i.e., $t+1 \notin \{r'_j(l)\}_{l=2}^{\delta'_j}$. Then*

$$\tilde{a}_{i',s';j',t'} = \begin{cases} 1_{Q_{j,l_{t+1}}} & \text{if } i' = j' = j,\ s' = t,\ t' = t+1 \\ 0_{Q_{i',k_{s'};j',l_{t'}}} & \text{otherwise.} \end{cases}$$

(2) *In the case $t+1 = r'_j(l_{t+1})$ (then $\delta'_j \geq 2$). From the definition of a left* QF*-well-indexed set, we have a Q-epimorphism*

$$\zeta_{l_{t+1}} : {}_Q Q f'_{j,\, l_{t+1}-1} \to {}_Q J(Q) f'_{j,l_{t+1}}$$

such that

$$\operatorname{Ker} \zeta_{l_{t+1}} = S({}_Q Q f'_{j,\, l_{t+1}-1}) = S(f'_{j,\, l_{t+1}-1} Q f'_{j,\, l_{t+1}-1})$$

Then

$$\tilde{a}_{i',s';j',t'} = \begin{cases} (f'_{j,\, l_{t+1}-1})\zeta_{l_{t+1}} & \text{if } i' = j' = j,\ s' = r'_j(l_{t+1}) - 1 \\ & \quad \text{and } t' = r'_j(l_{t+1}) \\ 0_{Q_{i',k_{s'};j',l_{t'}}} & \text{otherwise.} \end{cases}$$

We put

$$a_{j,t} \overset{put}{:=} \tilde{a}_{j,t} + \tilde{I} \in R$$

and define a left R-homomorphism

$$\zeta_{j,t} \overset{put}{:=} (a_{j,t})_R : Rf_{j,t} \to Jf_{j,\,t+1}\,.$$

Then $\zeta_{j,t}$ is a left H-epimorphism with

$$\operatorname{Ker} \zeta_{j,t} = \begin{cases} 0 & \text{if } t \notin \{p_j(u)\}_{u=1}^{\delta_j} \\ \bigoplus_{s=r_{\psi(j)}(u')}^{r_{\psi(j)}(u'+1)-1} S(f_{\psi(j),s} R f_{j,t}) & \text{if } t = p_j(u'), \text{ where } u' \in \\ & \quad \{1, 2, \ldots, \delta_j\}, \text{ and} \\ & \quad r'_j(l_{t+1}) \neq t+1 \\ \bigoplus_{s=r_j(y_{i,l_{t+1}-1})}^{r'_j(l_{t+1})-1} S(f_{j,s} R f_{j,t}) & \text{if } t = p'_j(l_t) \text{ and} \\ & \quad r'_j(l_{t+1}) = t+1. \end{cases}$$

Proof. We consider a left $\tilde{R}$-homomorphism $(\tilde{a}_{j,t})_R : \tilde{R}\tilde{f}_{j,t} \to \tilde{J}\tilde{f}_{j,t+1}$.

(1) In the case $t+1 \neq r'_j(l_{t+1})$, by the definition of $\tilde{R}$, $(\tilde{a}_{j,t})_R$ is a left $\tilde{R}$-isomorphism. So $(a_{j,t})_R$ is a well-defined left R-epimorphism by the definition of $\tilde{I}$.

As for H-epimorphism, it is enough to show that $\operatorname{Ker}\zeta_{j,t}$ satisfies the statement. We show it.

By the definition of $A_{i,s;j,t}$, $A_{i,s;j,t} \neq A_{i,s;j,t+1}$ if and only if $i = \psi(j)$, $k_s = l_t$ and $p_j(\tau^r_{\psi(j)}(s)) = t$. And $p_j(\tau^r_{\psi(j)}(s)) = t$ if and only if $r_{\psi(j)}(u') \leq s < r_{\psi(j)}(u'+1)$ by the definition of $\tau^r_{\psi(j)}(s)$ and $t = p_j(u')$. Further we see that, if $p_j(\tau^r_{\psi(j)}(s)) = t$, then $k_s = l_t$. In fact, in the case $\delta'_j = 1$, it is clear. We assume $\delta'_j \geq 2$. Then $\psi(j) = j$. And $k_s = k_{r_j(\tau^r_j(s))}$ by the definition of $\tau^r_j(s)$. And, since $(f_{j,r_j(\tau^r_j(s))}R, Rf_{j,p_j(\tau^r_j(s))})$ is an i-pair, $k_{r_j(\tau^r_j(s))} = l_{p_j(\tau^r_j(s))} = l_t$ by the assumption $p_j(\tau^r_{\psi(j)}(s)) = t$. Hence

$$\operatorname{Ker}\zeta_{j,t} = \begin{cases} 0 & \text{if } t \notin \{p_j(u)\}_{u=1}^{\delta_j} \\ \oplus_{s=r_{\psi(j)}(u')}^{r_{\psi(j)}(u'+1)-1} S(f_{\psi(j),s}Rf_{j,t}) & \text{if } t = p_j(u') \text{ for some} \\ & u' \in \{1,2,\ldots,\delta_j\}. \end{cases}$$

(2) In the case $t+1 = r'_j(l_{t+1})$. (Then $\delta'_j \geq 2$.) We put $l \overset{put}{:=} l_{t+1}$, i.e., $t+1 = r'_j(l)$, to read easy. From the definition of $\tilde{R}$, $(\tilde{a}_{j,t})_R$ is a left $\tilde{R}$-epimorphism such that $\operatorname{Ker}(\tilde{a}_{j,t})_R = \oplus_{s=r'_j(l_t)}^{r'_j(l)-1} S(\tilde{f}_{j,s}\tilde{R}\tilde{f}_{j,t})$. So $(a_{j,t})_R$ is a well-defined left R-epimorphism by the definition of $\tilde{I}$. As for H-epimorphism, it is enough to show that $\operatorname{Ker}\zeta_{j,t}$ satisfies the statement. But it is clear from the properties above of $\operatorname{Ker}(\tilde{a}_{j,t})_R$ and Lemma 3.1 (2). □

For each $i, j = 1, 2, \ldots, m'$, we consider the following sequences, where we let $p_j(0) = 0$ and $r_i(\delta_i + 1) = \gamma_i + 1$.

$(L\text{-}j\text{-}u)$ $Rf_{j,p_j(u-1)+1}$, $Rf_{j,p_j(u-1)+2}$, $\ldots$, $Rf_{j,p_j(u)}$ $(u = 1, 2, \ldots, \delta_j)$

$(L\text{-}j)$ $Rf_{j,1}$, $Rf_{j,2}$, $\ldots$, Rf_{j,γ_j}

$(R\text{-}i\text{-}u)$ $f_{i,r_i(u+1)-1}R$, $f_{i,r_i(u+1)-2}R$, $\ldots$, $f_{i,r_i(u)}R$ $(u = 1, 2, \ldots, \delta_i)$

$(R\text{-}i)$ $f_{i,\gamma_i}R$, $f_{i,\gamma_i-1}R$, $\ldots$, $f_{i,1}R$

Theorem 3.1. *Then R is a two-sided Harada ring which satisfy the following:*

(1) $(f_{\psi(j),r_{\psi(j)}(u)}R,\ Rf_{j,p_j(u)})$ *is an i-pair for any* $u = 1, 2, \ldots, \delta_j$.

(2) (*i*) (*L-j*) *is a left w-co-H-sequence for any* $j = 1, 2, \ldots, m'$.

(*ii*) (*R-i*) *is a right w-co-H-sequence for any* $i = 1, 2, \ldots, m'$.

(3) (*i*) (*L-j-u*) *is a left co-H-sequence for any* $j = 1, 2, \ldots, m'$ *and* $u = 1, 2, \ldots, \delta_j$.

(*ii*) (*R-i-u*) *is a right co-H-sequence for any* $i = 1, 2, \ldots, m'$ *and* $u = 1, 2, \ldots, \delta_i$.

Proof. It is clear that $\tilde{R}$ is an artinian ring and $\tilde{I}$ is its ideal. And R is a two-sided Harada ring from (3). So it is enough to show (1), (2), (3).

(1) By Lemma 2.2 (1) and the definition of $\tilde{R}$, $(\tilde{f}_{\psi(j),1}\tilde{R},\ \tilde{R}\tilde{f}_{j,\gamma_j})$ is an i-pair for $j = 1, 2, \ldots, m'$. So the statement holds by the definitions of R and $\tilde{I}$.

(2) (*i*) (WCHS1) holds by Lemma 3.2. We show (WCHS2) also holds.

We assume that there exists a left H-epimorphism $\zeta_1 : {}_RRf_{v,w} \to {}_RJf_{j,1}$ for some $v \neq j$ and $w \in \{1, 2, \ldots, \gamma_v\}$. Then $w = \gamma_v$ by $v \neq j$ and (WCHS1) which we already show.

In the case $r'_v(\delta'_v) = \gamma_v$, i.e., $\delta'_v \geq 2$ and $r'_v(\delta'_v) = \gamma_v$ or $\gamma_v = 1$, a left H-epimorphism $\zeta_1 : {}_RRf_{v,r'_v(\delta'_v)} \to {}_RJf_{j,1} = {}_RJf_{j,r'_j(1)}$ induces a left H-epimorphism $\zeta'_1 : {}_QQf'_{v,\delta'_v} \to {}_QJ(Q)f'_{j,1}$. This contradicts with the assumption $v \neq j$.

Next we consider the case $r'_v(\delta'_v) \neq \gamma_v$. Then

$$f_{\psi(v),w'}Rf_{v,\gamma_v} = \begin{cases} \overline{Q_{\psi(v),r'_{\psi(v)}(\delta'_v);v,r'_v(\delta'_v)}} & \text{if } r'_{\psi(v)}(\delta'_v) \leq w' < r_{\psi(v)}(\delta_v) \\ Q_{\psi(v),r'_{\psi(v)}(\delta'_v);v,r'_v(\delta'_v)} & \text{if } r_{\psi(v)}(\delta_v) \leq w' \leq \gamma_{\psi(v)}, \end{cases}$$

where $\overline{Q_{\psi(v),r'_{\psi(v)}(\delta'_v);v,r'_v(\delta'_v)}}$ means a factor module by its socle $S_{\psi(v),r'_{\psi(v)}(\delta'_i);v,r'_v(\delta'_v)}$. In fact, if $r_{\psi(v)}(\delta_v) \leq w' \leq \gamma_{\psi(v)}$, $\tau^r_{\psi(v)}(w') = \delta_v$. So $p_v(\tau^r_{\psi(v)}(w')) = \gamma_v$ by (†-1)(*i*). If $r'_{\psi(v)}(\delta'_v) \leq w' < r_{\psi(v)}(\delta_v)$, then $\tau^r_{\psi(v)}(w') < \delta_v$. So $p_v(\tau^r_{\psi(v)}(w')) < \gamma_v$. On the other hand, for all $w' \in \{r'_{\psi(v)}(\delta'_v), \ldots, \gamma_{\psi(v)}\}$,

$$f_{\psi(v),w'}Jf_{j,1} = \begin{cases} J_{j,1} & \text{if } \psi(v) = j \text{ and } \delta'_v = 1 \\ Q_{\psi(v),r'_{\psi(v)}(\delta'_v);j,1} & \text{otherwise.} \end{cases}$$

So

$$\oplus_{w'=r'_{\psi(v)}(\delta'_v)}^{r_{\psi(v)}(\delta_v)-1} S(f_{\psi(v),w'}Rf_{v,\gamma_v}) \bigcap \operatorname{Ker}\zeta_1 = 0$$

since ζ_1 is an H-epimorphism. Further we consider a composition mapping $\zeta \overset{put}{:=} \zeta_{v,r'_v(\delta'_v)}\,\zeta_{v,r_v(x_{v,\delta'_v})+1} \cdots \zeta_{v,\delta_v-1} : {}_RRf_{v,r'_v(\delta'_v)} \to$

${}_R J^{\delta_v - r'_v(\delta'_v)} f_{v,\delta_v}$, where mappings are given in Lemma 3.2. And $\operatorname{Ker}\zeta = \oplus_{s=r'_{\psi(v)}(\delta'_v)}^{r_{\psi(v)}(\delta_v)-1} S(f_{j,s} R f_{v,r'_v(\delta'_v)})$ by Lemma 3.2. So $\zeta\zeta_1 : {}_R R f_{v,r'_v(\delta'_v)} \to {}_R J^{\delta_v - r'_j(\delta'_v)+1} f_{j,1} = {}_R J^{\delta_v - r'_j(\delta'_v)+1} f_{v,r'_v(1)}$ satisfies $\operatorname{Ker}(\zeta\zeta_1)J = 0$. Therefore a left H-epimorphism $\zeta'_1 : {}_Q Q f'_{v,\delta'_v} \to {}_Q J(Q) f'_{j,1}$ is induced. This contradicts with the assumption $v \neq j$.

Next we assume that there exists a left H-epimorphism $\zeta_z : {}_R R f_{j,\gamma_j} \to {}_R J f_{v,w}$ for some $v \neq j$ and $w \in \{1, 2, \ldots, \gamma_v\}$. Then $w = 1$ from the assumption $v \neq j$ and (WCHS1) which we already show. And by the same argument as ζ_1, we see that such a ζ_2 does not exist.

(ii) We see by the same argument as in (i).

(3) (i) (CHS1) We see by Lemma 3.2.

(CHS2) We see from (1) and [7, Theorem 3.1].

(CHS3) For any $i' \in \{1, 2, \ldots, m'\}$ and $u' \in \{1, 2, \ldots, \delta_{i'}\}$ with $(i', u') \neq (j, u)$, $S({}_R R f_{i', p_{i'}(u')}) \not\cong S({}_R R f_{j, p_j(u)})$ from (1). So, for any $s \in \{1, 2, \ldots, \gamma_{i'}\}$, ${}_R R f_{i',s} \not\cong {}_R J f_{j, p_j(u-1)+1}$.

(ii) We see by the same way as in (i). □

Example 3.1. Let Q be an indecomposable basic QF ring such that

(i) its QF-well indexed set is $\{f'_{1,1}, f'_{1,2}, f'_{1,3}, f'_{2,1}, f'_{3,1}, f'_{4,1}\}$, and

(ii) $(f'_{1,1}Q, Qf'_{1,1})$, $(f'_{3,1}Q, Qf'_{2,1})$, $(f'_{2,1}Q, Qf'_{3,1})$, $(f'_{4,1}Q, Qf'_{4,1})$ are i-pairs.

The bijection $\psi : \{1, 2, 3, 4\} \to \{1, 2, 3, 4\}$ is defined by

$$\psi(1) = 1, \quad \psi(2) = 3, \quad \psi(3) = 2, \quad \psi(4) = 4$$

from (ii) above, and

$$\delta'_1 = 3, \qquad \delta'_2 = \delta'_3 = \delta'_4 = 1.$$

And for $i = 1, 2, 3, 4$, we let, for instance, δ_i, γ_i $p_i(u)$ and $r_i(u)$ $(u = 1, 2, \ldots, \delta_i)$ as follows.

- $\delta_1 = 5, \quad \delta_2 = \delta_3 = \delta_4 = 2.$
- $\gamma_1 = 9, \quad \gamma_2 = \gamma_3 = 2, \quad \gamma_4 = 3.$
- $p_1(1) = 2, \quad p_1(2) = 3, \quad p_1(3) = 5, \quad p_1(4) = 6, \quad p_1(5) = 9$
- $r_1(1) = 1, \quad r_1(2) = 2, \quad r_1(3) = 5, \quad r_1(4) = 8, \quad r_1(5) = 9$
- $r_2(1) = 1, \quad r_2(2) = 2, \quad p_2(1) = 1, \quad p_2(2) = 2$

- $r_3(1) = 1,\ \ r_3(2) = 2,\ \ p_3(1) = 1,\ \ p_3(2) = 2$
- $r_4(1) = 1,\ \ r_4(2) = 2,\ \ p_4(1) = 2,\ \ p_4(2) = 3$

Then $\delta'_i \le \delta_i \le \gamma_i$, ($\dagger$-1) and ($\dagger$-2) hold. Further, for $s = 1, 2, 3\,(= \delta'_1)$ and $t = 1, 2\,(= \delta'_1 - 1)$, we let $r'_1(s)$, $p'_1(t)$, $x_{1,s}$, $y_{1,t}$ as follows:

- $r'_1(1) = 1,\ \ r'_1(2) = 5,\ \ r'_1(3) = 9$
- $p'_1(1) = 3,\ \ p'_1(2) = 5$
- $x_{1,1} = 1,\ \ x_{1,2} = 3,\ \ x_{1,3} = 5$
- $y_{1,1} = 2,\ \ y_{1,2} = 3$

Then ($\dagger$-3) also holds. So, by Theorem 3.1, we can construct a two-sided Harada ring R with i-pairs

$$
\begin{array}{l}
(f_{1,1}R,\ Rf_{1,2}),\ \ (f_{1,2}R,\ Rf_{1,3}),\ \ (f_{1,5}R,\ Rf_{1,5}),\ \ (f_{1,8}R,\ Rf_{1,6}),\ \ (f_{1,9}R,\ Rf_{1,9}) \\
(f_{3,1}R,\ Rf_{2,1}),\ \ (f_{3,2}R,\ Rf_{2,2}), \\
(f_{2,1}R,\ Rf_{3,1}),\ \ (f_{2,2}R,\ Rf_{3,2}), \\
(f_{4,1}R,\ Rf_{4,2}),\ \ (f_{4,2}R,\ Rf_{4,3}).
\end{array}
$$

And, putting $Q_{i,k} \overset{put}{:=} Q_{i,k;i,k}$, $J_{i,k} \overset{put}{:=} J(Q_{i,k})$, $Q_{i,k;j,l} \overset{put}{:=} f'_{i,k} Q f'_{j,l}$ and $\overline{Q_{i,k;j,l}} \overset{put}{:=} Q_{i,k;j,l}/S(Q_{i,k;j,l})$, R is isomorphic to

$$
\left(\begin{array}{cccccccccccccccc}
Q_{11} & Q_{11} & \overline{Q_{11}} & \overline{Q_{11}} & Q_{11;12} & Q_{11;12} & Q_{11;12} & Q_{11;12} & Q_{11;13} & Q_{11;21} & Q_{11;21} & Q_{11;31} & Q_{11;31} & Q_{11;41} & Q_{11;41} & Q_{11;41} \\
J_{11} & Q_{11} & Q_{11} & \overline{Q_{11}} & Q_{11;12} & Q_{11;12} & Q_{11;12} & Q_{11;12} & Q_{11;13} & Q_{11;21} & Q_{11;21} & Q_{11;31} & Q_{11;31} & Q_{11;41} & Q_{11;41} & Q_{11;41} \\
J_{11} & J_{11} & Q_{11} & \overline{Q_{11}} & Q_{11;12} & Q_{11;12} & Q_{11;12} & Q_{11;12} & Q_{11;13} & Q_{11;21} & Q_{11;21} & Q_{11;31} & Q_{11;31} & Q_{11;41} & Q_{11;41} & Q_{11;41} \\
J_{11} & J_{11} & J_{11} & \overline{Q_{11}} & Q_{11;12} & Q_{11;12} & Q_{11;12} & Q_{11;12} & Q_{11;13} & Q_{11;21} & Q_{11;21} & Q_{11;31} & Q_{11;31} & Q_{11;41} & Q_{11;41} & Q_{11;41} \\
Q_{12;11} & Q_{12;11} & Q_{12;11} & Q_{12;11} & Q_{12} & \overline{Q_{12}} & \overline{Q_{12}} & \overline{Q_{12}} & Q_{12;13} & Q_{12;21} & Q_{12;21} & Q_{12;31} & Q_{12;31} & Q_{12;41} & Q_{12;41} & Q_{12;41} \\
Q_{12;11} & Q_{12;11} & Q_{12;11} & Q_{12;11} & J_{12} & \overline{Q_{12}} & \overline{Q_{12}} & \overline{Q_{12}} & Q_{12;13} & Q_{12;21} & Q_{12;21} & Q_{12;31} & Q_{12;31} & Q_{12;41} & Q_{12;41} & Q_{12;41} \\
Q_{12;11} & Q_{12;11} & Q_{12;11} & Q_{12;11} & J_{12} & \overline{J_{12}} & \overline{Q_{12}} & \overline{Q_{12}} & Q_{12;13} & Q_{12;21} & Q_{12;21} & Q_{12;31} & Q_{12;31} & Q_{12;41} & Q_{12;41} & Q_{12;41} \\
Q_{12;11} & Q_{12;11} & Q_{12;11} & Q_{12;11} & J_{12} & J_{12} & \overline{J_{12}} & \overline{Q_{12}} & Q_{12;13} & Q_{12;21} & Q_{12;21} & Q_{12;31} & Q_{12;31} & Q_{12;41} & Q_{12;41} & Q_{12;41} \\
Q_{13;11} & Q_{13;11} & Q_{13;11} & Q_{13;11} & Q_{13;12} & Q_{13;12} & Q_{13;12} & Q_{13;12} & Q_{13} & Q_{13;21} & Q_{13;21} & Q_{13;31} & Q_{13;31} & Q_{13;41} & Q_{13;41} & Q_{13;41} \\
Q_{21;11} & Q_{21;11} & Q_{21;11} & Q_{21;11} & Q_{21;12} & Q_{21;12} & Q_{21;12} & Q_{21;12} & Q_{21;13} & Q_{21} & Q_{21} & Q_{21;31} & \overline{Q_{21;31}} & Q_{21;41} & Q_{21;41} & Q_{21;41} \\
Q_{21;11} & Q_{21;11} & Q_{21;11} & Q_{21;11} & Q_{21;12} & Q_{21;12} & Q_{21;12} & Q_{21;12} & Q_{21;13} & J_{21} & Q_{21} & Q_{21;31} & Q_{21;31} & Q_{21;41} & Q_{21;41} & Q_{21;41} \\
Q_{31;11} & Q_{31;11} & Q_{31;11} & Q_{31;11} & Q_{31;12} & Q_{31;12} & Q_{31;12} & Q_{31;12} & Q_{31;13} & Q_{31;21} & \overline{Q_{31;21}} & Q_{31} & Q_{31} & Q_{31;41} & Q_{31;41} & Q_{31;41} \\
Q_{31;11} & Q_{31;11} & Q_{31;11} & Q_{31;11} & Q_{31;12} & Q_{31;12} & Q_{31;12} & Q_{31;12} & Q_{31;13} & Q_{41;21} & Q_{31;21} & J_{31} & Q_{31} & Q_{31;41} & Q_{31;41} & Q_{31;41} \\
Q_{41;11} & Q_{41;11} & Q_{41;11} & Q_{41;11} & Q_{41;12} & Q_{41;12} & Q_{41;12} & Q_{41;12} & Q_{41;13} & Q_{41;21} & Q_{41;21} & Q_{41;31} & Q_{41;31} & Q_{41} & Q_{41} & \overline{Q_{41}} \\
Q_{41;11} & Q_{41;11} & Q_{41;11} & Q_{41;11} & Q_{41;12} & Q_{41;12} & Q_{41;12} & Q_{41;12} & Q_{41;13} & Q_{41;21} & Q_{41;21} & Q_{41;31} & Q_{41;31} & J_{41} & Q_{41} & Q_{41} \\
Q_{41;11} & Q_{41;11} & Q_{41;11} & Q_{41;11} & Q_{41;12} & Q_{41;12} & Q_{41;12} & Q_{41;12} & Q_{41;13} & Q_{41;21} & Q_{41;21} & Q_{41;31} & Q_{41;31} & J_{41} & J_{41} & Q_{41}
\end{array}\right)
$$

In the forthcoming paper, we will show the importance of two-sided Harada rings in Theorem 3.1.

Acknowledgment

This work was supported by JSPS KAKENHI Grant Number JP17K05202.

References

[1] Y. Baba and K. Oshiro, On a theorem of Fuller, *J. Algebra* **154** (1) (1993), 86–94.

[2] Y. Baba, Injectivity of quasi-projective modules, projectivity of quasi-injective modules, and projective covers of injective modules, *J. Algebra* **155** (2) (1993), 415–434.

[3] Y. Baba and K. Iwase, On quasi-Harada rings, *J. Algebra* **185** (2) (1996), 544–570.

[4] Y. Baba and K. Oshiro, Classical artinian rings and related topics, World Scientific (2009).

[5] Y. Baba, On H-epimorphisms and co-H-sequences in two-sided Harada rings, to appear in Math. J. Okayama Univ.

[6] Y. Baba, On weak co-H-sequences in two-sided Harada rings, preprint.

[7] K. R. Fuller, On indecomposable injectives over artinian rings, *Pacific J. Math.* **29** (1969), 115–135.

[8] M. Harada, Non-small modules and non-cosmall modules, in "Ring Theory", Proceedings of 1978 Antwerp Conference (F. Van Oystaeyen, Ed.) Dekker, New York (1979), 669–690.

[9] M. Harada, Factor categories with applications to direct decomposition of modules, Lecture Note in Pure and Appl. Math., Vol. 88, Dekker, New York, (1983).

[10] K. Oshiro, Lifting modules, extending modules and their applications to QF-rings, Hokkaido Math. J. 13 (1984), 310–338.

[11] K. Oshiro, On Harada rings I, *Math. J. Okayama Univ.* **31** (1989), 161–178.

Morita Theory on f-representations of rings

N. Hijriati*, S. Wahyuni† and I. E. Wijayanti‡

*Department of Mathematics, Universitas Gadjah Mada,
Yogyakarta, Indonesia
Department of Mathematics, Universitas Lambung Mangkurat,
Banjarmasin, Indonesia
E-mail: naimah.hijriati@mail.ugm.ac.id; nh_hijriati@ulm.ac.id
www.ugm.ac.id; www.ulm.ac.id

†Department of Mathematics, Universitas Gadjah Mada,
Yogyakarta, Indonesia
E-mail: swahyuni@ugm.ac.id

‡Department of Mathematics, Universitas Gadjah Mada,
Yogyakarta, Indonesia
E-mail: ind_wijayanti@ugm.ac.id

The representation of rings on finite dimension vector spaces has been generalized to the representation of rings on modules over a commutative ring. Let S be a commutative ring with unity and M an S-module. A representation of ring R with unity on an S-module M is a ring homomorphism from R to the ring of endomorphisms of M. An S-module associated with a representation of R is called a representation module of R. For any ring homomorphism $f: R \to S$, we define a representation of ring R with unity on M via f, and it is called an f-representation of ring R which is a special case of the representation of ring R on an S-module. This S-module associated with the f-representation of ring R is called an f-representation module of R.

In case S is non-commutative, we give a sufficient condition for the S-module M to be a representation module of R. The category of f-representation modules of ring R is Abelian and Morita equivalent to the category of modules over an R-algebra. Thus, if the category of modules over the R-algebra which is equivalent to the category of f-representation modules of R satisfies the Krull-Schmidt Theorem, then the category of f-representation modules of R also satisfies Krull-Schmidt's Theorem.

Keywords: f-representation of ring; Morita Theory; Krull-Schmidt Theorem.

1. Introduction

Representation theory is a part of abstract algebras. One part of representation theory is a finite group representation that was first introduced by Frobenius in 1896. The representation of a finite group G on a finite dimensions vector space V over a field F is a group homomorphism $\varphi\colon G \to \mathrm{GL}_F(V)$, where a $\mathrm{GL}_F(V)$ is a group of all linear transformations from V to V that has an inverse. Furthermore, V is called a representation space of a group G, which is also an $F[G]$-module, where $F[G]$ is a group algebra. Noether first introduced the concept of space representation as $F[G]$-modules in 1929 [5].

Because of the formation of $F[G]$-modules from any representation of a finite group G on a finite dimension vector space over a field F, we can form a ring homomorphism $\rho\colon F[G] \to \mathrm{End}_F(V), a \mapsto \rho_a$, with $\rho_a \in \mathrm{End}_F(V)$ defined as $\rho_a(v) = av$ for every $v \in V$. Based on this, Burrow (1965) generalized $F[G]$ to any ring R with a unity, namely homomorphism ring $\rho\colon R \to \mathrm{End}_F(V)$. Based on this fact, comes the definition of ring representation on a finite dimension vector space of V over the field F.

From [2], if there is a non-zero ring homomorphism $f\colon R \to S$, then an S-module M is an R-module with a scalar multiplication over R defined by $r \cdot m = f(r)m$. Hence, if S is commutative then for any S-module M, it can always be constructed a ring homomorphism $\varphi\colon R \to \mathrm{End}_S(M), r \mapsto \varphi_r$ with $\varphi_r \in \mathrm{End}_S(M)$ is defined as $\varphi_r\colon M \to M, m \mapsto f(r)m$ for every $r \in R$ and $m \in M$. In other words, φ is a representation of ring R and M is a representation module of R. Later, in this study φ is called an f-representation of the ring R on S-modules M and M associated with φ is called an f-representation module of R [8].

In this study, it is assumed that the ring S is a commutative ring so that $Z(S) = S$, with $Z(S)$ is a center of S. If there is a non-zero ring homomorphism of $f\colon R \to S$, then based on the definition of R-algebra, ring S is an R-algebra. It is known that if the ring S is an Artin R-algebra, then the module category over S satisfy Krull-Schmidt's Theorem. Krull-Schmidt's theorem states that if a module over Artin R-algebra is decomposable into a finite number of indecomposable submodules, then the module decomposes uniquely [3]. Based on this fact and based on the definition of a completely reducible of ring representation, the Krull-Schmidt theorem guarantees the uniqueness of the decomposition of f-representation module into a finite direct sum of irreducible R-invariant submodules.

Problems arise when the ring S is not commutative. Although there is a homomorphism of the ring $f\colon R \to S$, it is not guaranteed that $Z(S)$ contains the image of f. As such, S is not necessarily an R-algebra. As a result, an S-module M not necessarily an f-representation module of R. The sufficient conditions for an S-module M is an f-representation module of R is $sf(r) - f(r)s \in \mathrm{Ann}_S(M)$ for any $s \in S$ and $r \in R$ [7]. Then, it is shown that the class of all of f-representation modules can still satisfy the properties that apply to the category of modules over an R-algebra. One way to solve this problem is by using the equivalence of two categories. Therefore, it is necessary to construct a category of modules over an R-algebra, so that the category of f-representation modules is equivalent to that category.

2. Equivalence of the Category of Representation Modules and the Category of Modules over an Algebra

Let S be a ring with unity. The set S^n, $n < \infty$ is a right free S-module with basis $B = \{e_1, e_2, ..., e_n \mid e_i \in S^n\}$, where e_i is an n-tuple with a 1 in the ith position and 0 in all other position. Since S^n is also a left S-module, for any $s \in S$, we can define the map $g_s\colon S^n \to S^n, a \mapsto sa$. The map $g\colon S \to \mathrm{End}_S(S^n), s \mapsto g_s$ is a ring homomorphism. If $f\colon R \to S$ is a ring homomorphism then a ring $\mathrm{End}_S(S^n)$ is an R-algebra.

Suppose that $\Lambda = \mathrm{End}_S(S^n)$ and $Q = \mathrm{Hom}_S(S^n, S)$. The additive Abelian group S^n and Q are (Λ, S)-bimodule and (S, Λ)-bimodule, respectively. So, we can construct a tensor product $Q \otimes_\Lambda S^n$ and $S^n \otimes_S Q$. Furthermore, there are a (Λ, Λ)-bimodule homomorphism and an (S, S)-bimodule homomorphism, i.e., $\delta\colon Q \otimes_\Lambda S^n \to S, (q, p) \mapsto qp$, and $\epsilon\colon S^n \otimes_S Q \to \Lambda, (p, q) \mapsto pq$, respectively.

There are $p = (s, 0, 0, ..., 0) \in S^n$ and $q\colon S^n \to S$, $(a_1, a_2, ..., a_n) \mapsto a_1$ in Q, such that $qp = q(p) = q(s, 0, ..., 0) = s$, for any $s \in S$. So the map δ is surjective. Based on this, the right S-module S^n is a generator associated with Morita context $(S, S^n, Q, \Lambda; \delta, \epsilon)$ and $Q \otimes_\Lambda S^n \cong S$. Moreover, the generator S^n is a free module, such that S^n is a finitely generated module and projective. Hence, S^n is a progenerator associated with Morita context $(S, S^n, Q, \Lambda; \delta, \epsilon)$ and $S^n \otimes_S Q \cong \Lambda$. Therefore, based on Morita Theory, we have the following proposition.

Proposition 2.1. *Rings S and Λ are Morita equivalent.*

Proof. Since S^n is a progenerator associate with Morita context $(S, S^n, Q, \Lambda; \delta, \epsilon)$, by Morita Theory (Theorem 18.24 [6]), there is a functor equivalence from S-Mod to Λ-Mod, i.e.,

$$S^n \otimes_S -\colon S\text{-Mod} \to \Lambda\text{-Mod},$$

with an inverse equivalence defined by functor

$$Q \otimes_\Lambda -\colon \Lambda\text{-Mod} \to S\text{-Mod}.$$

Hence, rings S and Λ are Morita equivalent. □

Based on Proposition 2.1, the category S-Mod equivalent with the category $\mathrm{End}_S(S^n)$-Mod where $\mathrm{End}_S(S^n)$ is an R-algebra. Recall that

$$\mathfrak{S}_f = \{M \in S\text{-Mod} \mid sf(r) - f(r)s \in \mathrm{Ann}_S(M)\}$$

is the category of f-representation modules with its morphism is an S-module homomorphism [7]. We construct a class of all Λ-modules, i.e.,

$$\Delta = \{S^n \otimes_S M \mid M \in \mathfrak{S}_f\}. \tag{1}$$

For any $S^n \otimes_S M, S^n \otimes_S N \in \Delta$ with $M, N \in \mathfrak{S}_f$ and $\alpha \in \mathrm{Hom}_{\mathfrak{S}_f}(M, N)$, a map $Id_{S^n} \otimes_S \alpha\colon S^n \otimes_S M \to S^n \otimes_S N, a \otimes m \mapsto a \otimes \alpha(m)$ is a Λ-module homomorphism. Next, $S^n \otimes_S M \in \Delta$ and the map $id_{S^n} \otimes_S \alpha$ are denoted by M° and α°, respectively. Furthermore, by Corollary 7.2.14 [1] we have

(1) For any $\alpha^\circ\colon M_1^\circ \to M_2^\circ$ and $\beta^\circ\colon M_2^\circ \to M_3^\circ$ with $M_1^\circ, M_2^\circ, M_3^\circ \in \Delta$ $\beta^\circ\alpha^\circ\colon M_1^\circ \to M_3^\circ$ is a Λ-module homomorphism.
(2) For any $M^\circ, N^\circ \in \Delta$, there are $id_M^\circ = id_{S^n} \otimes_S id_M\colon M^\circ \to M^\circ$ and $id_N^\circ = id_{S^n} \otimes_S id_N\colon N^\circ \to N^\circ$, such that for any $\alpha^\circ\colon M^\circ \to N^\circ$,

$$\begin{aligned}\alpha^\circ id_M^\circ = (id_{S^n} \otimes_S \alpha)(id_{S^n} \otimes_S id_M) &= id_{S^n} id_{S^n} \otimes_S \alpha id_M \\ &= id_{S^n} \otimes_S \alpha = \alpha^\circ,\end{aligned}$$

$$\begin{aligned}id_N^\circ \alpha^\circ = (id_{S^n} \otimes_S id_N)(id_{S^n} \otimes_S \alpha) &= id_{S^n} id_{S^n} \otimes_S id_N \alpha \\ &= id_{S^n} \otimes_S \alpha = \alpha^\circ.\end{aligned}$$

(3) For any $\alpha^\circ\colon M_1^\circ \to M_2^\circ$, $\beta^\circ\colon M_2^\circ \to M_3^\circ$, and $\gamma^\circ\colon M_3^\circ \to M_4^\circ$, with

$M_1^\circ, M_2^\circ, M_3^\circ, M_4^\circ \in \Delta$, we have

$$\begin{aligned}\gamma^\circ(\beta^\circ\alpha^\circ) &= (id_{S^n} \otimes_S \gamma)((id_{S^n} \otimes_S \beta)(id_{S^n} \otimes_S \alpha))\\ &= (id_{S^n} \otimes_S \gamma)(id_{S^n} \otimes_S \beta\alpha)\\ &= id_{S^n} \otimes_S \gamma(\beta\alpha)\\ &= id_{S^n} \otimes_S (\gamma\beta)\alpha\\ &= (id_{S^n} \otimes_S \gamma\beta)(id_{S^n} \otimes_S \alpha)\\ &= ((id_{S^n} \otimes_S \gamma)(id_{S^n} \otimes_S \beta))(id_{S^n} \otimes_S \alpha)\\ &= (\gamma^\circ\beta^\circ)\alpha^\circ.\end{aligned}$$

Definition 2.1. The category Γ is a class of objects in Δ ($Obj(\Gamma) = \Delta$) and for all M°, N° in Γ, a set of morphisms is all Λ-module homomorphisms

$$\mathrm{Hom}_\Gamma(M^\circ, N^\circ) = \{\alpha^\circ\colon M^\circ \to N^\circ \mid \alpha \in \mathrm{Hom}_{\mathfrak{S}_f}(M, N)\},$$

with its composition is a composition of the maps.

The category Γ is an additive category, because for any $\alpha^\circ \in \mathrm{Hom}_\Gamma(M_1^\circ, M_2^\circ)$, $\beta_1^\circ, \beta_2^\circ \in \mathrm{Hom}_\Gamma(M_2^\circ, M_3^\circ)$, and $\gamma^\circ \in \mathrm{Hom}_\Gamma(M_3^\circ, M_4^\circ)$, we have

$$\begin{aligned}((\beta_1^\circ + \beta_2^\circ)\alpha^\circ)(a \otimes m_1) &= (\beta_1 + \beta_2)^\circ\alpha^\circ(a \otimes m_1)\\ &= (\beta_1 + \beta_2)^\circ(a \otimes \alpha(m_1))\\ &= a \otimes (\beta_1 + \beta_2)\alpha(m_1)\\ &= a \otimes (\beta_1\alpha(m_1) + \beta_2\alpha(m_1))\\ &= a \otimes \beta_1\alpha(m_1) + a \otimes \beta_2\alpha(m_1)\\ &= \beta_1^\circ(a \otimes \alpha(m_1)) + \beta_2^\circ(a \otimes \alpha(m_1))\\ &= \beta_1^\circ\alpha^\circ(a \otimes m_1) + \beta_2^\circ\alpha^\circ(a \otimes m_1)\\ &= (\beta_1^\circ\alpha^\circ + \beta_2^\circ\alpha^\circ)(a \otimes m_1),\end{aligned}$$

$$\begin{aligned}(\gamma^\circ(\beta_1^\circ + \beta_2^\circ))(a \otimes m_2) &= \gamma^\circ((\beta_1 + \beta_2)^\circ(a \otimes m_2))\\ &= \gamma^\circ(a \otimes (\beta_1 + \beta_2)(m_2))\\ &= a \otimes \gamma(\beta_1 + \beta_2)(m_2)\\ &= a \otimes (\gamma\beta_1(m_2) + \gamma\beta_2(m_2))\\ &= a \otimes \gamma\beta_1(m_2) + a \otimes \gamma\beta_2(m_2)\\ &= \gamma^\circ(a \otimes \beta_1(m_2)) + \gamma^\circ(a \otimes \beta_2(m_2))\\ &= \gamma^\circ\beta_1^\circ(a \otimes m_2) + \gamma^\circ\beta_2^\circ(a \otimes m_2)\\ &= (\gamma^\circ\beta_1^\circ + \gamma^\circ\beta_2^\circ)(a \otimes m_2).\end{aligned}$$

Based on Definition 2.1, the Γ is a full subcategory Λ-Mod. A relation $S^n \otimes_S -: \mathfrak{S}_f \to \Gamma$ defined by $(S^n \otimes_S -)(M) = S^n \otimes_S M$ and $(S^n \otimes_S -)(\alpha) = \alpha^\circ$, and a relation $Q \otimes_\Lambda -: \Gamma \to \mathfrak{S}_f$ defined by $(Q \otimes_\Lambda -)(M^\circ) = M$ and $(Q \otimes_\Lambda -)(\alpha^\circ) = \alpha$, for all $M \in \mathfrak{S}_f$, $\alpha \in \mathrm{Hom}_{\mathfrak{S}_f}(M, N)$, $M^\circ \in \Gamma$ and $\alpha^\circ \in \mathrm{Hom}_\Gamma(M^\circ, N^\circ)$ are covariant functors. Therefore, by using covariant functors $S^n \otimes_S -$ and $Q \otimes_\Lambda -$, we have the following proposition.

Proposition 2.2. *The category $\mathfrak{S}_f$ equivalent to the category Γ.*

Proof. To prove that $\mathfrak{S}_f \approx \Gamma$, it is sufficient to show that $(S^n \otimes_S -)$ is a functor equivalence with an inverse equivalence $(Q \otimes_\Lambda -)$, i.e.,

$$(Q \otimes_\Lambda -)(S^n \otimes_S -) \cong Id_{\mathfrak{S}_f} \text{ and } (S^n \otimes_S -)(Q \otimes_\Lambda -) \cong Id_\Gamma.$$

Let M be any object in $\mathfrak{S}_f$ and $S^n \otimes_S N$ in Γ.

$$\begin{aligned}(Q \otimes_\Lambda -)(S^n \otimes_S -)(M) &= (Q \otimes_\Lambda -)(S^n \otimes_S M)\\ &= Q \otimes_\Lambda (S^n \otimes_S M)\\ &= (Q \otimes_\Lambda S^n) \otimes_S M\\ &= S \otimes_S M = M, \end{aligned} \tag{2}$$

$$\begin{aligned}(S^n \otimes_S -)(Q \otimes_\Lambda -)(S^n \otimes_S N) &= (S^n \otimes_S -)(Q \otimes_\Lambda (S^n \otimes_S N))\\ &= (S^n \otimes_S -)((Q \otimes_\Lambda S^n) \otimes_S N)\\ &= (S^n \otimes_S -)(S \otimes_S N)\\ &= (S^n \otimes_S -)(N)\\ &= S^n \otimes_S N. \end{aligned} \tag{3}$$

Furthermore, if we define a natural transformation $\eta: (Q \otimes_\Lambda -)(S^n \otimes_S -) \to Id_{\mathfrak{S}_f}$ and $\psi: (S^n \otimes_S -)(Q \otimes_\Lambda -) \to Id_\Gamma$ then based on (2) and (3), for any $\alpha: M \to N \in \mathrm{Hom}_{\mathfrak{S}_f}(M, N)$ and $\beta^\circ: M^\circ \to N^\circ \in \mathrm{Hom}_\Gamma(M^\circ, N^\circ)$, the diagrams

$$\begin{array}{ccc} GF(M) & \xrightarrow{GF(\alpha)} & GF(N) \\ \eta_M \downarrow & & \downarrow \eta_N \\ M & \xrightarrow[Id_{\mathfrak{S}_f}(\alpha)]{} & N \end{array}$$

$$\begin{array}{ccc} FG(M^\circ) & \xrightarrow{FG(\beta^\circ)} & FG(N^\circ) \\ \psi_{M^\circ} \downarrow & & \downarrow \psi_{N^\circ} \\ M^\circ & \xrightarrow[Id_\Gamma(\beta^\circ)]{} & N^\circ \end{array}$$

are commutative. Hence, the category $\mathfrak{S}_f$ is equivalent to Γ. □

By using Proposition 2.2 we have the following proposition.

Proposition 2.3. *If the category Γ satisfies Krull-Schmidt Theorem then the category $\mathfrak{S}_f$ also satisfies Krull-Schmidt Theorem.*

Proof. Based on Proposition 2.2, $\mathfrak{S}_f \approx \Gamma$. We define $F\colon \Gamma \to \mathfrak{S}_f$ as a functor equivalence with an inverse equivalence $G\colon \mathfrak{S}_f \to \Gamma$ such that $FG = Id_{\mathfrak{S}_f}$ and $GF = Id_\Gamma$. In another words G is a functor equivalence with F as an inverse equivalence. Next, we will prove that $\mathfrak{S}_f$ satisfies a Krull-Schmidt Theorem.

i. Let M be an indecomposable S-module in $\mathfrak{S}_f$. By Proposition 21.8 [2], $G(M)$ is indecomposable in Γ. Suppose that category Γ satisfies the Krull-Schmidt Theorem. So, $\mathrm{End}_S(G(M))$ is a local ring. Furthermore, by Proposition 21.2 [2], we have $\mathrm{End}_{\mathfrak{S}_f}(M) \cong \mathrm{End}_S(G(M))$. So $\mathrm{End}_{\mathfrak{S}_f}(M)$ is a local ring. Conversely, if $\mathrm{End}_{\mathfrak{S}_f}(M)$ is a local ring then $\mathrm{End}_S(G(M))$ is also a local ring, by Proposition 21.2 [2]. Since Γ satisfies the Krull-Schmidt Theorem, we have $G(M)$ is an indecomposable S-module. Hence, an S-module M is indecomposable.
ii. We will prove that if $\bigoplus_{i\in I} A_i \cong \bigoplus_{j\in J} B_j$ with $\{A_i\}_{i\in I}$ and $\{B_j\}_{j\in J}$ class of indecomposable finitely generated S-modules then there is a bijection $\sigma\colon I \to J$ such that $A_i \cong B_{\sigma(i)}$ for any $i \in I$.

   Suppose that $\{A_i\}_{i\in I}$ and $\{B_j\}_{j\in J}$ are two class of finitely generated indecomposable modules. Based on Proposition 21.8 [2], the classes $\{G(A_i)\}_{i\in I}$ and $\{G(B_j)\}_{j\in J}$ are the class of indecomposable finitely generated S-modules. If $\bigoplus_{i\in I} A_i \cong \bigoplus_{j\in J} B_j$, then by Proposition 21.2 [2]

$$\bigoplus_{i\in I} G(A_i) \cong \bigoplus_{j\in J} G(B_j).$$

   Since Γ satisfies the Krull-Schmidt Theorem, there is a bijection $\sigma : I \to J$ such that $G(A_i) \cong G(B_{\sigma(i)})$ for any $i \in I$. Hence by Proposition 21.2 [2] $A_i \cong B_{\sigma(i)}$ for any $i \in I$.

By (1) and (2), the category $\mathfrak{S}_f$ satisfies the Krull-Schmidt Theorem. □

3. Conclusions

In the case of ring S with unity (not necessarily commutative), the category of f-representation modules, i.e., $\mathfrak{S}_f = \{M \in S\text{-Mod} \mid sf(r) - f(r)s \in$

$\mathrm{Ann}_S(M)\}$ with its morphism is an S-module homomorphism, has enough injective objects and enough projective objects if $S \in \mathfrak{S}_f$ as an S-module. This causes the ring S to be an R-algebra and $\mathfrak{S}_f = S$-Mod. Therefore, if S-Mod satisfies Krull-Schmidt Theorem then $\mathfrak{S}_f$ also satisfies Krull-Schmidt Theorem.

If S as an S-module is not object in $\mathfrak{S}_f$, then there is a category of modules over an R-algebra $\mathrm{End}_S(S^n)$, i.e., the category Γ with objects in $\{S^n \otimes_S M \mid M \in \mathfrak{S}_f\}$, and its morphism is $\mathrm{End}_S(S^n)$-module homomorphism $\alpha^\circ\colon S^n \otimes_S M \to S^n \otimes_S M$, $a \otimes m \mapsto a \otimes \alpha(m)$ with $\alpha \in \mathrm{Hom}_{\mathfrak{S}_f}(M, N)$, such that $\mathfrak{S}_f$ equivalent to that category. So, Krull-Schmidt Theorem is still applies to the f-representation module, if the module over an R-algebra $\mathrm{End}_S(S^n)$ satisfies Krull-Schmidt Theorem.

Acknowledgements This paper is one of the results of the Doctoral Dissertation Research Grant (Hibah Penelitian Disertasi Doktor) from the Directorate General for Research and Development, Ministry of Research and Higher Education by Contract No. 2920/UN1.DITLIT/DIT-LIT/LT/2019.

References

[1] W. A. Adkins and S. H. Weintraub, *Algebra : An Approach Via Module Theory*, (Springer-Verlag, New York, 1992).

[2] F. W. Anderson, and K. R. Fuller, *Rings and Categories of Modules*, Second edition, (Springer-Verlag, 1992)

[3] M. Auslander, I. Reiten, and S. O. Smalo, *Representation Theory of Artin Algebras*, (Cambridge University Press, 1997).

[4] M. Burrow, *Representation Theory of Finite Groups*, (Academic Press, New York, 1965)

[5] C. W. Curtis, and I. Reiner, *Representation Theory Of Finite Groups And Associative Algebras*, (John Wiley & Sons. Inc, 1962).

[6] T. Y. Lam, *Lectures on Module and Rings*, (Springer-Verlag, New York, 1999).

[7] N. Hijriati, S. Wahyuni, and I. E. Wijayanti, Injectivity and projectivity Properties of The Category of Representation modules of Rings., *J. Phys.: Conf. Ser.* **1097**, 012078 (2018).

[8] N. Hijriati, S. Wahyuni, and I. E. Wijayanti, Generalization of Schur's Lemma in Ring Representations on Modules over a Commutative Ring, *European J. of Pure and Applied Math* **11**(3), 751 (2018).

An application of Hochschild cohomology to the moduli of subalgebras of the full matrix ring II

K. Nakamoto

Center for Medical Education and Sciences, Faculty of Medicine, University of Yamanashi, Yamanashi 409–3898, Japan
E-mail: nakamoto@yamanashi.ac.jp

T. Torii

Department of Mathematics, Okayama University, Okayama 700–8530, Japan
E-mail: torii@math.okayama-u.ac.jp

We show an application of Hochschild cohomology to the moduli of subalgebras of the full matrix ring without proofs. We also calculate Hochschild cohomology $H^i(\mathrm{S}_{11}(R), \mathrm{M}_3(R)/\mathrm{S}_{11}(R))$ for an R-subalgebra $\mathrm{S}_{11}(R)$ of $\mathrm{M}_3(R)$ over a commutative ring R. The calculation by using a spectral sequence will give us a useful technique for calculating Hochschild cohomology.

Keywords: Hochschild cohomology, subalgebra, matrix ring, moduli of molds.

1. Introduction

In this paper, we introduce another application of Hochschild cohomology to the moduli of subalgebras of the full matrix ring as a continuation of Ref. 5. First, let us define the moduli of subalgebras of the full matrix ring M_n as a closed subscheme of the Grassmann scheme $\mathrm{Grass}(d, n^2)$ (for details, see Ref. 3). For this, we need a notion of mold:

Definition 1.1. We say that a subsheaf $\mathcal{A}$ of $\mathcal{O}_X$-algebras of $\mathrm{M}_n(\mathcal{O}_X)$ is a *mold* of degree n on a scheme X if $\mathrm{M}_n(\mathcal{O}_X)/\mathcal{A}$ is a locally free sheaf, where $\mathcal{O}_X$ is the structure sheaf of X. Note that if $\mathrm{M}_n(\mathcal{O}_X)/\mathcal{A}$ is a locally free sheaf, then so is $\mathcal{A}$. We denote by $\mathrm{rank}\mathcal{A}$ the rank of $\mathcal{A}$ as a locally free sheaf.

Proposition 1.1. *Let* $(\mathbf{Sch})^{op}$ *be the opposite category of schemes and* $(\mathbf{Sets})$ *the category of sets. The following contravariant functor from the category of schemes to the category of sets is represented by a closed sub-*

scheme of the Grassmann scheme $\mathrm{Grass}(d, n^2)$*:*

$$\begin{array}{rl}\mathrm{Mold}_{n,d} : (\mathbf{Sch})^{op} \to (\mathbf{Sets}) & \\ X \quad \mapsto \{\mathcal{A} | \mathcal{A} \textit{ is a mold of degree } n \textit{ on } X \textit{ with } \mathrm{rank}\,\mathcal{A} = d\}. & \end{array}$$

Proof. For the proof, see Ref. 3 or Ref. 4. □

Definition 1.2. By Proposition 1.1 and the Yoneda lemma, the functor $\mathrm{Mold}_{n,d}$ can be regarded as a closed subscheme of $\mathrm{Grass}(d, n^2)$. Since $\mathrm{Mold}_{n,d}$ parametrizes rank d molds of M_n, we call $\mathrm{Mold}_{n,d}$ the *moduli of rank d subalgebras of* M_n or the *moduli of rank d molds of degree n*.

Next, let us define Hochschild cohomology groups.

Definition 1.3. Let A be an associative algebra over a commutative ring R. Assume that A is a projective module over R. Let $A^e := A \otimes_R A^{op}$ be the enveloping algebra of A. For A-bimodules A and M over R, we can regard them as A^e-modules. We define the i-th Hochschild cohomology group $H^i(A, M)$ as $\mathrm{Ext}^i_{A^e}(A, M)$. Note that $H^i(A, M)$ is an R-module. We can calculate $H^i(A, M)$ by taking the cohomology groups of the bar complex $(C^i(A, M), d^i)_{i\in\mathbb{Z}}$ which is given by

$$C^i(A, M) := \begin{cases} \mathrm{Hom}_R(A^{\otimes i}, M) & (i \geq 0) \\ 0 & (i < 0) \end{cases}$$

and $d^i : C^i(A, M) \to C^{i+1}(A, M)$ $(i \geq 0)$ defined by

$$\begin{aligned} & d^i(f)(a_1 \otimes a_2 \otimes \cdots \otimes a_{i+1}) \\ & = a_1 f(a_2 \otimes \cdots \otimes a_{i+1}) + \sum_{j=1}^{i} (-1)^j f(a_1 \otimes \cdots \otimes a_j a_{j+1} \otimes \cdots \otimes a_{i+1}) \\ & \qquad + (-1)^{i+1} f(a_1 \otimes a_2 \otimes \cdots \otimes a_i) a_{i+1} \end{aligned}$$

for $f \in C^i(A, M)$ $(i \geq 1)$ and

$$d^0(m)(a) = am - ma$$

for $m \in C^0(A, M) = M$, where $A^{\otimes i} = \overbrace{A \otimes_R \cdots \otimes_R A}^{i}$.

Remark 1.1. Suppose that the unit map $R \to A$ is a split monomorphism. We set $\overline{A} = A/RI$, where $I \in A$ is the image of $1 \in R$ under the unit map. For $p \geq 0$, let

$$\overline{B}_p(A, A, A) = A \otimes_R \overbrace{\overline{A} \otimes_R \cdots \otimes_R \overline{A}}^{p} \otimes_R A.$$

For an A-bimodule M over R, let us denote $\mathrm{Hom}_{A^e}(\overline{B}_*(A,A,A),M) \cong \mathrm{Hom}_R(\overline{A}^{\otimes i},M)$ by $\overline{C}^*(A,M)$. Then $\overline{C}^*(A,M)$ can be regarded as a subcomplex of $C^*(A,M)$. We can verify that $\overline{C}^*(A,M) \to C^*(A,M)$ induces an isomorphism

$$H^*(\overline{C}(A,M)) \cong H^*(A,M).$$

In Ref. 6, two important applications of Hochschild cohomology to the moduli of molds have been shown. Let $\mathcal{A}$ be the universal mold on $\mathrm{Mold}_{n,d}$. For $x \in \mathrm{Mold}_{n,d}$, denote by $\mathcal{A}(x) := \mathcal{A} \otimes_{\mathcal{O}_{\mathrm{Mold}_{n,d}}} k(x) \subset \mathrm{M}_n(k(x))$ the mold corresponding to x, where $\mathcal{O}_{\mathrm{Mold}_{n,d}}$ is the structure sheaf of $\mathrm{Mold}_{n,d}$ and $k(x)$ is the residue field of x. Let $T_{\mathrm{Mold}_{n,d}/\mathbb{Z},x}$ be the tangent space of $\mathrm{Mold}_{n,d}$ over $\mathbb{Z}$ at x. Then we have the following theorems:

Theorem 1.1. *For each point* $x \in \mathrm{Mold}_{n,d}$,

$$\begin{aligned}&\dim_{k(x)} T_{\mathrm{Mold}_{n,d}/\mathbb{Z},x}\\ &\qquad = \dim_{k(x)} H^1(\mathcal{A}(x), \mathrm{M}_n(k(x))/\mathcal{A}(x)) + n^2 - \dim_{k(x)} N(\mathcal{A}(x)),\end{aligned}$$

where $N(\mathcal{A}(x)) := \{b \in \mathrm{M}_n(k(x)) \mid [b,a] := ba - ab \in \mathcal{A}(x)$ *for any* $a \in \mathcal{A}(x)\}$.

Theorem 1.2. *Let* $x \in \mathrm{Mold}_{n,d}$. *If* $H^2(\mathcal{A}(x), \mathrm{M}_n(k(x))/\mathcal{A}(x)) = 0$, *then the canonical morphism* $\mathrm{Mold}_{n,d} \to \mathbb{Z}$ *is smooth at* x.

In Section 2, we introduce other applications (Theorems 2.1 and 2.2) without proofs. Let S be a locally noetherian scheme. Let $\mathcal{A}$ be a rank d mold of degree n on S. Set $\mathrm{PGL}_{n,S} := \mathrm{PGL}_n \otimes_{\mathbb{Z}} S$. Assume that $H^1(\mathcal{A}(x), \mathrm{M}_n(k(x))/\mathcal{A}(x)) = 0$ for each $x \in S$. By Theorem 2.2, the $\mathrm{PGL}_{n,S}$-orbit $\{P^{-1}\mathcal{A}P \mid P \in \mathrm{PGL}_{n,S}\}$ is open in $\mathrm{Mold}_{n,d} \otimes_{\mathbb{Z}} S$. For the long proof, see Ref. 6.

In this paper, we also show several calculations of $H^i(A, \mathrm{M}_n(R)/A)$ for R-subalgebras A of $\mathrm{M}_3(R)$ over a commutative ring R. In Section 3, we show a long proof of Theorem 3.1:

Theorem 1.3 (Theorem 3.1). *Let*

$$S_{11}(R) = \left\{ \begin{pmatrix} a & b & c \\ 0 & e & d \\ 0 & 0 & a \end{pmatrix} \in \mathrm{M}_3(R) \;\middle|\; a,b,c,d,e \in R \right\}$$

for a commutative ring R. *Then*

$$H^n(\mathrm{S}_{11}(R), \mathrm{M}_3(R)/\mathrm{S}_{11}(R)) \cong \begin{cases} R \ (n = 0,1), \\ 0 \ \ (n \geq 2). \end{cases}$$

The proof in Section 3 using a spectral sequence is different from that of Ref. 6. The technique for calculating Hochschild cohomology groups in Section 3 might be useful for readers. This is the reason why we describe another proof in this paper.

2. Several results

In this section, we describe an application of Hochschild cohomology to the moduli of subalgebras of the full matrix ring. We also show two examples of Hochschild cohomology $H^i(A, \mathrm{M}_3(R)/A)$ for R-subalgebras A of $\mathrm{M}_3(R)$ over a commutative ring R. For proofs, see Ref. 6.

The following theorem is one of applications of Hochschild cohomology.

Theorem 2.1. *Let S be a locally noetherian scheme. Let $\mathcal{A}$ be a rank d mold of degree n on S. Set $\mathcal{A}(x) := \mathcal{A} \otimes_{\mathcal{O}_S} k(x) \subseteq \mathrm{M}_n(k(x))$, where $k(x)$ is the residue field of a point $x \in S$. Put $\mathrm{PGL}_{n,S} := \mathrm{PGL}_n \otimes_{\mathbb{Z}} S$. Let us define the S-morphism $\phi_{\mathcal{A}} : \mathrm{PGL}_{n,S} \to \mathrm{Mold}_{n,d} \otimes_{\mathbb{Z}} S$ by $P \mapsto P^{-1}\mathcal{A}P$. Then $\phi_{\mathcal{A}}$ is smooth if and only if $H^1(\mathcal{A}(x), \mathrm{M}_n(k(x))/\mathcal{A}(x)) = 0$ for each $x \in S$.*

As a corollary of Theorem 2.1, we have:

Theorem 2.2. *Let S be a locally noetherian scheme. Let $\mathcal{A}$ be a rank d mold of degree n on S. Set $\mathrm{PGL}_{n,S} := \mathrm{PGL}_n \otimes_{\mathbb{Z}} S$. Assume that $H^1(\mathcal{A}(x), \mathrm{M}_n(k(x))/\mathcal{A}(x)) = 0$ for each $x \in S$. Then the $\mathrm{PGL}_{n,S}$-orbit $\{P^{-1}\mathcal{A}P \mid P \in \mathrm{PGL}_{n,S}\}$ is open in $\mathrm{Mold}_{n,d} \otimes_{\mathbb{Z}} S$.*

Example 2.1. For a commutative ring R, let

$$\mathrm{S}_8(R) = \left\{ \begin{pmatrix} a & c & d \\ 0 & b & 0 \\ 0 & 0 & b \end{pmatrix} \middle| \; a, b, c, d \in R \right\}.$$

Then

$$H^i(\mathrm{S}_8(R), \mathrm{M}_3(R)/\mathrm{S}_8(R)) \cong \begin{cases} R^3 & (i = 0) \\ 0 & (i > 0). \end{cases}$$

Hence $\phi_{\mathrm{S}_8} : \mathrm{PGL}_3 \to \mathrm{Mold}_{3,4}$ defined by $P \mapsto P^{-1}\mathrm{S}_8 P$ is smooth. In particular, the PGL_3-orbit of S_8 is open in $\mathrm{Mold}_{3,4}$. More precisely, we can see that the PGL_3-orbit of S_8 is isomorphic to $\mathbb{P}^2_{\mathbb{Z}}$. For details, see Ref. 7.

Let k be an algebraically closed field. There are 26 types of k-subalgebras of $\mathrm{M}_3(k)$ up to inner automorphisms of $\mathrm{M}_3(k)$. For all types of k-subalgebras A of $\mathrm{M}_3(k)$, we have calculated Hochschild cohomology $H^i(A, \mathrm{M}_n(k)/A)$. We introduce several results not only for an algebraically closed field k but also for any commutative ring R. For proofs, see Ref. 6.

Theorem 2.3. *Set* $\mathrm{N}_3(R) := \left\{ \begin{pmatrix} a & b & c \\ 0 & a & d \\ 0 & 0 & a \end{pmatrix} \middle| \; a, b, c, d \in R \right\} \subset \mathrm{M}_3(R)$ *for a commutative ring* R. *Then*

$$H^i(\mathrm{N}_3(R), \mathrm{M}_3(R)/\mathrm{N}_3(R)) \cong \begin{cases} R^2 & (i = 0) \\ R^{i+1} & (i > 0). \end{cases}$$

Theorem 2.4. *Set* $\mathrm{S}_4(R) := \left\{ \begin{pmatrix} a & b & c \\ 0 & a & 0 \\ 0 & 0 & a \end{pmatrix} \middle| \; a, b, c \in R \right\} \subset \mathrm{M}_3(R)$ *for a commutative ring* R. *Then*

$$H^i(\mathrm{S}_4(R), \mathrm{M}_3(R)/\mathrm{S}_4(R)) \cong \begin{cases} R^4 & (i = 0) \\ R^{3 \cdot 2^i} & (i > 0). \end{cases}$$

3. The case S_{11}

In this section, we calculate $H^i(\mathrm{S}_{11}(R), \mathrm{M}_3(R)/\mathrm{S}_{11}(R))$, where $\mathrm{S}_{11}(R)$ will be defined below. The long proof will give us a useful technique for calculating Hochschild cohomology.

For a commutative ring R, we define an R-subalgebra $\mathrm{S}_{11}(R)$ of $\mathrm{M}_3(R)$ by

$$S_{11}(R) = \left\{ \begin{pmatrix} a & b & c \\ 0 & e & d \\ 0 & 0 & a \end{pmatrix} \in \mathrm{M}_3(R) \;\middle|\; a, b, c, d, e \in R \right\}.$$

Set $A = \mathrm{S}_{11}(R)$ and $M = \mathrm{M}_3(R)$. Let us calculate the Hochschild cohomology $H^*(A, M/A)$.

We denote by $E_{ij} \in \mathrm{M}_3(R)$ the matrix with entry 1 in the (i, j)-component and 0 the other components. We set $I = E_{11} + E_{22} + E_{33}$, $F = E_{11} + E_{33}$, $U = E_{12}$, $V = E_{23}$ and $W = E_{13}$. The set $\{F, U, V, W\}$ forms a basis of the free R-module $\overline{A} = A/RI$.

We introduce a filtration on M/A. We set $F^0 = M/A$. Let L be the R-submodule of $M = \mathrm{M}_3(R)$ consisting of matrices in which the $(3, 1)$-entry is 0. We set $F^1 = L/A$ and $F^2 = B/A$, where $B = \mathrm{B}_3(R) := \{(a_{ij}) \in \mathrm{M}_3(R) \mid a_{ij} = 0 \text{ for } i > j\}$. We have obtained a filtration

$$0 = F^3 \subset F^2 \subset F^1 \subset F^0 = M/A$$

of A-bimodules over R. We denote the A-bimodule M/L over R by $T_{a,a}$. The A-bimodule L/B over R is isomorphic to the direct sum of $T_{a,e}$ and $T_{e,a}$, where $T_{a,e}$ is the submodule of L/B generated by $(E_{32} \mod B)$, and $T_{e,a}$ is the submodule of L/B generated by $(E_{21} \mod B)$. The A-bimodule B/A over R is isomorphic to $T_{a,a}$.

First, let us calculate the Hochschild cohomology $H^*(A, T_{a,a}) = \bigoplus_{i \geq 0} H^i(A, T_{a,a})$ of A with coefficients in $T_{a,a}$. We have an isomorphism

$$T_{a,a} \otimes_A T_{a,a} \cong T_{a,a}$$

of A-bimodules over R. This implies that $T_{a,a}$ is a monoid object in the category of A-bimodules over R, where the unit $u : A \to T_{a,a}$ is given by $u(I) = (E_{31} \mod L)$. For A-bimodules N_1 and N_2, the cup product

$$\cup : C^p(A, N_1) \otimes_R C^q(A, N_2) \to C^{p+q}(A, N_1 \otimes_R N_2)$$

defined by

$$(f \cup g)(a_1 \otimes \cdots a_p \otimes b_1 \otimes \cdots \otimes b_q) = f(a_1 \otimes \cdots \otimes a_p) \otimes g(b_1 \otimes \cdots \otimes b_q)$$

is R-linear and satisfies

$$d^{p+q}(f \cup g) = d^p(f) \cup g + (-1)^p f \cup d^q(g)$$

for $f \in C^p(A, N_1)$ and $g \in C^q(A, N_2)$. Note that the cup product induces an R-linear map $\cup : \overline{C}^p(A, N_1) \otimes_R \overline{C}^q(A, N_2) \to \overline{C}^{p+q}(A, N_1 \otimes_R N_2)$. See, for examples, Ref. 1 or Ref. 8 for the cup product. Since $T_{a,a}$ is a monoid object in the category of A-bimodules over R, $\overline{C}^*(A, T_{a,a}) = \bigoplus_{i \geq 0} \overline{C}^i(A, T_{a,a})$ has the structure of a differential graded algebra over R, which induces on $H^*(A, T_{a,a})$ the structure of a graded associative algebra over R.

Let $e_{a,a} \in T_{a,a}$ be the image of I under the unit $u : A \to T_{a,a}$. We denote by $F^*, U^*, V^*, W^* \in \overline{C}^1(A, T_{a,a})$ the R-homomorphisms $\overline{A} \to T_{a,a}$ given by

$$F^*(n) = \begin{cases} e_{a,a} & \text{if } n = F \\ 0 & \text{if } n = U, V, W, \end{cases}$$

$$U^*(n) = \begin{cases} e_{a,a} & \text{if } n = U \\ 0 & \text{if } n = F, V, W, \end{cases}$$

$$V^*(n) = \begin{cases} e_{a,a} & \text{if } n = V \\ 0 & \text{if } n = F, U, W, \end{cases}$$

$$W^*(n) = \begin{cases} e_{a,a} & \text{if } n = W \\ 0 & \text{if } n = F, U, V, \end{cases}$$

respectively. We observe that $\overline{C}^*(A, T_{a,a})$ is a differential graded algebra which is isomorphic to the free graded associative algebra $R\langle F^*, U^*, V^*, W^*\rangle$ over R generated by F^*, U^*, V^*, W^* with differential

$$\begin{aligned}\delta(F^*) &= F^*F^*,\\ \delta(U^*) &= U^*F^*,\\ \delta(V^*) &= F^*V^*,\\ \delta(W^*) &= -U^*V^*.\end{aligned}$$

We consider the set M of all monomials of U^*, V^*, W^* and denote by $R\mathsf{M}$ the free R-module generated by M. We have an isomorphism of R-modules between $R\mathsf{M}$ and $R\langle U^*, V^*, W^*\rangle$. Let $\mathsf{M}_{V^*\cdots U^*\cdots}$ be a subset of M given by

$$\mathsf{M}_{V^*\cdots U^*\cdots} = \{\overbrace{V^*\cdots V^*}^{i}\overbrace{U^*\cdots U^*}^{j} \in \mathsf{M} \mid i, j \in \mathbb{Z}_{\geq 0}\},$$

and let M_{W^*} be a subset of M given by

$$\mathsf{M}_{W^*} = \{W^*, U^*V^*\}.$$

There is a bijection between M and

$$\mathsf{M}_{V^*\cdots U^*\cdots} \times \left(\coprod_{r\geq 0} \overbrace{(\mathsf{M}_{W^*} \times \mathsf{M}_{V^*\cdots U^*\cdots}) \times \cdots \times (\mathsf{M}_{W^*} \times \mathsf{M}_{V^*\cdots U^*\cdots})}^{r}\right). \quad (1)$$

Hence we have an isomorphism of R-modules between $R\langle U^*, V^*, W^*\rangle$ and

$$R\mathsf{M}_{V^*\cdots U^*\cdots} \otimes_R \left(\bigoplus_{r\geq 0} (R\mathsf{M}_{W^*} \otimes_R R\mathsf{M}_{V^*\cdots U^*\cdots})^{\otimes r}\right),$$

where $R\mathsf{M}_{V^*\cdots U^*\cdots}$ and $R\mathsf{M}_{W^*}$ are the free R-modules generated by $\mathsf{M}_{V^*\cdots U^*\cdots}$ and M_{W^*}, respectively.

Let C_{F^*} be the R-submodule of $\overline{C}^*(A, T_{a,a})$ given by

$$C_{F^*} = \bigoplus_{r\geq 0} R\,\overbrace{F^*\cdots F^*}^{r}.$$

It is easy to see that C_{F^*} is a subcomplex of $\overline{C}^*(A, T_{a,a})$ and that $H^*(C_{F^*}) \cong R$.

We set

$$C_{U^*} = RU^* \otimes_R C_{F^*},$$
$$C_{V^*} = C_{F^*} \otimes_R RV^*,$$
$$C_{W^*} = RW^* \oplus (RU^* \otimes_R C_{F^*} \otimes_R RV^*).$$

We can regard C_{U^*}, C_{V^*} and C_{W^*} as subcomplexes of $\overline{C}^*(A, T_{a,a})$ and we easily obtain that $H^*(C_{U^*}) = H^*(C_{V^*}) = H^*(C_{W^*}) = 0$.

We define a subcomplex $C_{V^*\cdots U^*\cdots}$ of $\overline{C}^*(A, T_{a,a})$ by

$$C_{V^*\cdots U^*\cdots} = \bigoplus_{i,j\geq 0} \overbrace{C_{V^*} \otimes_R \cdots \otimes_R C_{V^*}}^{i} \otimes_R C_{F^*} \otimes_R \overbrace{C_{U^*} \otimes_R \cdots \otimes_R C_{U^*}}^{j}.$$

Using the Künneth isomorphism and $H^*(C_{U^*}) = H^*(C_{V^*}) = 0$, we obtain that $H^*(C_{V^*\cdots U^*\cdots}) \cong H^*(C_{F^*}) \cong R$.

Using the bijection between M and (1), we see that there is an isomorphism of differential graded algebras between $\overline{C}^*(A, T_{a,a})$ and

$$C_{V^*\cdots U^*\cdots} \otimes_R \left(\bigoplus_{r\geq 0} (C_{W^*} \otimes_R C_{V^*\cdots U^*\cdots})^{\otimes r} \right),$$

because monomials of F^*, U^*, V^*, W^* are obtained from monomials in M by inserting $\overbrace{F^* \cdots F^*}^{s}$ for $s \geq 0$. Using the Künneth isomorphism and $H^*(C_{V^*\cdots U^*\cdots}) \cong R$, we obtain that $H^*(\overline{C}(A, T_{a,a})) \cong H^*(C_{V^*\cdots U^*\cdots}) \cong R$. Hence we obtain the following lemma.

Lemma 3.1. *We have an isomorphism*

$$H^*(A, T_{a,a}) \cong R$$

of graded associative algebras over R.

Next, we calculate the Hochschild cohomology $H^*(A, T_{a,e})$ of A with coefficients in $T_{a,e}$. Recall that $T_{a,e}$ is a submodule of the A-bimodule L/B over R generated by E_{32}. Since there is an isomorphism

$$T_{a,a} \otimes_A T_{a,e} \cong T_{a,e}$$

of A-bimodules over R, we see that $T_{a,e}$ is a left module object over $T_{a,a}$ in the category of A-bimodules over R. This implies that the cochain complex $\overline{C}^*(A, T_{a,e})$ is a differential graded left module over the differential graded algebra $\overline{C}^*(A, T_{a,a})$. We denote by $e_{a,e}$ the element of $T_{a,e} \subset L/B$

represented by E_{32}. We regard $e_{a,e}$ as a 0-cochain of $\overline{C}^*(A, T_{a,e})$ and we have

$$\delta(e_{a,e}) = F^* e_{a,e}.$$

We notice that $\overline{C}^*(A, T_{a,e})$ is a free module over $\overline{C}^*(A, T_{a,a})$ generated by $e_{a,e}$.

We consider R-submodules $C_{F^*} e_{a,e}$ and $C_{U^*} e_{a,e}$ of $\overline{C}^*(A, T_{a,e})$. We easily see that they are subcomplexes of $\overline{C}^*(A, T_{a,e})$ and obtain that $H^*(C_{F^*} e_{a,e}) = 0$ and $H^*(C_{U^*} e_{a,e}) = R[U^* e_{a,e}] \cong R$, where $[U^* e_{a,e}] \in H^1(C_{U^* e_{a,e}})$ is represented by $U^* e_{a,e}$. Combining these results with the Künneth isomorphism, $H^*(C_{F^*}) \cong R$, and $H^*(C_{U^*}) = H^*(C_{V^*}) = 0$, we see that $H^*(C_{V^* \dots U^* \dots} e_{a,e}) \cong H^*(C_{F^*} \otimes_R C_{U^*} e_{a,e}) \cong R[U^* e_{a,e}]$. Since $\overline{C}^*(A, T_{a,e})$ is a free left $\overline{C}^*(A, T_{a,a})$-module generated by $e_{a,e}$, we have an isomorphism of differential graded modules between $\overline{C}^*(A, T_{a,e})$ and

$$\left(\bigoplus_{r \geq 0} (C_{V^* \dots U^* \dots} \otimes_R C_{W^*})^{\otimes r} \right) \otimes_R C_{V^* \dots U^* \dots} e_{a,e}.$$

Using $H^*(C_{W^*}) = 0$, $H^*(C_{V^* \dots U^* \dots} e_{a,e}) \cong R[U^* e_{a,e}]$ and the Künneth isomorphism again, we obtain the following lemma.

Lemma 3.2. *We have an isomorphism*

$$H^*(A, T_{a,e}) \cong R[U^* e_{a,e}]$$

of graded modules over R, where the degree of $[U^* e_{a,e}]$ is 1.

By the similar argument, we obtain the following lemma.

Lemma 3.3. *We have an isomorphism*

$$H^*(A, T_{e,a}) \cong R[e_{e,a} V^*]$$

of graded modules over R, where $e_{e,a}$ is the element of $T_{e,a}$ represented by E_{21} and the degree of $[e_{e,a} V^*]$ is 1.

We recall that we have defined a filtration of A-bimodules over R on M/A:

$$0 = F^3 \subset F^2 \subset F^1 \subset F^0 = M/A,$$

where $F^0 = M/A$, $F^1 = L/A$ and $F^2 = B/A$. We denote by $\mathrm{Gr}^p(M/A)$ the p-th associated graded module F^p/F^{p+1}. The filtration induces a long exact sequence

$$\cdots \to H^{p+q}(A, F^{p+1}) \to H^{p+q}(A, F^p) \to H^{p+q}(A, \mathrm{Gr}^p(M/A))$$
$$\to H^{p+q+1}(A, F^{p+1}) \to \cdots .$$

We set

$$D^{p,q} = H^{p+q}(A, F^p),$$
$$E^{p,q} = H^{p+q}(A, \mathrm{Gr}^p(M/A)).$$

We obtain an exact couple

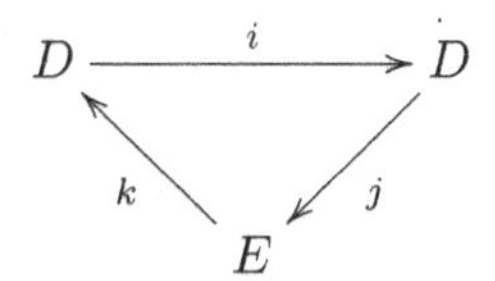

where $D = \oplus_{p,q} D^{p,q}$ and $E = \oplus_{p,q} E^{p,q}$. By standard construction, we obtain a spectral sequence

$$E_1^{p,q} = H^{p+q}(A, \mathrm{Gr}^p(M/A)) \Longrightarrow H^{p+q}(A, M/A)$$

with

$$d_r : E_r^{p,q} \longrightarrow E_r^{p+r,q-r+1}$$

for $r \geq 1$. See, for examples, Ref. 2 for construction of spectral sequences. Note that $E_1^{p,q} = 0$ unless $0 \leq p \leq 2$ and $p + q \geq 0$. Thus, the spectral sequence collapses at the E_3-page.

We recall that $M/L = T_{a,a}$. The A-bimodule L/B over R is isomorphic to the direct sum of $T_{a,e}$ and $T_{e,a}$, and the A-bimodule B/A over R is isomorphic to $T_{a,a}$. By Lemma 3.1, we have

$$E_1^{0,q} \cong \begin{cases} Re_{a,a} & (q = 0), \\ 0 & (q \neq 0), \end{cases}$$

and

$$E_1^{2,q-2} \cong \begin{cases} R & (q = 0), \\ 0 & (q \neq 0). \end{cases}$$

By Lemmas 3.2 and 3.3, we have

$$E_1^{1,q-1} \cong \begin{cases} R[U^* e_{a,e}] \oplus R[e_{e,a} V^*] & (q = 1), \\ 0 & (q \neq 1). \end{cases}$$

We calculate $d_1 : E_1^{0,0} \to E_1^{1,0}$. The differential d_1 is identified with the connecting homomorphism $\delta : H^0(A, M/L) \to H^1(A, L/B)$. From this observation, we see that

$$d_1(e_{a,a}) = -[U^* e_{a,e}] + [e_{e,a} V^*].$$

Hence we obtain that

$$E_2^{p,q} \cong \begin{cases} R \ ((p,q) = (1,0), (2,-2)), \\ 0 \ \text{(otherwise)}. \end{cases}$$

Thus, the spectral sequence collapses at the E_2-page and there is no extension problem.

Theorem 3.1. *The R-module $H^n(A, M/A)$ is free for all $n \geq 0$. The rank of $H^n(A, M/A)$ over R is given by*

$$\operatorname{rank}_R H^n(A, M/A) = \begin{cases} 1 \ (n = 0, 1), \\ 0 \ (n \geq 2). \end{cases}$$

References

[1] V. Ginzburg, Lectures on Noncommutative Geometry, preprint, 2005, arXiv:math/0506603.

[2] J. McCleary, *A user's guide to spectral sequences*, Second edition. Cambridge Studies in Advanced Mathematics, 58 (Cambridge University Press, Cambridge, 2001).

[3] K. Nakamoto, The moduli of representations with Borel mold, *Internat. J. Math.* **25**, 1450067, 31 pp (2014).

[4] K. Nakamoto and T. Torii, The moduli of subalgebras of the full matrix ring of degree 3, Proceedings of the 50th Symposium on Ring Theory and Representation Theory, 137–149, Symp. Ring Theory Represent. Theory Organ. Comm., Yamanashi, (2018).

[5] K. Nakamoto and T. Torii, An application of Hochschild cohomology to the moduli of subalgebras of the full matrix ring, Proceedings of the 51st Symposium on Ring Theory and Representation Theory, 110–118, Symp. Ring Theory Represent. Theory Organ. Comm., Shizuoka, (2019).

[6] K. Nakamoto and T. Torii, Applications of Hochschild cohomology to the moduli of subalgebras of the full matrix ring, preprint, 2020, arXiv:2006.07870.

[7] K. Nakamoto and T. Torii, On the classification of subalgebras of the full matrix ring of degree 3, in preparation.

[8] S. J. Witherspoon, *Hochschild cohomology for algebras*, Graduate Studies in Mathematics, 204, American Mathematical Society, Providence, RI, 2019.

On generalized Nakayama-Azumaya Lemma and NAS-modules

Masahisa Sato

Center for Regional Policy Studies, Aichi University,
1-1 Machihata-cho, Toyohashi, Aichi, 441-8522, Japan
E-mail: msato@yamanashi.ac.jp

Nakayama-Azumaya Lemma asserts that $MJ(R) = M$ implies $M = 0$ for a finitely generated R-module M, here $J(R)$ is the Jacobson radical of a ring R. Also this lemma holds for any projective modules by H. Bass [2]. To unify and generalize these two lemmas, we consider an R-module M which is isomorphic to a direct summand of a direct sum of finitely generated modules and satisfies $MJ(R) = M$. Generalized Nakayama-Azumaya Lemma (GNA Lemma) is the assertion that $M = 0$ for a module satisfying the above property. We are concern whether the GNA Lemma holds or not. In this paper, we introduce Nakayama-Azumaya special module (NAS-module) and show that GNA Lemma holds if and only if there are no non-zero NAS modules. Also we investigate NAS modules.

Keywords: Nakaya-Azumaya Lemma; Nakayama-Azumaya special module; existence of maximal submodule.

1. Introduction

Let R be a ring with identity and $J(R)$ the Jacobson radical of R. Throughout this paper, we always consider right R-modules.

The following statement is well known as Nakayama-Azumaya Lemma.

$MJ(R) = M$ implies $M = 0$ for a finitely generated R-module M.

This statement holds for any projective modules [1. Proposition 17.14, 5. Lemma 1.53] originally first proved by H. Bass [2].

As a generalization of these two statements, we consider an R-module M satisfying two properties;
(1) M is isomorphic to a direct summand of a direct sum of finitely generated modules.
(2) $MJ(R) = M$.
Generalized Nakayama-Azumaya Lemma (GNA Lemma) is the assertion $M = 0$ for a module M with the properties (1) and (2).

In this paper, we concern GNA Lemma holds or not. To study this problem, we introduce the following NAS-modules.

Definition A module M is called Nakayama-Azumaya special module (NAS-module) if M satisfies the following conditions;

(1) There is a subsequence $M_1 \subset M_2 \subset \cdots$ of finitely generated submodules of M such that $M_i \subset M_{i+1}J(R)$ for any $i \in \mathbb{N}$ and $M = \bigcup_{i\in\mathbb{N}} M_i$. Consequently, $M = MJ(R)$.

(2) Let $f : \sum_{i\in\mathbb{N}} \oplus M_i \to M$ be the homomorphism given by $f((x_i)) = \sum_{i\in\mathbb{N}} x_i$ for $(x_i) \in \sum_{i\in\mathbb{N}} \oplus M_i$. There is a homomorphism $g : M \to \sum_{i\in\mathbb{N}} \oplus M_i$ such that $fg = 1_M$

(3) $M_i \cap g(M) = M_i \cap N = 0$ for any $i \in \mathbb{N}$. Here $N = \ker f$.

One of the main theorems in this paper is the reduction of GNA Lemma to the non-existence of non-zero NAS-modules.

Theorem The following properties are equivalent.

(1) GNA Lemma holds.

(2) There are no non-zero NAS-modules.

We remark that even indecomposable module M which is a direct summand of a direct sum of finitely generated modules is not necessary finitely generated [3, 4], but it is a direct sum of countably generated modules [1. Theorem 26.1] originally proved by I. Kaplansky [4] and C. Walker [7].

In case of projective modules or finitely generated modules, Nakayama-Azumaya Lemma is equivalent to the existence of a maximal submodule. So in section 5, we investigate this property for a module isomorphic to a direct summand of a direct sum of finitely generated modules.

2. Generalized Nakayama-Azumaya Lemma

In this section, we generalize the following fact;
$MJ(R) = M$ implies $M = 0$ for a projective module or a finitely generated module M. The latter case is the original Nakayama-Azumaya Lemma.

Before we generalize and unify these theorems, we show the following lemma to use in the proof of the main theorem.

Lemma 2.1. *For a module with two direct decompositions* $\sum_{\delta\in\Delta} \oplus F_\delta = M \oplus N$, *there is a direct decomposition* $\sum_{\delta\in\Delta} \oplus (F_\delta/((F_\delta \cap M) \oplus (F_\delta \cap N))) = (M/\sum_{\delta\in\Delta} \oplus (F_\delta \cap M)) \oplus (N/\sum_{\delta\in\Delta} \oplus (F_\delta \cap N))$ *such that, for each* $\gamma \in \Delta$,

$$(F_\gamma/((F_\gamma \cap M) \oplus (F_\gamma \cap N))) \cap (M/\sum_{\delta\in\Delta} \oplus(F_\delta \cap M)) = 0$$

and

$$(F_\gamma/((F_\gamma \cap M) \oplus (F_\gamma \cap N))) \cap (N/\sum_{\delta\in\Delta} \oplus(F_\delta \cap N)) = 0.$$

Proof. We remark that the direct decomposition is shown by

$$\begin{aligned}
&(\sum_{\delta\in\Delta} \oplus F_\delta)/(\sum_{\delta\in\Delta} \oplus((F_\delta \cap M) \oplus (F_\delta \cap N))) \\
&= (M \oplus N)/\sum_{\delta\in\Delta} \oplus((F_\delta \cap M) \oplus (F_\delta \cap N)) \\
&= \left(M/\sum_{\delta\in\Delta} \oplus(F_\delta \cap M)\right) \oplus \left(N/\sum_{\delta\in\Delta} \oplus(F_\delta \cap N)\right).
\end{aligned}$$

But to clarify the meaning of the intersection property, we use the following canonical epimorphism;

$\alpha : \sum_{\delta\in\Delta} \oplus F_\delta = M \oplus N \to \left(M/\sum_{\delta\in\Delta} \oplus(F_\delta \cap M)\right) \oplus \left(N/\sum_{\delta\in\Delta} \oplus(F_\delta \cap N)\right)$
and show $\ker(\alpha) = \sum_{\delta\in\Delta} \oplus((F_\delta \cap M) \oplus (F_\delta \cap N))$ though we know this fact itself from the above remark.

We describe an element of $M \oplus N$ by $\binom{x}{y}$ for $x \in M$ and $y \in N$.

Take any $a = \sum_{\delta\in\Delta} a_\delta \in \sum_{\delta\in\Delta} \oplus F_\delta$, here $a_\delta \in F_\delta$. We have unique expression $a_\delta = \binom{x_\delta}{y_\delta} \in M \oplus N$ for some $x_\delta \in M$ and $y_\delta \in N$.

Assume $\alpha(a) = 0$, then $\sum_{\delta\in\Delta} x_\delta \in \sum_{\delta\in\Delta} \oplus(F_\delta \cap M)$ and $\sum_{\delta\in\Delta} y_\delta \in \sum_{\delta\in\Delta} \oplus(F_\delta \cap N)$. So there are $x'_\delta \in F_\delta \cap M$ and $y'_\delta \in F_\delta \cap N$ such that $\sum_{\delta\in\Delta} x_\delta = \sum_{\delta\in\Delta} x'_\delta \in \sum_{\delta\in\Delta} \oplus(F_\delta \cap M)$ and $\sum_{\delta\in\Delta} y_\delta = \sum_{\delta\in\Delta} y'_\delta \in \sum_{\delta\in\Delta} \oplus(F_\delta \cap N)$.

Hence $a = \sum_{\delta\in\Delta} a_\delta = \sum_{\delta\in\Delta} \binom{x'_\delta}{0} + \sum_{\delta\in\Delta} \binom{0}{y'_\delta} \in \sum_{\delta\in\Delta} ((F_\delta \cap M) \oplus (F_\delta \cap N))$. Thus $\ker \alpha \subset \sum_{\delta\in\Delta} ((F_\delta \cap M) \oplus (F_\delta \cap N))$. The converse is clear, so they coincide.

Next we show $(F_\gamma/((F_\gamma \cap M) \oplus (F_\gamma \cap N))) \cap (N/\sum_{\delta\in\Delta} \oplus(F_\delta \cap N)) = 0$.

Take any $a \in F_\gamma$. We denote $a = \binom{x}{y}$, here $x \in M$ and $y \in N$, then $\alpha(a) = \overline{x} + \overline{y}$, here $\overline{x} \in M/\sum_{\delta\in\Delta} \oplus(F_\delta \cap M)$ and $\overline{y} \in N/\sum_{\delta\in\Delta} \oplus(F_\delta \cap N)$ are residue classes of x and y, respectively.

Assume $\alpha(a) \in N/\sum_{\delta\in\Delta} \oplus(F_\delta \cap N)$, then $\overline{x} = \overline{0}$, which means that x has the expression $x = \sum_{\delta\in\Delta} \binom{z_\delta}{0}$ such that $z_\delta \in F_\delta \cap M$. Since $\overline{y} \in N/\sum_{\delta\in\Delta} \oplus(F_\delta \cap N)$, y has the expression $y = \sum_{\delta\in\Delta} \binom{x_\delta}{y_\delta}$ such that $\sum_{\delta\in\Delta} x_\delta = 0$.

Since $a \in F_\gamma$ and the above summation is direct sum, $x_\delta = y_\delta = z_\delta = 0$

if $\delta \neq \gamma$. Also $\sum_{\delta \in \Delta} x_\delta = 0$ implies $x_\gamma = 0$, hence $y = \binom{0}{y_\gamma} \in F_\gamma \cap N$.

Thus we have $a = \binom{z_\gamma}{0} + \binom{0}{y_\gamma} \in (F_\gamma \cap M) \oplus (F_\gamma \cap N)$. That is, $\overline{a} = \overline{0}$.

By replacing N as M, we have $(F_\gamma/((F_\gamma \cap M) \oplus (F_\gamma \cap N))) \cap (M/\sum_{\delta \in \Delta} \oplus(F_\delta \cap M)) = 0$. □

Definition 2.1. A module M is called Nakayama-Azumaya special module (NAS-module) if M satisfies the following conditions;

(1) There is a subsequence $M_1 \subset M_2 \subset \cdots$ of finitely generated submodules of M such that $M_i \subset M_{i+1}J(R)$ for any $i \in \mathbb{N}$ and $M = \bigcup_{i \in \mathbb{N}} M_i$. Consequently, $M = MJ(R)$.

(2) Let $f : \sum_{i \in \mathbb{N}} \oplus M_i \to M$ be the homomorphism given by $f((x_i)) = \sum_{i \in \mathbb{N}} x_i$ for $(x_i) \in \sum_{i \in \mathbb{N}} \oplus M_i$. There is a homomorphism $g : M \to \sum_{i \in \mathbb{N}} \oplus M_i$ such that $fg = 1_M$.

(3) $M_i \cap g(M) = M_i \cap N = 0$ for any $i \in \mathbb{N}$. Here $N = \ker f$.

We remark that there are some subsequences which make the same NAS-module. In fact, take any finitely generated submodule M_0 of $M_1J(R)$ in Definition 2.1, then a subsequence $M_0 \subset M_1 \subset M_2 \subset \cdots$ gives the same NSA-module M as easily shown. But there is, in some sense, a unique minimal subsequence, so called *reduced*, which is defined and shown in the following.

Definition 2.2. A subsequence $M_1 \subset M_2 \subset \cdots$ in Definition 2.1 is said to be reduced if $g(M) \not\subset \sum_{i \in (\mathbb{N} \setminus S)} \oplus M_i$ for any non-empty subset S of $\mathbb{N}$.

Lemma 2.2. *Let $M_1 \subset M_2 \subset \cdots$ be a subsequence which gives NAS-module M. Then there is a unique minimal subset $T = \{t_1 < t_2 < \cdots\}$ of $\mathbb{N}$ such that $M_{t_1} \subset M_{t_2} \subset \cdots$ is a reduced subsequence of NAS-module M.*

Proof. Let Γ be a set of a subset S of $\mathbb{N}$ such that $g(M) \subset \sum_{i \in S} \oplus M_i$. Then $\sum_{i \in S} \oplus M_i = g(M) \oplus (N \bigcap \sum_{i \in S} \oplus M_i)$ by modular law. We set $T = \bigcap_{S \in \Gamma} S = \{t_1 < t_2 < \cdots\}$. Since $g(M) \subset \bigcap_{S \in \Gamma} \sum_{i \in S} \oplus M_i = \sum_{t_i \in T} \oplus M_{t_i}$, by applying modular law again, we have $\sum_{t_i \in T} \oplus M_{t_i} = g(M) \oplus (N \bigcap \sum_{t_i \in T} \oplus M_{t_i})$.

Hence $M_{t_1} \subset M_{t_2} \subset \cdots$ is a subsequence which gives NAS-module M. By the construction of T, $g(M) \not\subset \sum_{t_i \in (T \setminus S)} \oplus M_{t_i}$ for any non-empty subset S of T. Thus $M_{t_1} \subset M_{t_2} \subset \cdots$ is a reduced subsequence. □

Theorem 2.1. *The following properties are equivalent.*

(1) *GNA Lemma holds.*
(2) *There are no non-zero NAS-modules given by reduced subsequences.*

Proof. If GNA Lemma holds, then clearly $M = 0$ for any NAS-module M.

Let M be a module such that $MJ(R) = M$. Assume that there are finitely generated modules F_δ $(\delta \in \Delta)$ such that M is a direct summand of $F = \sum_{\delta\in\Delta} \oplus F_\delta$. Since M is a direct sum of countably generated modules [1, 4, 7], we may assume M is countably generated since, for $M = \sum_{\gamma\in\Gamma} \oplus M_\gamma$, $MJ(R) = M$ is equivalent to $M_\gamma J(R) = M_\gamma$ for every $\gamma \in \Gamma$ and the property that $MJ(R) = M$ implies $M = 0$ is equivalent to the property that $M_\gamma J(R) = M_\gamma$ implies $M_\gamma = 0$ for every $\gamma \in \Gamma$. Under this assumption, also we may assume an index set Δ is a countable set since we can take a countable generator set of M and each generator is included in a finite direct sum of F_δ's. So we can set $\Delta = \mathbb{N}$ the set of natural numbers and $F_n = \sum_{s=1}^{i_n} a_{ns}R$ $(a_{ns} \in F_n)$ for each $n \in \mathbb{N}$ since F_n's are finitely generated.

There is some module N such that $F = M \oplus N$ by assumption. Let $f : F \to M, \overline{g} : F \to N$ and $g : M \to F, \overline{f} : N \to F$ be the canonical projections and the canonical injections, respectively. Then $\ker f = N$ and it holds $1_F = gf + \overline{f}\,\overline{g}, fg = 1_M, \overline{g}\,\overline{f} = 1_N, f\overline{f} = 0, \overline{g}g = 0$. So we have $a_{ns} = gf(a_{ns}) + \overline{f}\,\overline{g}(a_{ns})$. We set $x_{ns} = f(a_{ns}) \in M$ and $y_{ns} = \overline{g}(a_{ns}) \in N$, then $a_{ns} = g(x_{ns}) + \overline{f}(y_{ns})$. Also we set $p_{ns} = g(x_{ns})$, then we have $f(a_{ns}) = g(p_{ns}) = x_{ns}$. So we denote $a_{ns} = \binom{x_{ns}}{y_{ns}}$. We define a set $G_n = \{p_{ns} \mid s = 1, 2, \cdots, i_n\}$ for each $n \in \mathbb{N}$, then $\bigcup_{n\in\mathbb{N}} G_n$ is a generator set of $g(M)$.

Each p_{ns} is expressed uniquely by a finite sum of the form $\sum_{j=1}^{t_{ns}} m_{(ns)_j}$ for each n and s, here $(ns)_j \in \mathbb{N}$ and $m_{(ns)_j} \in F_{(ns)_j}$. Also each $m_{(ns)_j}$ has an expression $m_{(ns)_j} = \sum_{k=1}^{i_{(ns)_j}} a_{(ns)_j k} r_{(ns)_j k}$ for $r_{(ns)_j k} \in R$ as an element of $F_{(ns)_j}$. We remark that $g(M) \subset \sum_{n\in\mathbb{N}} \oplus F_n J(R)$ since $MJ(R) = M$. So we can take $r_{(ns)_j k}$ from $J(R)$.

Now we have the equation,

$$x_{ns} = f(p_{ns}) = f(\sum_{j=1}^{t_{ns}} m_{(ns)_j}) = \sum_{j=1}^{t_{ns}} f(m_{(ns)_j})$$

$$= \sum_{j=1}^{t_{ns}} \sum_{k=1}^{i_{(ns)_j}} f(a_{(ns)_j k}) r_{(ns)_j k} = \sum_{j=1}^{t_{ns}} \sum_{k=1}^{i_{(ns)_j}} x_{(ns)_j k} r_{(ns)_j k}.$$

By applying g to the above equation, we have the equation

$$p_{ns} = \sum_{j=1}^{t_{ns}} \sum_{k=1}^{i_{(ns)_j}} p_{(ns)_j k} r_{(ns)_j k}.$$

We take $n = s = 1$, then we may assume $(1,1)_1 = 1$. Hence we have

$$p_{1,1} = \sum_{j=1}^{t_{1,1}} \sum_{k=1}^{i_{(1,1)_j}} p_{(1,1)_j k} r_{(1,1)_j k}.$$

If $r_{(1,1)_1 1} = r_{1,1} = 0$, then we can remove $p_{1,1}$ from the set G_1.

If $r_{(1,1)_1 1} = r_{1,1} \neq 0$, then we have

$$p_{1,1}(1 - r_{1,1}) = \sum_{k=2}^{i_1} p_{1k} r_{1k} + \sum_{j=2}^{t_{1,1}} \sum_{k=1}^{i_{(1,1)_j}} p_{(1,1)_j k} r_{(1,1)_j k}.$$

Since $r_{1,1} \in J(R)$, $(1 - r_{1,1})$ is invertible, we have the equations

$$p_{1,1} = \sum_{k=2}^{i_1} p_{1k} r_{1k}(1 - r_{1,1})^{-1} + \sum_{j=2}^{t_{1,1}} \sum_{k=1}^{i_{(1,1)_j}} p_{(1,1)_j k} r_{(1,1)_j k}(1 - r_{1,1})^{-1},$$

$$f(p_{1,1}) = \sum_{k=2}^{i_1} f(p_{1k}) r_{1k}(1 - r_{1,1})^{-1} + \sum_{j=2}^{t_{1,1}} \sum_{k=1}^{i_{(1,1)_j}} f(p_{(1,1)_j k}) r_{(1,1)_j k}(1 - r_{1,1})^{-1}$$

$$= \sum_{k=2}^{i_1} x_{1k} r_{1k}(1 - r_{1,1})^{-1} + \sum_{j=2}^{t_{1,1}} \sum_{k=1}^{i_{(1,1)_j}} x_{(1,1)_j k} r_{(1,1)_j k}(1 - r_{1,1})^{-1}.$$

Hence we can remove $p_{1,1}$ from the set G_1 and

$$x_{11} \in \sum_{k=2}^{i_1} x_{1k} R + \sum_{n \geq 2} \sum_{k=1}^{i_n} x_{nk} R.$$

We repeat this process for $\{p_{12}, \cdots, p_{1i_1}\}$ and $\{x_{12}, \cdots, x_{1i_1}\}$, we have $p_{11}, p_{12}, \cdots, p_{1i_1} \in \sum_{n=2}^{s} \sum_{k=1}^{i_n} p_{nk} R$ and $x_{11}, x_{12}, \cdots, x_{1i_1} \in \sum_{n=2}^{s} \sum_{k=1}^{i_n} x_{nk} R$ for some $s \in \mathbb{N}$.

Since the set $\bigcup_{1 \leq n \leq s} f(G_n) = \{x_{nk} \mid n = 1, \cdots s,\ k = 1, \cdots, i_n\}$ is a finite set, we have $\bigcup_{1 \leq n \leq s} G_n \subset \sum_{n=1}^{s'} \oplus F_n$ for some $s' \in \mathbb{N}$. We remark

$$\{\overline{f}(y_{nk}) \mid n = 1, \cdots s,\ k = 1, \cdots, i_n\} \subset \sum_{n=1}^{s'} \oplus F_n$$

since $\overline{f}(y_{mk}) = a_{mk} - g(x_{mk}) \in \sum_{n=1}^{s'} \oplus F_n$ for $m \leq s$. From now on, when we write $F = M \oplus N$, we use the expression $g(x) = x \in F$ for any $x \in M$ and $\overline{f}(y) = y \in F$ for any $y \in N$. Hence $N = \overline{f}(N)$ and $p_{ns} = x_{ns}$.

Since $\sum_{n=1}^{s'} \oplus F_n$ is finitely generated, we can replace $\sum_{n=2}^{s'} \oplus F_n$ with F_2.

Then for any $\binom{x}{y} \in F_1$, here $x \in M, y \in N$, there is some $y' \in N$ such that $\binom{x}{y'} \in F_2$.

We repeat this process, we may assume the following property (*), that is, for any $i \in \mathbb{N}$ and $\binom{x}{y} \in F_i$, here $x \in M, y \in N$, there is some $y' \in N$ such that $\binom{x}{y'} \in F_{i+1}$.

Set $N' = \sum_{i\geq 1} \oplus(F_i \cap N)$ the submodule of N, then we have $\sum_{i\geq 1} \oplus (F_i/(F_i \cap N)) = M \oplus (N/N')$. We replace $F_i/(F_i \cap N)$ and N/N' with F_i and N, respectively. By this setting, we remark if $\binom{0}{y} \in F_i$ for some $y \in N$, then $y = 0$.

We set $f_i = f|F_i : F_i \to M$, that is, $f_i\binom{x}{y} = x$ and $M_i = \mathrm{Im} f_i$. We claim if $\binom{x}{y} \in F_i$ and $\binom{x}{y'} \in F_i$, then $y = y'$, which means f_i gives an isomorphism $F_i \to M_i$. In fact, $\binom{0}{y-y'} \in F_i$ and this implies $y = y'$ by the above remark.

From the above fact, for $x \in M$ such that $\binom{x}{y} \in F_i$ for some $y \in N$, this y is determined uniquely in F_i, so we denote this y by $y_x^{(i)}$. Also we have the following properties by the uniqueness;

(1) $y_x^{(i)} r = y_{xr}^{(i)}$ and (2) $y_x^{(i)} + y_{x'}^{(i)} = y_{x+x'}^{(i)}$ for any $x, x' \in M, r \in R$.

Now we set $\overline{g_i} = \overline{g}|F_i : F_i \to N$ and $N_i = \mathrm{Im}\overline{g_i}$. Then the above properties (1) and (2) mean that the map $\alpha_i : M_i \to N_i$ defined by $\alpha_i(x) = y_x^{(i)}$ is an R-homomorphism.

The assumption M is a direct summand of F means $F = g(M) \oplus N$ exactly, but we simply write $F = M \oplus N$ same as stated before.

We consider the factor module $\sum_{i\geq 1} \oplus(F_i/(F_i \cap M)) = (M/\sum_{i\geq 1} \oplus(F_i \cap M)) \oplus N$, then clearly $(M/\sum_{i\geq 1} \oplus(F_i \cap M))J(R) = M/\sum_{i\geq 1} \oplus(F_i \cap M)$.

We consider the case $M = \sum_{i\geq 1} \oplus(F_i \cap M)$. Then $F = (\sum_{i\geq 1} \oplus(F_i \cap M)) \oplus N$. For each $n \in \mathbb{N}$, by modular law, we have $F_n = F \cap F_n = (F_n \cap M) \oplus (((\sum_{i\neq n} \oplus(F_i \cap M)) \oplus N) \cap F_n)$. Hence each $F_n \cap M$ is finitely generated and $(F_n \cap M)J(R) = F_n \cap M$ by $MJ(R) = M$. Thus $F_n \cap M = 0$ by the original Nakayama-Azumaya Lemma and we have $M = 0$.

So it suffices to show that $M/\sum_{i\geq 1} \oplus(F_i \cap M) = 0$.

By Lemma 2.1, we have the direct decomposition;

$$\sum_{j\in\mathbb{N}} \oplus (F_j/((F_j \cap M) \oplus (F_j \cap N)))$$
$$= (M/\sum_{j\in\mathbb{N}} \oplus(F_j \cap M)) \oplus (N/\sum_{j\in\mathbb{N}} \oplus(F_j \cap N))$$

such that, for each $i \in \mathbb{N}$,

$$(F_i/(F_i \cap M)) \cap (M/\sum_{j\in\mathbb{N}} \oplus(F_j \cap M)) = 0$$

and

$$(F_i/(F_i \cap M) \cap (N/\sum_{j\in\mathbb{N}} \oplus(F_j \cap N)) = 0.$$

Since $F_j \cap N = 0$, for each $i \in \mathbb{N}$, we have the direct decomposition $\sum_{j\in\mathbb{N}} \oplus (F_j/(F_j \cap M)) = (M/\sum_{j\in\mathbb{N}} \oplus(F_j \cap M)) \oplus N$ such that $(F_i/(F_i \cap M)) \cap (M/\sum_{j\in\mathbb{N}} \oplus(F_j\ capM)) = 0$ and $(F_i/(F_i \cap M)) \cap N = 0$.

Now we replace $M/\sum_{i\geq 1} \oplus(F_i \cap M)$ with M and $F_i/(F_i \cap M)$ with F_i, then it satisfies $F_i \cap N = 0$ and $F_i \cap M = 0$ for all $i \in \mathbb{N}$ in this setting. By applying the same discussion as above, we know α_i is a monomorphism, hence an isomorphism.

There is a canonical inclusion $\beta_i : M_i \to M_{i+1}$ by the property (*). Also we define a map $\gamma_i : N_i \to N_{i+1}$ by $\gamma_i(y_x^{(i)}) = y_x^{(i+1)}$ for each $i \geq 1$, then γ_i is an R-homomorphism, in fact, monomorphism by the uniqueness. These maps induce a monomorphism $\delta_i = \begin{pmatrix} \beta_i f_i \\ \gamma_i \alpha_i f_i \end{pmatrix} : F_i \to F_{i+1}$ and an isomorphism $\epsilon_i : M_i \to L_i$ given by $\epsilon_i(x) = \begin{pmatrix} x \\ y_x^{(i)} \end{pmatrix}$ for each $i \in \mathbb{N}$.
We remark $\delta_i \begin{pmatrix} x \\ y_x^{(i)} \end{pmatrix} = \begin{pmatrix} x \\ y_x^{(i+1)} \end{pmatrix}$ for each $i \in \mathbb{N}$.

We summarize the above facts. For each $i \in \mathbb{N}$, we have;
(1) M_i is finitely generated and $M_i \subset M_{i+1}J(R)$ and $\bigcup_{j\in\mathbb{N}} M_j = M$.
(2) The map $\epsilon_i : M_i \to L_i$ is an isomorphism. So F is isomorphic to $\sum_{j\in\mathbb{N}} \oplus M_j$. Thus the map $f : \sum_{j\in\mathbb{N}} \oplus M_j \to M$ given by $f((x_j)) = \sum_{j\in\mathbb{N}} x_j$ has its retraction map $g : M \to \sum_{j\in\mathbb{N}} \oplus M_j$ such that $fg = 1_M$.
(3) $F_i \bigcap M = F_i \bigcap N = 0$ means $M_i \bigcap g(M) = M_i \bigcap N = 0$.
Thus M is NAS-module. Hence $M = 0$ by assumption. □

3. Nakayama-Azumaya Special Modules

In this section, we study what properties NAS-modules have by continuing the discussion of NAS-module M constructed in the last part of Theorem 2.1. In the following, we use the same notations used in Theorem 2.1.

In the discussion of the proof of Theorem 2.1, we have the following commutative diagrams for each i;

$$\begin{array}{ccc} M_i & \xrightarrow{\beta_i} & M_{i+1} \\ {\scriptstyle\alpha_i}\downarrow & & \downarrow{\scriptstyle\alpha_{i+1}} \\ N_i & \xrightarrow{\gamma_i} & N_{i+1}, \end{array} \qquad \begin{array}{ccc} L_i & \xrightarrow{\delta_i} & L_{i+1} \\ {\scriptstyle\epsilon_i}\downarrow & & \downarrow{\scriptstyle\epsilon_{i+1}} \\ M_i & \xrightarrow{\beta_i} & M_{i+1} \end{array}$$

Take direct limits of $\{\beta_i, M_i\}$ and $\{\gamma_i, N_i\}$, then both are isomorphic since they are direct system of isomorphic modules.

It is clear that $\varinjlim_{i\in\mathbb{N}}\{\beta_i, M_i\} = \cup M_i = M$. So $\varinjlim_{i\in\mathbb{N}}\{\gamma_i, N_i\}$ is isomorphic to M. Also we remark $\beta_i(M_i) \subset M_{i+1}J(R)$ and $\delta_i(L_i) \subset L_{i+1}J(R)$.

We denote $\tilde{F}_i = \sum_{j\geq i} \oplus F_j$, $\tilde{\delta}_i = (\delta_i, 1_{\tilde{F}_{(i+1)}}) : \tilde{F}_i \to \tilde{F}_{i+1}$ and $\tilde{f}_i = f|\tilde{F}_i : \tilde{F}_i \to M$, $\tilde{N}_i = \ker \tilde{f}_i$ for each $i \in \mathbb{N}$.

Then we have the commutative diagram with exact rows, which is a direct system of short exact sequences.

$$
\begin{array}{ccccccccc}
0 & \longrightarrow & N = \tilde{N}_1 & \xrightarrow{\bar{f}} & F = \tilde{F}_1 & \xrightarrow{f} & M & \longrightarrow & 0 \\
 & & \zeta_1 \big\downarrow & & \tilde{\delta}_1 \big\downarrow & & \big\downarrow 1_M & & \\
0 & \longrightarrow & \tilde{N}_2 & \xrightarrow{\bar{f}_2} & \tilde{F}_2 & \xrightarrow{\tilde{f}_2} & M & \longrightarrow & 0 \\
 & & \zeta_2 \big\downarrow & & \tilde{\delta}_2 \big\downarrow & & \big\downarrow 1_M & & \\
0 & \longrightarrow & \tilde{N}_3 & \xrightarrow{\bar{f}_3} & \tilde{F}_3 & \xrightarrow{\tilde{f}_3} & M & \longrightarrow & 0 \\
 & & \big\downarrow & & \tilde{\delta}_3 \big\downarrow & & \big\downarrow 1_M & & \\
 & & \cdots & & \cdots & & \cdots & &
\end{array}
$$

Here $\zeta_i : \tilde{N}_i \to \tilde{N}_{i+1}$ is the homomorphism induced from $\tilde{\delta}_i$ for each $i \geq 1$ and 1_M is the identity map over M.

We remark that elements of $\tilde{N}_i$ as elements of $\tilde{F}_i$ consist of $\begin{pmatrix} x_i \\ y^{(i)}_{x_i} \end{pmatrix} + \cdots + \begin{pmatrix} x_n \\ y^{(n)}_{x_n} \end{pmatrix} \in F_i \oplus \cdots \oplus F_n$ such that $x_i + \cdots + x_n = 0$ for some $n \geq i$, in fact, $\begin{pmatrix} 0 \\ y^{(i)}_{x_i} + \cdots + y^{(n)}_{x_n} \end{pmatrix} \in \tilde{N}_i$. Also $\bar{f}_i$ is given by $\bar{f}_i \begin{pmatrix} 0 \\ y \end{pmatrix} = \begin{pmatrix} x_i \\ y^{(i)}_{x_i} \end{pmatrix} + \cdots + \begin{pmatrix} x_n \\ y^{(n)}_{x_n} \end{pmatrix}$.

We denote $g_1 = g$ and $g_{i+1} = \tilde{\delta}_i g_i$ for each $i \in \mathbb{N}$. Then $\tilde{f}_i g_i = 1_M$ since $\tilde{f}_i g_i(x) = \tilde{f}_i(\sum_{j=1}^{i} \begin{pmatrix} x_j \\ y^{(i)}_{x_j} \end{pmatrix} + \sum_{j=i+1}^{n} \begin{pmatrix} x_j \\ y^{(j)}_{x_j} \end{pmatrix}) = x_1 + \cdots + x_n = x$ for any $x \in M$.

Hence there is a split exact sequence $0 \leftarrow \tilde{N}_i \xleftarrow{\tilde{g}_i} \tilde{F}_i \xleftarrow{g_i} M \leftarrow 0$ and we have $\tilde{F}_i = g_i(M) \oplus \bar{f}_i(\tilde{N}_i)$ for each $i \in \mathbb{N}$.

We summarize the above discussion.

Theorem 3.1. *Let $M_1 \subset M_2 \subset \cdots$ be a subsequence to make NAS-module M and $f : \sum_{i\in\mathbb{N}} \oplus M_i \to M$ a homomorphism given by $f((x_i)) = \sum_{i\in\mathbb{N}} x_i$ for $(x_i) \in \sum_{i\in\mathbb{N}} \oplus M_i$.*

(1) *There is a map* $g_i : M \to \sum_{j \geq i} \oplus M_j$ *such that* $fg_i = 1_M$ *for each* $i \in \mathbb{N}$.
(2) $\varinjlim_{i\in\mathbb{N}}(\tilde{\delta}_i, \sum_{j\geq i} \oplus M_i) = M$ *and* $\varinjlim_{i\in\mathbb{N}}(\zeta_i, \tilde{N}_i) = 0$.
(3) *These exact sequences make a following commutative diagram;*

$$\begin{array}{ccccccccc} 0 & \longleftarrow & \tilde{N}_i & \xleftarrow{\tilde{g}_i} & \sum_{j\geq i} \oplus M_j & \xleftarrow{g_i} & M & \longleftarrow & 0 \\ & & \zeta_i \downarrow & & \tilde{\delta}_i \downarrow & & 1_M \downarrow & & \\ 0 & \longleftarrow & \tilde{N}_{i+1} & \xleftarrow{\tilde{g}_{i+1}} & \sum_{j\geq i+1} \oplus M_j & \xleftarrow{g_{i+1}} & M & \longleftarrow & 0. \end{array}$$

(4) M *is isomorphic to a direct summand of* $\sum_{i\in\mathbb{N}\setminus S} \oplus M_i$ *for any finite subset* S *of* $\mathbb{N}$.
(5) *There are infinitely many NAS-modules as a factor module of* M.

We go back to the former form $F = \sum_{i\in\mathbb{N}} \oplus F_i = M \oplus N$, where each F_i is finitely generated and $MJ(R) = M$. We may assume $f(F_1) = M_1 \subset f(F_2) = M_2 \subset \cdots$ is a reduced subsequence of M. We know the following facts.

Proposition 3.1.

(1) $\tilde{F}_n = g_n(M) \oplus \bar{f}_n(\tilde{N}_n)$.
(2) $\delta_n(\tilde{F}_n) = \delta_n(g_n(M)) \oplus \delta_n(\bar{f}_n(\tilde{N}_n))$.
(3) $\tilde{F}_{n+1} = \delta_n(\tilde{F}_n)$, $g_{n+1}(M) = \delta_n(g_n(M))$, $\bar{f}_{n+1}(\tilde{N}_n) = \delta_n(\bar{f}_n(\tilde{N}_n))$.
(4) $F_n \cap g_n(M) = g_n(M \bigcap (\sum_{i=1}^{n} \oplus F_i))$.
(5) $(F_n/(F_n \bigcap g_n(M))) \oplus \tilde{F}_{n+1} = (g(M)/g_n(M \bigcap (\sum_{i=1}^{n} \oplus F_i))) \oplus \bar{f}_n(N)$.
(6) *There is an infinite series* $M/L_1, M/L_2, \cdots$ *of factor modules of* M *such that* $L_1 \subset L_2 \subset \cdots$, $\bigcup_{i\in\mathbb{N}} L_i = M$ *and each* M/L_i *is a non-zero NAS-module with a reduced subsequence.*

Proof. The statements (1), (2) and (3) are already shown. We show the statement (4). We have $g_n((F_1 \oplus \cdots \oplus F_n) \bigcap M) \subset F_n \bigcap g_n(M)$ by definition of g_n. If $\left(\begin{smallmatrix} x \\ y_x^{(n)} \end{smallmatrix}\right) = g_n(z)$ for some $z \in M$ and we express $\left(\begin{smallmatrix} z \\ 0 \end{smallmatrix}\right) = \sum_{i=1}^{m} \left(\begin{smallmatrix} z_i \\ y_{z_i}^{(i)} \end{smallmatrix}\right)$, then $g_n(z) = \sum_{i=1}^{n} \left(\begin{smallmatrix} z_i \\ y_{z_i}^{(n)} \end{smallmatrix}\right) + \sum_{i=n+1}^{m} \left(\begin{smallmatrix} z_i \\ y_{z_i}^{(i)} \end{smallmatrix}\right) \in F_n$, so $\left(\begin{smallmatrix} z_i \\ y_{z_i}^{(i)} \end{smallmatrix}\right) = \left(\begin{smallmatrix} 0 \\ 0 \end{smallmatrix}\right)$ for any $i > n$. Hence $z \in M \bigcap (\sum_{i=1}^{n} \oplus F_i)$. Thus $F_n \cap g_n(M) = g_n(M \bigcap (\sum_{i=1}^{n} \oplus F_i))$.

(5) is clear from (4).
From the statement (5), we can construct non-zero NAS-modules as a factor module of M for each $i \in \mathbb{N}$ by the reduction methods used in Theorem 2.1 (more precisely, see the last part of this section). By applying the above process, for example, for $n > 2, 2n, \cdots$, we have a series of NAS-modules required in (6). $\square$

By these observation, we suspect the existence of non-zero NAS-modules. So we give the following conjecture.

Conjecture 3.1. *There exists no non-zero NAS-modules.*

Before we conclude this section, we summarize the reduction process given in our former discussion to make our target module NAS-module with a reduced subsequence.

Reduction Process
Let $\{F_i | i \in \mathbb{N}\}$ be a set of finitely generated modules and M a direct summand of $\sum_{i \in \mathbb{N}} \oplus F_i$ such that $MJ(R) = M$.

(1) **Rearrangement Process** (Process used in Theorem 2.1)
Rearrange $\{F_i\}$ such that $M_i \subset M_{i+1}J(R)$ for all $i \in \mathbb{N}$.
(2) **Non-redundant Process** (Process used in Lemma 2.1)
Replace M with $M/\sum_{i \in \mathbb{N}} \oplus(M \cap F_i)$ and N with $N/\sum_{i \in \mathbb{N}} \oplus(N \cap F_i)$.
Replace F_i with $F_i/\sum_{i \in \mathbb{N}} \oplus((M \cap F_i) \oplus (N \cap F_i))$ for all $i \in \mathbb{N}$.
(3) **Correspondence Process** (Process used in Lemma 2.2)
Take a unique minimal subset $T \subset \mathbb{N}$ such that $\sum_{t \in T} \oplus F_t = M \oplus N'$.
Replace F with $\sum_{t \in T} \oplus F_t$.

4. Example

We explain one example to imagine that the conjecture in the former section 3 seems affirmative. Also this example will help readers understand what the observation in the former section 3 means. The author believes the method used in the following example will be effective in general to prove GNA Lemma.

The following example appears in M. Sato [6, Example 2].

Example 4.1. Let F be a field and Z a commutative F-algebra with the basis $\{v_x \mid 0 < x \leq 1\}$ equipped with multiplication $v_x \cdot v_y = v_{xy}$. Let

J_i $(0 < i \le 1)$ be an open ideal of Z with the basis $\{v_x \mid 0 < x < i\}$ and $\overline{J_i}$ a closure ideal of J_i of Z with the basis $\{v_x \mid 0 < x \le i\}$.

We consider the factor algebra $S = Z/J_{\frac{1}{2}}$ of Z, then S is a local ring and its radical $J(S)$ is $U = J_1/J_{\frac{1}{2}}$.

We use, for simplicity, the same notation v_t instead of its residue class $\overline{v_t} = v_t + J_{\frac{1}{2}}$ $(\frac{1}{2} \le t \le 1)$ in order to denote the induced basis of S.

We know that U is a uniserial module satisfying $UJ(S) = U$ and its submodules have the one of the following forms for $\frac{1}{2} < j \le 1$;

(1) $J_i/J_{\frac{1}{2}} = \{\alpha v_j \mid \alpha \text{ is 0 or a unit in } S, \frac{1}{2} \le j < i\}$,

(2) $\overline{J_i}/J_{\frac{1}{2}} = \{\alpha v_j \mid \alpha \text{ is 0 or a unit in } S, \frac{1}{2} \le j \le i\}$.

We use the notions J_i and $\overline{J_i}$ instead of (1) and (2), respectively.

Take $M = U$ and for each $i \in \mathbb{N}$, consider a cyclic module $M_i = \overline{J_{(1-\frac{1}{2^i})}}$, then $M_i \subset M_{i+1}J(S)$ and $\sum_{i\in\mathbb{N}} M_i = M$.

Define an S-epimorphism $f : \sum_{i\in\mathbb{N}} \oplus M_i \to M$ by $f((w_i)) = \sum_{i\in\mathbb{N}} w_i$ for $(w_i) \in \sum_{i\in\mathbb{N}} \oplus M_i$. We show that f is not splittable.

Assume that there is a retraction $g : M \to \sum_{i\in\mathbb{N}} \oplus M_i$. Take one of bases $v_x \in M$ and express

$$g(v_x) = (\alpha_{x_1}v_{x_1}, \alpha_{x_2}v_{x_2}, \cdots, \alpha_{x_m}v_{x_m}, 0, \cdots)$$

such that $v_{x_i} \in M_i$ and $v_x = \alpha_{x_1}v_{x_1} + \alpha_{x_2}v_{x_2} + \cdots + \alpha_{x_m}v_{x_m}$, here, each α_{x_i} is 0 or a unit element in S. We remark that $x = \max\{x_i \mid \alpha_{x_i} \ne 0\}$.

For any t $(\frac{1}{2} < t \le x)$, there is some s $(\frac{1}{2} < s \le 1)$ such that $t = xs$. Since $v_t = v_x v_s$, we have $g(v_t) = g(v_x)v_s$, so $g(\overline{J_x}) \subset \sum_{i=1}^{m} \oplus M_i$. Since $\sum_{i=1}^{m} M_i = M_m \ne M$, there is some v_z such that $g(v_z) \notin \sum_{i=1}^{m} \oplus M_i$.

Set $g(v_z) = (\alpha_{z_1}v_{z_1}, \cdots, \alpha_{z_n}v_{z_n}, 0, \cdots)$. Assume $z_i \le x$ for all $i = 1, \cdots, n$, then $v_z \in \overline{J_x}$. Hence $g(v_z) \in \sum_{i=1}^{m} \oplus M_i$ by the above fact, which is a contradiction. Thus there is some k such that $z_k > x$ and $\alpha_{z_k} \ne 0$.

Since $z = \max\{z_i \mid \alpha_{z_i} \ne 0\}$, we can take z_k such that $z = z_k$. Choose s $(\frac{1}{2} < s < 1)$ such that $x = zs = z_k s$. Then $g(v_x) = g(v_z)v_s \in \sum_{i=1}^{m} \oplus M_i$, so z_k-component of $g(v_x)$ is $\alpha_{z_k}v_{z_k}v_s = \alpha_{z_k}v_{z_k s} = 0$. On the other hand, $\alpha_{z_k}v_{z_k s} = \alpha_{z_k}v_{zs} = \alpha_{z_k}v_x \ne 0$, which is a contradiction.

5. On Existence of a Maximal Submodule

It holds $MJ(R) = \text{rad } M$ in many cases like R is a right artinian ring or M is a projective module. In this case, Nakayama-Azumaya Lemma is equivalent to the existence of a maximal submodule. But in general, $MJ(R) \subset \text{rad } M$, but they do not coincide. So in this section, we study about existence of a maximal submodule for a module which is a direct summand or a submodule of a direct sum of finitely generated modules.

Remark 5.1. A non-zero projective module has a maximal submodule [1, Proposition 17.14]. On the other hand, there is a submodule of a projective module which is not included in any maximal submodules. For example, take $\mathbb{Z}$-module $\mathbb{Q}$ the set of rational numbers and consider an epimorphism $f : \sum \oplus \mathbb{Z} \to \mathbb{Q}$, then $\ker f$ is not included in any maximal submodule of $\sum \oplus \mathbb{Z}$ since there is no non-zero $\mathbb{Z}$-homomorphism from $\mathbb{Q}$ to any simple $\mathbb{Z}$-modules. But the following fact holds for a finitely generated module.

Lemma 5.1. *For any proper submodule L of a finitely generated module M, there is some maximal submodule of M including L.*

Proof. We can give the same proof as the existence of a maximal ideal by using Zorn's Lemma. □

As a consequence of Lemma 5.1, we have the following theorem.

Theorem 5.1. *Let F be a direct sum of finitely generated modules $\{F_\delta\}_{\delta\in\Delta}$ and M a non-zero submodule of F, then M has a maximal submodule if $M \not\subset \text{rad } F$.*

Proof. Let $f_\gamma : \sum_{\delta\in\Delta} \oplus F_\delta \to F_\gamma$ be the canonical projection. Since $M \not\subset \text{rad } F$ by assumption, there is some $\gamma \in \Delta$, $a \in M$ and maximal submodule T of F_γ such that $f_\gamma(a) \notin T$. Hence $f_\gamma(M) + T = F_\gamma$. We set a simple module $S = F_\gamma/T = (f_\gamma(M) + T)/T \cong f_\gamma(M)/(f_\gamma(M) \cap T)$. Thus there is an epimorphism $M \to S$ as a composition map $M \to \sum_{\delta\in\Delta} \oplus F_\delta \to F_\gamma \to S$. Its kernel is a maximal submodule of M. Thus M has a maximal submodule. □

A projective module is a direct summand of a direct sum of copies of R and a projective module has a maximal submodule [1, Proposition 17.14]. We have more general result as a corollary of Theorem 5.1.

Corollary 5.1. *Any non-zero submodule M of a free module $F = \sum_{\delta \in \Delta} \oplus R$ has a maximal submodule if $M \not\subset FJ(R) = \sum_{\delta \in \Delta} \oplus J(R)$. Particularly a non-zero projective module has a maximal submodule.*

Proof. The first statement holds directly by Theorem 5.1. Assume that M is a projective module and $F = \sum_{\delta \in \Delta} \oplus R = M \oplus N$. Then rad $F = FJ(R) = \sum_{\delta \in \Delta} \oplus J(R) = MJ(R) \oplus NJ(R)$. If $M \subset$ rad F, then $M = MJ(R)$. Hence $M = 0$ by Nakayama-Azumaya Lemma for projective modules. This contradicts the fact that $M \neq 0$. □

Theorem 5.2. *Assume GNA Lemma holds. Let M be a non-zero direct summand of a direct sum of finitely generated modules. Then M has a maximal submodule if $MJ(R) =$ rad M.*

Particularly, let F_δ ($\delta \in \Delta$) be finitely generated modules satisfying $F_\delta J(R) =$ rad F_δ for every $\delta \in \Delta$, then a non-zero direct summand of $F = \sum_{\delta \in \Delta} \oplus F_\delta$ has a maximal submodule.

Proof. Assume $MJ(R) =$ rad M. If M does not have any maximal submodules, then rad $M = M$, therefore $MJ(R) = M$. Hence $M = 0$ by GNA Lemma. Thus M has a maximal submodule.

Next we show $MJ(R) =$ rad M for any direct summand M of F if $F_\delta J(R) =$ rad F_δ for every $\delta \in \Delta$.

Let $F = \sum_{\delta \in \Delta} \oplus F_\delta = M \oplus N$ be direct decompositions. Then rad$F = \sum_{\delta \in \Delta} \oplus$rad $F_\delta =$ rad $M \oplus$ rad N (see for example [1, Proposition 9.19]). On the other hand, $\sum_{\delta \in \Delta} \oplus$rad $F_\delta = \sum_{\delta \in \Delta} \oplus F_\delta J(R) = (\sum_{\delta \in \Delta} \oplus F_\delta)J(R) = FJ(R) = MJ(R) \oplus NJ(R)$. Hence rad $M = MJ(R)$ and rad $N = NJ(R)$ since $MJ(R) \subset$ rad M and $NJ(R) \subset$ rad N.
Thus M has a maximal submodule. □

Acknowledgment

The author expresses great thanks to Professor Y. Kuratomi for the useful suggestions during the preparation of this paper. Also the author thanks to the referees for their careful reading of the manuscript and useful comments.

References

[1] F.W. Anderson, K.R. Fuller, *Rings and Categories of Modules*, GTM **13**, Springer-Verlag (1992).

[2] H. Bass, *Finitistic dimension and a homological generalization of semiprimary rings*, Trans. of the American Math. Soc. **95**, 466–488 (1960).

[3] S. Hinohara, *Projective modules II*, The sixth proceeding of Japan algebraic symposium (Homological algebra and its applications) **6**, 24–28 (1964).

[4] I. Kaplansky, *Projective modules*, Ann. of Math. **68**, 372–377 (1958).

[5] W.K. Nicholson, M.F. Yousif, *Quasi-Frobenius Rings*, Cambridge University Press (2002).

[6] M. Sato, *Projective modules with unique maximal submodules are cyclic*, Proc. of the American Math. Soc. **148**, 3673–3684 (2020).

[7] C.P. Walker, *Relative homological algebra and abelian groups*, Illinois J. Math. **10**, 263–276 (1966).

A Baer-Kaplansky Theorem in additive categories

S. Crivei

Department of Mathematics, Babeş-Bolyai University,
Str. M. Kogălniceanu 1, 400084 Cluj-Napoca, Romania
E-mail: crivei@math.ubbcluj.ro

D. Keskin Tütüncü

Department of Mathematics, Hacettepe University,
06800 Beytepe, Ankara, Turkey
E-mail: keskin@hacettepe.edu.tr

R. Tribak

Centre Régional des Métiers de L'Education et de la Formation (CRMEF)-Tanger,
Avenue My Abdelaziz, Souani, B.P.:3117, Tangier 90000, Morocco
E-mail: tribak12@yahoo.com

We prove the following Baer-Kaplansky Theorem in additive categories. If M and N are objects of an additive category $\mathcal{C}$ such that: (1) $M = A \oplus X$ and $N = B \oplus X$ for some objects A, B and some indecomposable object X; (2) N is internally cancellable with respect to X; (3) the functor $H = \mathrm{Hom}_{\mathcal{C}}(X, -) : \mathcal{C} \to \mathrm{Ab}$ has the property that $H(A) \cong H(B)$ implies $A \cong B$; (4) there exists an IP-isomorphism $\Phi : \mathrm{End}_{\mathcal{C}}(M) \to \mathrm{End}_{\mathcal{C}}(N)$; then $M \cong N$. We also deduce consequences in Grothendieck and finitely accessible additive categories.

Keywords: Baer-Kaplansky Theorem, additive category, indecomposable object, (internally) cancellable object, IP-isomorphism, Grothendieck category, finitely accessible additive category.

1. Introduction

A class of objects of an additive category $\mathcal{C}$ is called *Baer-Kaplansky* if a Baer-Kaplansky Theorem holds for its objects, that is, any of its objects are isomorphic whenever they have isomorphic endomorphism rings (see [7, 14]). Some examples of Baer-Kaplansky classes are the following ones: torsion abelian groups [9, Theorem 108.1], finitely generated abelian groups [15, Example 1.3], modules over a primary artinian uniserial ring [17, Lemma 7.4], modules over a non-singular artinian serial ring [13, Theorem 9], finitely generated modules over an indecomposable FGC commutative ring R each of which has a copy of R as a direct summand [14, Theorem 4], projective (respectively injective, semisimple) modules over a local artinian ring with radical W such that $W^2 = 0$, $Q = R/W$ is commutative, $\dim_Q W = 1$ and $\dim W_Q = 2$ [15, Example 2.3].

Further examples may be obtained by using the natural concept of IP-isomorphism (i.e., indecomposable-preserving ring isomorphism), proposed by Ivanov in the study of Baer-Kaplansky classes [13]. Such results have been recently obtained in [7] in the context of Grothendieck categories, and further extended to certain types of additive categories.

In the present paper we establish a Baer-Kaplansky Theorem in the general setting of additive categories. We first prove it in idempotent complete additive categories, and then we transfer it to arbitrary additive categories by using the important result that every additive category has an idempotent completion (or Karoubian completion). Our Baer-Kaplansky Theorem states as follows: *If M and N are objects of an additive category $\mathcal{C}$ such that: (1) $M = A \oplus X$ and $N = B \oplus X$ for some objects A, B and some indecomposable object X; (2) N is internally cancellable with respect to X; (3) the functor $H = \mathrm{Hom}_{\mathcal{C}}(X, -) : \mathcal{C} \to \mathrm{Ab}$ has the property that $H(A) \cong H(B)$ implies $A \cong B$; (4) there exists an IP-isomorphism $\Phi : \mathrm{End}_{\mathcal{C}}(M) \to \mathrm{End}_{\mathcal{C}}(N)$; then $M \cong N$.* We also deduce consequences in Grothendieck categories and finitely accessible additive categories.

2. A Baer-Kaplansky Theorem

We recall some needed terminology in additive categories. Let $A_1, \ldots, A_n$ be objects of an additive category $\mathcal{C}$. Then a *direct sum* $A = A_1 \oplus \cdots \oplus A_n$ is an object A together with morphisms $j_i : A_i \to A$ (called canonical injections) and $p_i : A \to A_i$ (called canonical projections) for every $i \in \{1, \ldots, n\}$ such that $\sum_{i=1}^n j_i p_i = 1_A$ and $p_i j_i = 1_{A_i}$ for every $i \in \{1, \ldots, n\}$. It is well known that a finite direct sum of objects is isomorphic to both their product and coproduct. A non-zero object A of $\mathcal{C}$ is called *indecomposable* if whenever $A = A_1 \oplus A_2$ for some objects A_1 and A_2, one has either $A_1 = 0$ or $A_2 = 0$.

Following [16], an object N of an additive category $\mathcal{C}$ is called *internally cancellable* if whenever $N = A \oplus C = B \oplus D$ such that $C \cong D$, one has $A \cong B$. We say that an object N of $\mathcal{C}$ is *internally cancellable with respect to a subobject X of N* if whenever $N = A \oplus X = B \oplus Y$ such that $X \cong Y$, one has $A \cong B$. Hence an object is internally cancellable if and only if it is internally cancellable with respect to any of its subobjects.

Let M and N be objects of an additive category $\mathcal{C}$. Recall that a ring isomorphism $\Phi : \mathrm{End}_{\mathcal{C}}(M) \to \mathrm{End}_{\mathcal{C}}(N)$ is called an *IP-isomorphism* if for every primitive idempotent $e \in \mathrm{End}_{\mathcal{C}}(M)$, one has $\Phi(e)N \cong eM$ [13].

An additive category $\mathcal{C}$ is called *idempotent complete* if all idempotents split in $\mathcal{C}$, in the sense that for every object A of $\mathcal{C}$ and every idempotent $e \in \mathrm{End}_{\mathcal{C}}(A)$ there is an object B and morphisms $p : A \to B$ and $i : B \to A$ in $\mathcal{C}$ such that $pi = 1_B$ and $ip = e$ (e.g., see [3, Remark 6.2] and [18, p. 6]).

Now we may establish the following proposition, which is one of the key ingredients for the main result of the paper. We denote by Ab the category of abelian groups.

Proposition 2.1. *Let M and N be objects of an idempotent complete additive category $\mathcal{C}$ such that:*

(1) $M = A \oplus X$ and $N = B \oplus X$ for some objects A, B and some indecomposable object X;
(2) N is internally cancellable with respect to X;
(3) the functor $H = \mathrm{Hom}_{\mathcal{C}}(X, -) : \mathcal{C} \to \mathrm{Ab}$ has the property that $H(A) \cong H(B)$ implies $A \cong B$;
(4) there exists an IP-isomorphism $\Phi : \mathrm{End}_{\mathcal{C}}(M) \to \mathrm{End}_{\mathcal{C}}(N)$.

Then $M \cong N$.

Proof. Let $i : A \to M$ and $j : X \to M$ be the canonical injections, and let $p : M \to A$ and $q : M \to X$ be the canonical projections. Consider the idempotent $e = jq \in \mathrm{End}_{\mathcal{C}}(M)$. Then $1 - e = ip$. Since $\mathcal{C}$ is idempotent complete, e has a kernel and an image, and we have the isomorphisms

$$M \cong \mathrm{Ker}(e) \oplus \mathrm{Im}(e) \cong \mathrm{Im}(1 - e) \oplus \mathrm{Im}(e)$$

(e.g., see [3, Remark 6.2] and [18, p. 6]). Note that $X \cong \mathrm{Im}(e)$ and $A \cong \mathrm{Im}(1 - e)$.

Since X is indecomposable, e is a primitive idempotent in $\mathrm{End}_{\mathcal{C}}(M)$. Then $\Phi(e)$ is a primitive idempotent in $\mathrm{End}_{\mathcal{C}}(N)$. Hence $C = \mathrm{Im}(\Phi(e))$ is an indecomposable direct summand of N. Then $N = D \oplus C$, where $D \cong \mathrm{Im}(\Phi(1-e))$. Since Φ is an IP-isomorphism, we have $C = \mathrm{Im}(\Phi(e)) \cong \mathrm{Im}(e) \cong X$. Since $N = B \oplus X$ and N is internally cancellable with respect to X, we deduce that $B \cong D$.

Note that the morphism

$$\varphi : \mathrm{Hom}_{\mathcal{C}}(\mathrm{Im}(e), \mathrm{Im}(1 - e)) \to (1 - e)\mathrm{End}_{\mathcal{C}}(M)e$$

defined by $\varphi(f) = (1-e)i\beta f\alpha q e$ for every $f \in \mathrm{Hom}_{\mathcal{C}}(\mathrm{Im}(e), \mathrm{Im}(1-e))$, where $\alpha : X \to \mathrm{Im}(e)$ and $\beta : \mathrm{Im}(1-e) \to A$ are isomorphisms, is an abelian group isomorphism with inverse

$$\psi : (1-e)\mathrm{End}_{\mathcal{C}}(M)e \to \mathrm{Hom}_{\mathcal{C}}(\mathrm{Im}(e), \mathrm{Im}(1-e))$$

defined by $\psi((1-e)ge) = \beta^{-1}pgj\alpha^{-1}$ for every $g \in \mathrm{End}_{\mathcal{C}}(M)$.

Then we have the following abelian group isomorphisms:

$$\begin{aligned}\mathrm{Hom}_{\mathcal{C}}(X, A) &\cong \mathrm{Hom}_{\mathcal{C}}(\mathrm{Im}(e), \mathrm{Im}(1-e)) \cong (1-e)\mathrm{End}_{\mathcal{C}}(M)e \\ &\cong \Phi(1-e)\mathrm{End}_{\mathcal{C}}(N)\Phi(e) \cong \mathrm{Hom}_{\mathcal{C}}(\mathrm{Im}(\Phi(e)), \mathrm{Im}(\Phi(1-e))) \\ &\cong \mathrm{Hom}_{\mathcal{C}}(C, D) \cong \mathrm{Hom}_{\mathcal{C}}(X, D) \cong \mathrm{Hom}_{\mathcal{C}}(X, B).\end{aligned}$$

Now the hypothesis implies that $A \cong B$. Hence we have $M \cong N$. □

We need the following lemma.

Lemma 2.2. *Let $F : \mathcal{A} \to \mathcal{B}$ be a fully faithful additive covariant functor between additive categories $\mathcal{A}$ and $\mathcal{B}$. Then F preserves the following conditions:*

(1) $M = A \oplus X$ and $N = B \oplus X$ for some objects A, B and some indecomposable object X;
(2) N is internally cancellable with respect to X;
(3) the functor $H = \mathrm{Hom}_{\mathcal{A}}(X, -) : \mathcal{A} \to \mathrm{Ab}$ has the property that $H(A) \cong H(B)$ implies $A \cong B$;
(4) there exists an IP-isomorphism $\Phi : \mathrm{End}_{\mathcal{A}}(M) \to \mathrm{End}_{\mathcal{A}}(N)$.

Proof. Since F preserves finite direct sums and indecomposables, we have $F(M) = F(A) \oplus F(X)$ and $F(N) = F(B) \oplus F(X)$ with $F(X)$ indecomposable, and $F(N)$ is internally cancellable with respect to $F(X)$.

Now consider the functor $H' = \mathrm{Hom}_{\mathcal{B}}(F(X), -) : \mathcal{B} \to \mathrm{Ab}$ and assume that $H'(F(A)) \cong H'(F(B))$. Since F is fully faithful, we have the following abelian group isomorphisms:

$$\mathrm{Hom}_{\mathcal{A}}(X, A) \cong \mathrm{Hom}_{\mathcal{B}}(F(X), F(A)) \cong \mathrm{Hom}_{\mathcal{B}}(F(X), F(B)) \cong \mathrm{Hom}_{\mathcal{A}}(X, B).$$

Then we have $A \cong B$ by hypothesis, and so $F(A) \cong F(B)$.

Finally, since $\Phi : \mathrm{End}_{\mathcal{A}}(M) \to \mathrm{End}_{\mathcal{A}}(N)$ is an IP-isomorphism and F is fully faithful, we have the composed induced ring isomorphism:

$$\mathrm{End}_{\mathcal{B}}(F(M)) \cong \mathrm{End}_{\mathcal{A}}(M) \overset{\Phi}{\cong} \mathrm{End}_{\mathcal{A}}(N) \cong \mathrm{End}_{\mathcal{B}}(F(N)),$$

which is an IP-isomorphism, because F preserves indecomposables and $F(e)F(M) = F(eM)$ for any idempotent $e \in \mathrm{End}_{\mathcal{A}}(M)$. □

Let us note that every additive category has an *idempotent completion* (or *Karoubian completion*). More precisely, for every additive category $\mathcal{C}$, there exists an idempotent complete additive category $\widehat{\mathcal{C}}$ and a fully faithful additive covariant functor $F : \mathcal{C} \to \widehat{\mathcal{C}}$. The reader is referred to [3, Section 6] for further details on the idempotent completion of an additive category.

Now we are in a position to give the main result of the paper, which is the following Baer-Kaplansky Theorem.

Theorem 2.3. *Let M and N be objects of an additive category $\mathcal{C}$ such that:*

(1) $M = A \oplus X$ and $N = B \oplus X$ for some objects A, B and some indecomposable object X;
(2) N is internally cancellable with respect to X;
(3) the functor $H = \mathrm{Hom}_{\mathcal{C}}(X, -) : \mathcal{C} \to \mathrm{Ab}$ has the property that $H(A) \cong H(B)$ implies $A \cong B$;
(4) there exists an IP-isomorphism $\Phi : \mathrm{End}_{\mathcal{C}}(M) \to \mathrm{End}_{\mathcal{C}}(N)$.

Then $M \cong N$.

Proof. Consider the idempotent completion $\widehat{\mathcal{C}}$ of $\mathcal{C}$ and the associated fully faithful additive covariant functor $F : \mathcal{C} \to \widehat{\mathcal{C}}$. Then F preserves properties (1)–(4) by Lemma 2.2. We have $F(M) \cong F(N)$ by Proposition 2.1. Finally, we deduce that $M \cong N$ because F is fully faithful. $\square$

Remark 2.4. (a) Condition (2) from Theorem 2.3 is clearly fulfilled when N is internally cancellable and, in particular, when X is *cancellable* in the sense that whenever A and B are objects such that $A \oplus X \cong B \oplus X$, one has $A \cong B$ (e.g., see [16, Proposition 3.4], whose proof is categorical).

(b) Condition (3) from Theorem 2.3 clearly holds true when the functor H is injective on objects and, in particular, when H is fully faithful.

3. Applications

In this section we illustrate our general Baer-Kaplansky Theorem in certain classes of additive categories, namely Grothendieck categories and finitely accessible additive categories. We consider more particular conditions that ensure the validity of our Baer-Kaplansky Theorem.

We first recall some concepts, mainly following [16].

A ring R *has stable range one* if for every $a, b \in R$ such that $aR+bR = R$, there exists $y \in R$ such that $a+by$ is invertible in R. By Evans' Cancellation Theorem [16, Theorem 5.1], a large class of cancellable modules consists of modules whose endomorphism ring has stable range one.

A ring R is called *semilocal* if the ring $R/J(R)$ is semisimple, where $J(R)$ is the Jacobson radical of R. A ring R is called *unit-regular* if for every $a \in R$, one has $a = aua$ for some invertible $u \in R$. A ring R is called *strongly π-regular* if for every $a \in R$, one has $a^n \in a^{n+1}R$ for some positive integer n. Note that strongly π-regular rings, unit-regular rings and semilocal rings all have stable range one by [1, Theorem 4], [16, Theorem 2.9] and [16, Corollary 2.10], respectively.

Let (T, H) be an adjoint pair of covariant functors $T : \mathcal{A} \to \mathcal{B}$ and $H : \mathcal{B} \to \mathcal{A}$ between arbitrary categories $\mathcal{A}$ and $\mathcal{B}$ with counit of the adjunction $\varepsilon : TH \to 1_{\mathcal{B}}$. Recall that an object $B \in \mathcal{B}$ is called *H-static* if $\varepsilon_B : TH(B) \to B$ is an isomorphism (e.g., see [4, p. 382]). We will denote by $\mathrm{Stat}(H)$ the full subcategory of $\mathcal{B}$ consisting of H-static objects.

Theorem 3.1. *Let M and N be objects of a Grothendieck category $\mathcal{C}$ such that:*

(1) *$M = A \oplus X$ and $N = B \oplus X$ for some objects A, B and some indecomposable object X;*
(2) *$\mathrm{End}_{\mathcal{C}}(X)$ has stable range one (in particular, $\mathrm{End}_{\mathcal{C}}(X)$ is semilocal, unit-regular or strongly π-regular);*
(3) *$A, B \in \mathrm{Stat}(\mathrm{Hom}_{\mathcal{C}}(X, -))$ (in particular, X is a generator of $\mathcal{C}$);*
(4) *there exists an IP-isomorphism $\Phi : \mathrm{End}_{\mathcal{C}}(M) \to \mathrm{End}_{\mathcal{C}}(N)$.*

Then $M \cong N$.

Proof. We check conditions (2) and (3) from Theorem 2.3.

First, let U be a generator of $\mathcal{C}$ with $R = \mathrm{End}_{\mathcal{C}}(U)$, and consider the additive functor $F = \mathrm{Hom}_{\mathcal{C}}(U, -) : \mathcal{C} \to \mathrm{Mod}(R)$, where $\mathrm{Mod}(R)$ is the category of right R-modules. By the Gabriel-Popescu Theorem [20, Theorem 4.1, p. 220], F is fully faithful, hence $\mathrm{End}_R(F(X)) \cong \mathrm{End}_{\mathcal{C}}(X)$ has stable range one. By Evans' Cancellation Theorem [16, Theorem 5.1], $F(X)$ is cancellable in $\mathrm{Mod}(R)$. Since cancellability is clearly reflected by fully faithful additive functors, X is cancellable in $\mathcal{C}$, and so N is internally cancellable with respect to X.

Now consider the additive functor $H = \mathrm{Hom}_{\mathcal{C}}(X, -) : \mathcal{C} \to \mathrm{Ab}$. In fact, we have $H : \mathcal{C} \to \mathrm{Mod}(S)$, where $S = \mathrm{End}_{\mathcal{C}}(X)$. Then H has a left adjoint $T : \mathrm{Mod}(S) \to \mathcal{C}$. Assume that $H(A) \cong H(B)$. Since $A, B \in \mathrm{Stat}(H)$, we have $A \cong TH(A) \cong TH(B) \cong B$. If X is a generator of $\mathcal{C}$, then H is fully faithful by the Gabriel-Popescu Theorem [20, Theorem 4.1, p. 220], so $\mathcal{C} = \mathrm{Stat}(H)$. Now Theorem 2.3 completes the proof. $\square$

As a first consequence we recover the following result of Ivanov and Vámos, which includes the classical one that finitely generated abelian groups form a Baer-Kaplansky class. Recall that a ring is called an *FGC ring* if every finitely generated module is a direct sum of cyclic modules.

Corollary 3.2. *[14, Theorem 4] Let R be an indecomposable FGC commutative ring. Let M and N be finitely generated R-modules such that $M = A \oplus R$, $N = B \oplus R$ and there exists a ring isomorphism $\Phi : \mathrm{End}_R(M) \to \mathrm{End}_R(N)$. Then $M \cong N$.*

Proof. Every finitely generated module over a commutative FGC ring R has a unique (up to isomorphism) indecomposable decomposition into cyclic modules [2, Theorem 9.2]. The uniqueness of such a decomposition implies immediately that every finitely generated R-module is internally cancellable with respect to R. From the proof of [14, Theorem 4], it follows that every ring isomorphism $\Phi : \mathrm{End}_R(M) \to \mathrm{End}_R(N)$ is an IP-isomorphism. Noting that R is a generator of $\mathrm{Mod}(R)$, one concludes that $M \cong N$ by using Theorem 2.3 and the proof of Theorem 3.1. □

Corollary 3.3. *Let R be an indecomposable ring. Let M and N be free right R-modules such that N is internally cancellable with respect to R, and there exists an IP-isomorphism $\Phi : \mathrm{End}_R(M) \to \mathrm{End}_R(N)$. Then $M \cong N$.*

Proof. This follows by Theorem 2.3 and the proof of Theorem 3.1. □

Example 3.4. Let R be a local artinian ring with radical W such that $W^2 = 0$, $Q = R/W$ is commutative, $\dim_Q W = 1$ and $\dim W_Q = 2$. Let P and P' be projective right R-modules. Then R is indecomposable and cancellable because it has stable range one. Also, every ring isomorphism $\mathrm{End}_R(P) \to \mathrm{End}_R(P')$ is an IP-isomorphism [15, Example 2.3]. Since projective right R-modules coincide with free right R-modules, Corollary 3.3 implies that the class of projective right R-modules is a Baer-Kaplansky class in $\mathrm{Mod}(R)$.

Recall that a right R-module X is called *automorphism invariant* (resp. *automorphism coinvariant*) if for any automorphism $\varphi : E \to E$ (resp. $\varphi : P \to P$), there exists an endomorphism $f : X \to X$ such that $uf = \varphi u$ (resp. $fp = p\varphi$), where $u : X \to E$ is an injective envelope (resp. $p : P \to X$ is a projective cover) of X [12]. Note that every module over a right perfect ring has a projective cover.

Corollary 3.5. *Let M and N be right R-modules such that:*

(1) $M = A \oplus X$ *and* $N = B \oplus X$ *for some right* R*-modules* A*,* B *and some indecomposable right* R*-module* X*;*

(2) X *is automorphism invariant (or automorphism coinvariant when* R *is right perfect);*

(3) $A, B \in \mathrm{Stat}(\mathrm{Hom}_R(X, -))$*;*

(4) there exists an IP-isomorphism $\Phi : \mathrm{End}_R(M) \to \mathrm{End}_R(N)$*.*

Then $M \cong N$*.*

Proof. With the above notation, it is well-known that $S = \mathrm{End}_R(E)$ satisfies that $S/J(S)$ is a von Neumann regular, right self-injective ring and idempotents lift modulo $J(S)$ [21, 22.1].

Now let us consider the case of a perfect ring R. Recall that P is a projective cover of X and $S = \mathrm{End}_R(P)$. In this case every right R-module is cotorsion [22, Proposition 3.3.1], hence P is a flat cotorsion right R-module. Then S is a right cotorsion ring [11, Example 1]. By [11, Corollary 4, Theorem 6, Corollary 9], it follows that $S/J(S)$ is a von Neumann regular, right self-injective ring and idempotents lift modulo $J(S)$.

Since X is also indecomposable, it is directly finite, in the sense that X is not isomorphic to a proper direct summand of itself. Therefore X is cancellable by [12, Theorems 3.3, 3.6]. Now use Theorem 2.3 and the proof of Theorem 3.1. □

We continue with an application to finitely accessible additive categories. Let us recall some needed terminology, mainly following [18]. An additive category $\mathcal{C}$ is called *finitely accessible* if it has direct limits, the class of finitely presented objects is skeletally small, and every object is a direct limit of finitely presented objects. Examples of finitely accessible additive categories include the category of unitary modules over a ring with enough idempotents, any locally finitely presented Grothendieck category, the category of torsion abelian groups [18, Example 10.2] and the category of torsion-free abelian groups [18, Example 10.5]. Note that Grothendieck categories are not finitely accessible in general. For instance, a module category of the form $\sigma[M]$ (of modules subgenerated by a module M) is Grothendieck, but it may not be finitely accessible [19, Example 1.7].

For every finitely accessible additive category $\mathcal{C}$, there exists a Grothendieck category $\mathcal{A}(\mathcal{C})$ and a fully faithful additive functor $F : \mathcal{C} \to \mathcal{A}(\mathcal{C})$, which induces an equivalence between $\mathcal{C}$ and the full subcategory of flat objects of $\mathcal{A}(\mathcal{C})$ (see [5, Theorem 1.4]). Note that $\mathcal{C}$ is equivalent to the category $\mathrm{Mod}(R)$ of unitary right R-modules, where R is the functor ring (with enough idempotents) of $\mathcal{C}$ (see [8, Theorem 1.1]). Also, finitely accessible additive categories admit a natural notion of purity, and a short exact sequence in $\mathcal{C}$ is pure if and only if the functor F takes it into a short exact sequence in $\mathcal{A}(\mathcal{C})$. Following [6], for an object X of $\mathcal{C}$, we denote by $\mathrm{PAdd}(X)$ the class of objects Z of $\mathcal{C}$ for which there is a pure epimorphism $X^{(I)} \to Z$ for some set I.

Corollary 3.6. *Let M and N be objects of a finitely accessible additive category $\mathcal{C}$ such that:*

(1) $M = A \oplus X$ and $N = B \oplus X$ for some objects A, B and some indecomposable object X;
(2) $\mathrm{End}_{\mathcal{C}}(X)$ has stable range one (in particular, $\mathrm{End}_{\mathcal{C}}(X)$ is semilocal, unit-regular or strongly π-regular);
(3) X is finitely presented and $A, B \in \mathrm{PAdd}(X)$;
(4) there exists an IP-isomorphism $\Phi : \mathrm{End}_{\mathcal{C}}(M) \to \mathrm{End}_{\mathcal{C}}(N)$.

Then $M \cong N$.

Proof. Note that the functor $F : \mathcal{C} \to \mathrm{Mod}(R)$ is fully faithful, where R is the functor ring of $\mathcal{C}$. It follows that $\mathrm{End}_R(F(X))$ has stable range one because so has $\mathrm{End}_{\mathcal{C}}(X)$. Since X is finitely presented, $F(X)$ is finitely generated projective [8, Theorem 1.1], and so finitely presented. Since $A, B \in \mathrm{PAdd}(X)$, we have $F(A), F(B) \in \mathrm{PAdd}(F(X))$ because $F(A)$ and $F(B)$ are flat right R-modules. Then $F(A), F(B) \in \mathrm{Stat}(\mathrm{Hom}_R(F(X), -))$ by [10, Lemma 2.4]. Hence $F(M) \cong F(N)$ by Lemma 2.2 and Theorem 3.1, and so $M \cong N$. □

Acknowledgments

This work was supported by Hacettepe University Scientific Research Projects Coordination Unit. Project Number: FBA-2017-16200. The second author would like to thank Hacettepe University for the financial support which facilitated the visit to Nagoya University (Japan) between August 26–31, 2019.

References

[1] P. Ara, Strongly π-regular rings have stable range one, *Proc. Amer. Math. Soc.* **124** (1996), 3293–3298.

[2] W. Brandal, Commutative Rings whose Finitely Generated Modules Decompose, *Lecture Notes in Mathematics*, Vol. **723** (Springer, Berlin, Heidelberg, New York, 1979).

[3] T. Bühler, Exact categories, *Expo. Math.* **28** (2010), 1–69.

[4] F. Castaño Iglesias, J. Gómez-Torrecillas and R. Wisbauer, Adjoint functors and equivalences of subcategories, *Bull. Sci. Math.* **127** (2003), 379–395.

[5] W. Crawley-Boevey, Locally finitely presented additive categories, *Comm. Algebra* **22** (1994), 1641–1674.

[6] S. Crivei, Σ-extending modules, Σ-lifting modules, and proper classes, *Comm. Algebra* **36** (2008), 529–545.

[7] S. Crivei and D. Keskin Tütüncü, Baer-Kaplansky classes in Grothendieck categories and applications, *Mediterr. J. Math.* **16:90** (2019).

[8] N. V. Dung and J. L. García, Additive categories of locally finite representation type, *J. Algebra* **238** (2001), 200–238.

[9] L. Fuchs, Infinite Abelian Groups, Pure and Applied Mathematics, 36-II (Academic Press, New York, 1973).

[10] J. L. Gómez Pardo and P. A. Guil Asensio, Indecomposable decompositions of finitely presented pure-injective modules, *J. Algebra* **192** (1997), 200–208.

[11] P. A. Guil Asensio and I. Herzog, *Left cotorsion rings*, Bull. London. Math. Soc. **36** (2004), 303–309.

[12] P. A. Guil Asensio, D. Keskin Tütüncü and A. K. Srivastava, Modules invariant under monomorphisms of their envelopes, *Contemporary Mathematics*, Vol. **715** (2018), pp. 171–179.

[13] G. Ivanov, Generalizing the Baer-Kaplansky theorem, *J. Pure Appl. Algebra* **133** (1998), 107–115.

[14] G. Ivanov and P. Vámos, A characterization of FGC rings, *Rocky Mountain J. Math.* **32** (2002), 1485–1492.

[15] D. Keskin Tütüncü and R. Tribak, On Baer-Kaplansky classes of modules, *Algebra Colloq.* **24** (2017), 603–610.

[16] T. Y. Lam, A crash course on stable range, cancellation, substitution and exchange, *J. Algebra Appl.* **3** (2004), 301–343.

[17] K. Morita, Category-isomorphisms and endomorphism rings of modules, *Trans. Amer. Math. Soc.* **103** (1962), 451–469.

[18] M. Prest, Definable additive categories: purity and model theory, *Mem. Amer. Math. Soc.* **210** (2011), No. 987.

[19] M. Prest and R. Wisbauer, Finite presentation and purity in categories $\sigma[M]$, *Colloq. Math.* **99** (2004), 189–202.
[20] B. Stenström, Rings of Quotients, Grundlehren der Math., Vol. **217** (Springer, Berlin, Heidelberg, New York, 1975).
[21] R. Wisbauer, Foundations of Module and Ring Theory (Gordon and Breach, Reading, 1991).
[22] J. Xu, Flat covers of modules, Lecture Notes in Math., **1634**, Springer, Berlin, 1996.

Leibniz conformal algebras of rank three

Zhixiang Wu
School of Mathematical Sciences, Zhejiang University, Hangzhou, 310027, P. R. China
E-mail: wzx@zju.edu.cn

In this article, we classify the Leibniz conformal algebras of rank three.

Keywords: Leibniz pseudoalgebra; Lie conformal algebra; Hopf algebra.

1. Introduction

Conformal algebras provide an axiomatic description of the singular part of the operator product expansion. A synonym of a Lie conformal algebra is a Lie H-pseudoalgebra, where $H = \mathbf{k}[s]$ is a polynomial algebra over a field $\mathbf{k}$ of characteristic zero with variable s. In [9], we have classified Leibniz H-pseudoalgebras of rank two and some Lie conformal algebras of rank three and rank four. In the present paper, we will classify Leibniz H-pseudoalgebras of rank three.

Any finite Leibniz H-pseudoalgebra L has a solvable radical I such that L/I is a Lie semisimple H-pseudoalgebra. The classification of finite semisimple H-pseudoalgebras was achieved in [1]. For any Leibniz H-pseudoalgebra L, $Z_l(L) := \{a \in L|[a, b] = 0, \forall b \in L\}$ is an ideal of L such that $L/Z_l(L)$ is a Lie H-pseudoalgebra. It is obvious that $Z_l(L)$ is contained in I, the solvable radical of L. In the remainder of this paper, we always use L to denote a Leibniz H-pseudoalgebra of rank three with a nonzero solvable radical I. We can classify all Leibniz H-pseudoalgebras of rank three according to the rank of I. If the rank of I is three, then this Leibniz H-pseudoalgebra is solvable. Solvable Leibniz H-pseudoalgebras are classified in Section 3. They can be classified into three classes: (1) abelian Leibniz H-pseudoalgebras, (2) solvable H-pseudoalgebras with maximal derived series, (3) non-abelian solvable Leibniz H-pseudoalgebras which are not of maximal derived series. The third class (3) is classified as (3i) Leibniz H-pseudoalgebras L where the rank of $L^{(1)}$ is one, and (3ii) Leibniz H-pseudoalgebras where the rank of $L^{(1)}$ is two. The second class (3ii) is classified into four classes: (3iia) Leibniz H-pseudoalgebras such that the rank of $Z_l(L) \cap L^{(1)}$ is one and the rank of $Z_l(L)$ is two, (3iib) Leibniz H-pseudoalgebras such that the rank of $Z_l(L) \cap L^{(1)}$ and the rank of $Z_l(L)$ are one, (3iic) Leibniz H-pseudoalgebras such that the rank of $Z_l(L) \cap L^{(1)}$ is two, and (3iid) Leibniz H-pseudoalgebras such that $Z_l(L) = 0$. The Leibniz H-pseudoalgebras in class (2), (3i), (3iia), (3iib), (3iic) and (3iid) are described by Theorem 3.1, Theorem 3.2, Theorem 3.3, Theorem 3.4, Theorem 3.5 and Theorem 3.6 respectively. The non-solvable H-pseudoalgebras are classified in Section 4. Non-semisimple non-solvable Leibniz H-pseudoalgebras have solvable radicals I. According to the rank of I, we can classify them into two types: (1) the rank of I is one, (2) the rank of I is two. We further classify Leibniz H-pseudoalgebras with a rank two solvable radical into two kinds: (2i) I is abelian and (2ii) I is not abelian. The class (2i) can be classified into three subclasses: (2ia) the rank of $Z_l(L)$ is two, (2ib) the rank of $Z_l(L)$ is one, and (2ic) $Z_l(L) = 0$. These three classes (2ia), (2ib), (2ic) Leibniz H-pseudoalgebras are described by Theorem 4.1, Theorem 4.2 and Theorem 4.3 respectively. The class (2ii) can be classified into three subclasses (2iia), (2iib) and (2iic) according to the structure constants of the solvable ideal I. These structure constants are given by Lemma 2.1 in Section 2. These three subclasses (2iia), (2iib) and (2iic) are described by Theorem 4.4, Theorem 4.5 and Theorem 4.6 respectively. Finally, we describe the Leibniz H-pseudoalgebras with a solvable radical of rank one in Theorem 4.7 and Theorem 4.8. In Section 2, we recall some basic concepts and fix some notation.

Throughout this paper, $\mathbf{k}$ is an algebraically closed field of characteristic zero. The unadorned $\otimes$ means the tensor product over $\mathbf{k}$. For the other undefined terms, we refer to [5], [6] and [7].

2. Preliminaries

In the sequel, we always assume that $H = \mathbf{k}[s]$ is a polynomial algebra over a field $\mathbf{k}$ of characteristic zero with a variable s. A left H-module A is called a Leibniz *H-pseudoalgebra* (or a *Leibniz conformal algebra*) if there is a mapping $\rho : A \otimes A \to H^{\otimes 2} \otimes_H A$ such that

$$\rho(h_1 a_1 \otimes h_2 a_2) = (h_1 \otimes h_2 \otimes_H 1)\rho(a_1 \otimes a_2) \tag{1}$$

$$\begin{aligned}\rho(\rho(a_1 \otimes a_2) \otimes a_3) = {} & \rho(a_1 \otimes \rho(a_2 \otimes a_3)) \\ & -((12) \otimes_H 1)\rho(a_2 \otimes \rho(a_1 \otimes a_3))\end{aligned} \tag{2}$$

for any $h_1, h_2 \in H$ and $a_1, a_2, a_3 \in A$. $\rho(a \otimes b)$, denoted by $[a, b]$, is usually called *pseudoproduct*, and Equation (2) is called Jacobi Identity. An asymmetric Leibniz H-pseudoalgebra is a *Lie H-pseudoalgebra*.

Suppose that A is an H-pseudoalgebra. Then A becomes a conformal algebra introduced in [4], where the mapping $\rho'({}_\lambda) : A \otimes A \to \mathbf{k}[\partial] \otimes A$ is given by $\rho'(a_\lambda b) = \sum_i p_i(-\lambda)c_i$ if $\rho(a, b) = \sum_i p_i(-s) \otimes 1 \otimes_H c_i$ for any $a, b \in A$. Conversely, if A is a conformal algebra, then A becomes an H-pseudoalgebra with the mapping $\rho' : A \otimes A \to H^{\otimes 2} \otimes_H A$, where $\rho'(a, b) = \sum_i Q_i(-s \otimes 1, \Delta(s)) \otimes_H c_i$ if $\rho(a, b) = \sum_i Q_i(\lambda, \partial) \otimes c_i$ for any $a, b \in A$.

For any Leibniz H-pseudoalgebra A, let $A^{(1)} = [A, A]$ and $A^{(n)} = [A^{(n-1)}, A^{(n-1)}]$ for any $n \geq 2$. Then $A, A^{(1)}, \cdots, A^{(n)}, \cdots$ is called the *derived series* of A. Suppose that the rank of a solvable Leibniz H-pseudoalgebra A is n. Then A is said to be a *solvable Leibniz H-pseudoalgebra with maximal derived series* if $A^{(n-1)} \neq 0$. The following lemma was proved in [9].

Lemma 2.1. *Let A be a rank two solvable Leibniz H-pseudoalgebra with maximal derived series. Suppose $B(s) = \sum_{i=0}^{n} b_i s^{(i)}, C(s) = \sum_{j=0}^{n'} c_j s^{(j)}, Y(s) = \sum_{i=0}^{m} y_i s^{(i)} \in H$. Then A has a basis $\{e_1, e_2\}$ such that $[e_1, e_1] = \alpha' \otimes_H e_2$, $[e_1, e_2] = \eta_1 \otimes_H e_2$, $[e_2, e_1] = \eta_2 \otimes_H e_2$ and $[e_2, e_2] = 0$, where $\eta_1 = B(s) \otimes 1$. In addition, one of the following cases holds:*

(i) $\eta_1 = \eta_2 = 0$, α' is an arbitrary nonzero element in $H \otimes H$.

(ii) $\eta_2 = -(12)\eta_1 \neq 0$ and $\alpha' = B(s) \otimes Y(s) - Y(s) \otimes B(s)$.

(iii) $\eta_1 = B(s) \otimes 1 \neq 0$, $\eta_2 = 0$, $\alpha' = B(s) \otimes C(s)$.

The next three lemmas play a key role in classifying of Leibniz H-pseudoalgebras of rank three.

Lemma 2.2. *Let $\alpha = s \otimes 1 - 1 \otimes s$, $\eta = \lambda s \otimes 1 - 1 \otimes s + \kappa \otimes 1$, where $\lambda, \kappa \in \mathbf{k}$. Suppose $\beta \in H^{\otimes 2}$ is subject to*

$$(\alpha\Delta \otimes 1)\beta = (1 \otimes \eta\Delta)\beta - (12)(1 \otimes \eta\Delta)\beta. \tag{3}$$

Then $\beta = x_{00} \otimes 1 + \eta\Delta(A(s))$ for some $x_{00} \in \mathbf{k}$ and $A(s) \in H$, where $(\lambda + 1)x_{00} = 0$.

Lemma 2.3. *With the same α and η as Lemma 2.2. Suppose $\beta \in H^{\otimes 2}$ is subject to*

$$(\alpha\Delta \otimes 1)\beta = (1 \otimes \beta\Delta)\eta - (12)(1 \otimes \beta\Delta)\eta. \tag{4}$$

Then $\beta = y_m \otimes s^{(m)} + \eta(1 \otimes E(s))$ *for some* $y_m \in \mathbf{k}$ *and some* $E(s) \in H$, *where* $(\lambda - 2)y_m = 0$ *if* $\kappa \neq 0$, $y_m = 0$ *if* $\kappa = 0$.

Lemma 2.4. *With the same* α *and* η *as Lemma 2.2. Let* $\xi = \lambda' s \otimes 1 - 1 \otimes s + \kappa' \otimes 1$ *and* $t = 1 + \lambda - \lambda'$ *for some* $\lambda, \kappa \in \mathbf{k}$. *Suppose* $\beta = \sum_{ij} x_{ij} s^{(i)} \otimes s^{(j)}$ *for some* $x_{ij} \in \mathbf{k}$ *is subject to*

$$\begin{aligned}(\alpha\Delta \otimes 1)\beta = (1 \otimes \eta\Delta)\beta - (12)(1 \otimes \eta\Delta)\beta \\ +(1 \otimes \beta\Delta)\xi - (12)(1 \otimes \beta\Delta)\xi.\end{aligned} \tag{5}$$

Then $\beta = \beta_t + \eta\Delta(A(s)) - \xi(1 \otimes A(s))$ *for some* $A(s) \in H$, *where*

$$\beta_t = \begin{cases} x_{10} s \otimes 1 + x_{00} \otimes 1, & t = 1, \kappa = \kappa' \\ x_{20} s^{(2)} \otimes 1 + x_{11} s \otimes s + x_{01}(1 \otimes s - \kappa \otimes 1), & \begin{cases} t = 2, (\lambda, \lambda') = (0, -1) \\ \kappa = \kappa' \end{cases} \\ x_{20} s^{(2)} \otimes 1, \quad if \quad \kappa = \kappa', & t = 2, (\lambda, \lambda') \neq (0, -1) \\ \sum_{p=0}^{t-2} \sum_{i=2}^{t-p} \frac{(-\kappa)^p}{p!} x_{i\ t-i} s^{(i)} \otimes s^{(t-p-i)}, & 3 \leq t \leq 5, \kappa = \kappa' \\ \sum_{p=0}^{4} \sum_{i=2}^{6-p} \frac{(-\kappa)^p}{p!} x_{i\ 6-i} s^{(i)} \otimes s^{(6-p-i)}, & t = 6, \kappa = \kappa', \lambda \in \{0, 4\} \\ \sum_{p=0}^{5} \sum_{i=2}^{7-p} \frac{(-\kappa)^p}{p!} x_{i\ 7-i} s^{(i)} \otimes s^{(7-p-i)}, & \begin{cases} t = 7, \kappa = \kappa', \\ \lambda = \frac{1}{2}(5 \pm \sqrt{19}) \end{cases} \\ 0, & otherwise. \end{cases}$$

Moreover, $x_{i\ t-i}$ *is determined by* $x_{3\ t-3}$ *and* $x_{2\ t-2}$ *as follows.* $x_{40} = \lambda(2x_{31} - 3x_{22})$; $x_{41} = (2\lambda - 1)x_{32} - (3\lambda - 1)x_{23}$, $x_{50} = \lambda(2\lambda - 1)x_{32} - \lambda(3\lambda + 1)x_{23}$; $x_{60} = 2\lambda(3\lambda - 5)x_{33} - 3\lambda(3\lambda - 2)x_{24}$, $x_{51} = (\lambda - 1)(2\lambda - 1)x_{33} - \frac{1}{2}(6\lambda^2 - 3\lambda + 1)x_{24}$, $x_{42} = 2(\lambda - 1)x_{33} - (3\lambda - 2)x_{24}$; $x_{70} = \frac{3}{2}\lambda x_{52} - \frac{5}{2}\lambda x_{34}$ *for* $\lambda = \frac{1}{2}(5 \pm \sqrt{19})$, $x_{43} = (2\lambda - 3)x_{34} - 3(\lambda - 1)x_{25}$, $x_{52} = (\lambda - 1)(2\lambda - 3)x_{34} - (3\lambda^2 - 4\lambda + 2)x_{25}$ *and* $x_{61} = (6\lambda^2 - 15\lambda + 4)x_{34} - (3\lambda - 1)(3\lambda - 3)x_{25}$.

From Lemma 2.4, we obtain the following

Corollary 2.1. *Let* $\alpha = s \otimes 1 - 1 \otimes s$, $\xi = \lambda' s \otimes 1 - 1 \otimes s + \kappa' \otimes 1$ *and* $n = 2 - \lambda'$. *Suppose* β *is subject to*

$$\begin{aligned}(\alpha\Delta \otimes 1)\beta = (1 \otimes \alpha\Delta)\beta - (12)(1 \otimes \alpha\Delta)\beta \\ +(1 \otimes \beta\Delta)\xi - (12)(1 \otimes \beta\Delta)\xi.\end{aligned} \tag{6}$$

If $\kappa' \neq 0$, *then* $\beta = \alpha\Delta(A(s)) - \xi(1 \otimes A(s))$ *for some* $A(s) \in H$. *If* $\kappa' = 0$ *and* $\lambda' \notin \{1, -1, -2, -3\}$, *then* $\beta = \alpha\Delta(A(s)) - \xi(1 \otimes A(s))$.

(1) If $\kappa' = 0$ *and* $\lambda' = 1$, *then* $\beta = x_{00} \otimes 1 + x_{10} s \otimes 1 + \alpha\Delta(A(s)) - \xi(1 \otimes A(s))$.

(2) If $\kappa' = 0$ *and* $\lambda' = -1$, *then* $\beta = (x_{30} - 3x_{21})s^{(3)} \otimes 1 + \alpha\Delta(A(s)) - \xi(1 \otimes A(s))$.

(3) If $\kappa' = 0$ *and* $\lambda' = -2$, *then* $\beta = -3x_{22}s^{(4)} \otimes 1 + x_{22}s^{(2)} \otimes s^{(2)} + \alpha\Delta(A(s)) - \xi(1 \otimes A(s))$.

(4) If $\kappa' = 0$ *and* $\lambda' = -3$, *then* $\beta = (x_{32} + x_{23})(s^{(5)} \otimes 1 + s^{(4)} \otimes s + s^{(3)} \otimes s^{(2)}) + \alpha\Delta(A(s)) - \xi(1 \otimes A(s))$.

3. Solvable Leibniz H-pseudoalgebras of rank three

In this section, we classify solvable Leibniz H-pseudoalgebras of rank three.

3.1. *Leibniz H-pseudoalgebras with maximal derived series*

Suppose that L is a solvable Leibniz H-pseudoalgebra of rank three with maximal derived series.

Theorem 3.1. *Suppose the rank three Leibniz H-pseudoalgebra L is a solvable Leibniz H-pseudoalgebra with maximal derived series. Then L is isomorphic to one of the following H-pseudoalgebras:*

(1) L has a basis $\{e_0, e_1, e_2\}$ such that the only nonzero pseudobrackets with this basis are $[e_0, e_1] = -((12) \otimes_H 1)[e_1, e_0] = A(s) \otimes 1 \otimes_H e_1$, $[e_0, e_2] = 2A(s) \otimes 1 \otimes_H e_2$, $[e_1, e_1] = y'_{00} \otimes 1 \otimes_H e_2 \neq 0$, *for some nonzero* $A(s) \in H$ *and for some nonzero* $y'_{00} \in \mathbf{k}$.

(2) L has a basis $\{e_0, e_1, e_2\}$ such that the only nonzero pseudobrackets with this basis are $[e_0, e_1] = -((12) \otimes_H 1)[e_1, e_0] = A(s) \otimes 1 \otimes_H e_1 + z'A(s) \otimes 1 \otimes_H e_2$, $[e_0, e_2] = -((12) \otimes_H 1)[e_2, e_0] = -2A(s) \otimes 1 \otimes_H e_2$, $[e_1, e_1] = y'_{10}(s \otimes 1 - 1 \otimes s) \otimes_H e_2$ *for some nonzero* $A(s) \in H$, *nonzero* $y'_{10} \in \mathbf{k}$ *and any* $z' \in \mathbf{k}$. *Moreover, L is a Lie H-pseudoalgebra.*

Next suppose that L is a nonabelian Leibniz H-pseudoalgebra of rank three such that $L^{(2)} = 0$. Suppose that the rank of $L^{(1)}$ is one. We have the following result.

Theorem 3.2. *Suppose that L is a Leibniz H-pseudoalgebra of rank three such that $L^{(2)} = 0$ and the rank of $L^{(1)}$ is one. Then L is isomorphic to one of the following Leibniz H-pseudoalgebras.*

(1) L has a basis $\{e_0, e_1, e_2\}$ such that the only nonzero pseudobrackets with this basis are $[e_0, e_1] = -((12) \otimes_H 1)[e_1, e_0] = A(s) \otimes B(s) \otimes_H e_2$, $[e_1, e_2] = -((12) \otimes_H 1)[e_2, e_1] = B(s) \otimes 1 \otimes_H e_2$, $[e_0, e_2] = -((12) \otimes_H 1)[e_2, e_0] = A(s) \otimes 1 \otimes_H e_2$ *for some nonzero* $A(s), B(s) \in H$.

(2) L has a basis $\{e_0, e_1, e_2\}$ such that the only nonzero pseudobrackets with this basis are $[e_1, e_2] = -((12) \otimes_H 1)[e_2, e_1] = B(s) \otimes 1 \otimes_H e_2$ for some nonzero $B(s) \in H$.

(3) L has a basis $\{e_0, e_1, e_2\}$ such that the only nonzero pseudobrackets with this basis are $[e_1, e_2] = B(s) \otimes 1 \otimes_H e_2$, $[e_1, e_0] = B(s) \otimes C(s) \otimes_H e_2$, $[e_0, e_2] = A(s) \otimes 1 \otimes_H e_2$ for some nonzero $A(s)$, $B(s)$, $C(s) \in H$.

(4) L has a basis $\{e_0, e_1, e_2\}$ such that the only nonzero pseudobracket with this basis is $[e_1, e_2] = B(s) \otimes 1 \otimes_H e_2$ for some nonzero $B(s) \in H$.

(5) L has a basis $\{e_0, e_1, e_2\}$ such that the only possibly nonzero pseudobrackets with this basis are $[e_0, e_1] = \eta_{01} \otimes_H e_2$, $[e_1, e_0] = \eta_{10} \otimes_H e_2$, $[e_0, e_0] = \eta_{00} \otimes_H e_2$, $[e_1, e_1] = \eta_{11} \otimes_H e_2$, where $\eta_{01}, \eta_{10}, \eta_{00}, \eta_{11}$ are arbitrary elements in $H^{\otimes 2}$ and are not all zero.

Next we assume that L is a Leibniz H-pseudoalgebra of rank three such that the rank of $L^{(1)}$ is equal to two and $L^{(2)} = 0$. Suppose that L is a Leibniz H-pseudoalgebra such that the rank of $L^{(1)}$ is two, $L^{(2)} = 0$ and the rank of $Z_l(L) \cap L^{(1)}$ is one.

Theorem 3.3. *Let L be a Leibniz H-pseudoalgebra of rank three such that the rank of both $L^{(1)}$ and $Z_l(L)$ is two, $L^{(2)} = 0$ and the rank of $Z_l(L) \cap L^{(1)}$ is one. Then L has a basis $\{e_0, e_1, e_2\}$ such that the only nonzero pseudobracket with this basis is*

$$[e_1, e_0] = \sum_{i=0}^{n} s^{(i)} \otimes 1 \otimes_H (a_i e_1 + b_i e_2)$$

for some $a_i, b_i \in H$, where the rank of the left H-module generated by $\{a_i e_1 + b_i e_2 | 0 \leq i \leq n\}$ is two.

Next, we determine the solvable rank three Leibniz H-pseudoalgebras L such that the rank of $L^{(1)}$ is two, the rank of $Z_l(L)$ and the rank of $Z_l(L) \cap L^{(1)}$ are one.

Theorem 3.4. *Suppose L is a solvable H-pseudoalgebra of rank three, the rank of both $Z_l(L)$ and $Z_l(L) \cap L^{(1)}$ is one and the rank of $L^{(1)}$ is two. Then L has a basis $\{e_0, e_1, e_2\}$ such that $L^{(1)} \subseteq He_1 + He_2$ and $Z_l(L) = He_2$. Furthermore, L is isomorphic to one of the following Leibniz H-pseudoalgebras.*

(I) A Leibniz H-pseudoalgebra with a basis $\{e_0, e_1, e_2\}$ such that the only nonzero pseudobrackets with this basis are $[e_0, e_0] = \eta_{00} \otimes_H e_1 + \xi_{00} \otimes_H e_2$, $[e_0, e_1] = \xi_{10} \otimes_H e_2$ and $[e_1, e_0] = \xi_{10} \otimes_H e_2$, where $\eta_{00} = -(12)\eta_{00} \neq 0$ and $(\eta_{00}\Delta \otimes 1)\xi_{10} = (1 \otimes \eta_{00}\Delta)\xi_{01} - (12)(1 \otimes \eta_{00}\Delta)\xi_{01}$.

(II) A Leibniz H-pseudoalgebra with a basis $\{e_0, e_1, e_2\}$ such that the only nonzero pseudobrackets with this basis are $[e_0, e_1] = -((12) \otimes_H 1)[e_1, e_0] = A(s) \otimes 1 \otimes_H e_1$, $[e_0, e_0] = \xi_{00} \otimes_H e_2$ for some nonzero element $A(s) \in H$ and some nonzero $\xi_{00} \in H \otimes H$.

(III) A Leibniz H-pseudoalgebra with a basis $\{e_0, e_1, e_2\}$ such that the only nonzero pseudobrackets with this basis are $[e_0, e_1] = -((12) \otimes_H 1)[e_1, e_0] = A(s) \otimes 1 \otimes_H e_1$, $[e_0, e_2] = B(s) \otimes 1 \otimes_H e_2$ for some nonzero elements $A(s), B(s) \in H$.

(IV) A Leibniz H-pseudoalgebra with a basis $\{e_0, e_1, e_2\}$ such that the only nonzero pseudobrackets with this basis are $[e_0, e_1] = A(s) \otimes 1 \otimes_H e_1$, $[e_1, e_0] = -1 \otimes A(s) \otimes_H e_1 + 1 \otimes B(s) \otimes_H e_2$ and $[e_0, e_2] = A(s) \otimes 1 \otimes_H e_2$ for some nonzero $A(s), B(s) \in H$.

Remark 3.1. If $\eta_{00} \in H \otimes H$ satisfies $(12)\eta_{00} = -\eta_{00}$, then $\eta_{00} = \alpha\eta'_{00}$, where $(12)\eta'_{00} = \eta'_{00}$. Let $\eta_{00} = \alpha\Delta(A(s))$ and $\xi_{10} = \xi_{01} = \Delta(A(s)B(s))\alpha$ for any $A(s), B(s) \in H$. Then $(\eta_{00}\Delta \otimes 1)\xi_{10} = (1 \otimes \eta_{00}\Delta)\xi_{01} - (12)(1 \otimes \eta_{00}\Delta)\xi_{01}$. This means there are many Leibniz H-pseudoalgebras of type (I) in Theorem 3.4.

Suppose the rank of $L^{(1)}$, $Z_l(L)$ and $Z_l(L) \cap L^{(1)}$ is equal to two. Then we have the following

Theorem 3.5. *Suppose L is a solvable Leibniz H-pseudoalgebra of rank three such that the rank of $L^{(1)} \cap Z_l(L)$, the rank of $L^{(1)}$ and the rank of $Z_l(L)$ are two. Then L has a basis $\{e_0, e_1, e_2\}$ satisfying $Z_l(L) = He_1 \oplus He_2$. Thus L is isomorphic to one of the following Leibniz H-pseudoalgebras.*

(1) L has a basis $\{e_0, e_1, e_2\}$ such that the only nonzero pseudobrackets with this basis are $[e_0, e_1] = B(s) \otimes 1 \otimes_H e_1$, $[e_0, e_2] = A(s) \otimes 1 \otimes_H e_2$ for some nonzero $A(s), B(s) \in H$.

(2) L has a basis $\{e_0, e_1, e_2\}$ such that the only possibly nonzero pseudobrackets with this basis are $[e_0, e_0] = \xi_{00} \otimes_H e_2$, $[e_0, e_1] = B(s) \otimes 1 \otimes_H e_1$ for some nonzero $B(s) \in H$, and some nonzero $\xi_{00} \in H^{\otimes 2}$.

(3) L has a basis $\{e_0, e_1, e_2\}$ such that the only nonzero pseudobracket with this basis is $[e_0, e_0] = a \otimes 1 \otimes_H be_1 + c \otimes 1 \otimes_H de_2$ for some $a, b, c, d \in H$, where a, c are linearly independent over $\mathbf{k}$ and $abcd \neq 0$.

(4) L has a basis $\{e_0, e_1, e_2\}$ such that the only nonzero pseudobrackets with this basis are $[e_0, e_0] = \eta_{00} \otimes_H e_1 + \xi_{00} \otimes_H e_2$, $[e_0, e_1] = \xi_{01} \otimes_H e_2$ for some nonzero $\eta_{00}, \xi_{01} \in H^{\otimes 2}$ and some $\xi_{00} \in H^{\otimes 2}$, where η_{00} and ξ_{01} satisfy $(1 \otimes \eta_{00}\Delta)\xi_{01} = (12)(1 \otimes \eta_{00}\Delta)\xi_{01}$.

Theorem 3.6. *Suppose L is a solvable Lie H-pseudoalgebra such that the rank of $L^{(1)}$ is two and $Z_l(L) = L^{(2)} = 0$. Then L has a basis $\{e_0, e_1, e_2\}$ such that $L^{(1)} \subseteq He_1 + He_2$ and $[e_i, e_j] = 0$ for $i, j = 1, 2$. Thus $He_1 + He_2$ is a representation of the Lie H-pseudoalgebra $L/He_1 + He_2$. Moreover, L is isomorphic to one of the following types.*

(1) L has a basis $\{e_0, e_1, e_2\}$ such that only nonzero pseudobrackets with this basis are $[e_0, e_1] = -((12) \otimes_H 1)[e_1, e_0] = B(s) \otimes 1 \otimes_H e_1$, $[e_0, e_2] = -((12) \otimes_H 1)[e_2, e_0] = A(s) \otimes 1 \otimes_H e_2$ for some nonzero $A(s), B(s) \in H$.

(2) L has a basis $\{e_0, e_1, e_2\}$ such that only nonzero pseudobrackets with this basis are $[e_0, e_0] = \eta_{00} \otimes_H e_1$, $[e_0, e_2] = -((12) \otimes_H 1)[e_2, e_0] = A(s) \otimes 1 \otimes_H e_2$ for some nonzero $A(s) \in H$ and some nonzero asymmetric element in $H^{\otimes 2}$.

Remark 3.2. The final solvable Leibniz H-pseudoalgebra of rank three is an abelian Leibniz H-pseudoalgebra.

4. Nonsolvable Leibniz H-pseudoalgebras of rank three

In this section, we classify nonsolvable Leibniz H-pseudoalgebras of rank three. First, let us assume that the rank of the solvable radical I is equal to two, $\mathcal{O}_1 = \{(-1,-4), (-1,-5), (-2,-4), (-2,-5), (-3,-5)\}$ and $\mathcal{O}_2 = \{(1,1), (1,-1), (1,-2), (1,-3), (-1,-1), (-2,-2), (-2,-5), (-3,-3), (-3,-6), (-3,-7)\}$ in the next theorem.

Theorem 4.1. *Suppose L is a nonsolvable Leibniz H-pseudoalgebra of rank three, and the rank of $Z_l(L)$ is two. Then L is isomorphic to one of the following types.*

(1) L is a direct sum of the Virasoro Lie conformal algebra He_0 and the abelian Lie H-pseudoalgebra $He_1 \oplus He_2$.

(2) L has a basis $\{e_0, e_1, e_2\}$ such that the only nonzero pseudobrackets with this basis are $[e_0, e_0] = \alpha \otimes_H e_0 + \xi_{00} \otimes_H e_2$, and $[e_0, e_2] = (\lambda_2 s \otimes 1 - 1 \otimes s) \otimes_H e_2$ for some $\lambda_2 \in \mathbf{k}$, where

$$\xi_{00} = \begin{cases} d'_{00} \otimes 1 + d'_{10} s \otimes 1, & \lambda_2 = 1 \\ d'_{30} s^{(3)} \otimes 1, & \lambda_2 = -1 \\ d'_{22}(s^{(2)} \otimes s^{(2)} - s^{(4)} \otimes 1), & \lambda_2 = -2 \\ d'_{23}(s^{(5)} \otimes 1 + s^{(4)} \otimes s + s^{(3)} \otimes s^{(2)}), & \lambda_2 = -3 \\ 0, & \text{otherwise}, \end{cases} \quad \text{for some } d'_{ij} \in \mathbf{k}.$$

(2') L has a basis $\{e_0, e_1, e_2\}$ such that the only nonzero pseudobrackets with this basis are $[e_0, e_0] = \alpha \otimes_H e_0$, $[e_0, e_1] = y_m \otimes s^{(m)} \otimes_H e_2$ and

$[e_0, e_2] = (\lambda_2 s \otimes 1 - 1 \otimes s + \kappa_2 \otimes 1) \otimes_H e_2$, *where for some nonzero* $\kappa_2 \in \mathbf{k}$ *and some* $\lambda_2, y_m \in \mathbf{k}$, *where* $(\lambda_2 - 2)y_m = 0$.

(3) L has a basis $\{e_0, e_1, e_2\}$ *such that the only nonzero pseudobrackets with this basis are* $[e_0, e_0] = \alpha \otimes_H e_0 + \eta_{00} \otimes_H e_1$, $[e_0, e_1] = \eta_{01} \otimes_H e_1 + \xi_{01} \otimes_H e_2$, *where* $\eta_{01} = \lambda_1 s \otimes 1 - 1 \otimes s + \kappa_1 \otimes 1$ *for some* $\lambda_1, \kappa_1 \in \mathbf{k}$, $\xi_{01} = x_{00} \otimes 1$ *for some* $x_{00} \in \mathbf{k}$ *satisfying* $(\lambda_1 + 1)x_{00} = 0$,

$$\eta_{00} = \begin{cases} d_{00} \otimes 1 + d_{10} s \otimes 1, & (\kappa_1, \lambda_1) = (0, 1) \\ d_{30} s^{(3)} \otimes 1, & (\kappa_1, \lambda_1) = (0, -1) \\ d_{22}(s^{(2)} \otimes s^{(2)} - 3s^{(4)} \otimes 1), & (\kappa_1, \lambda_1) = (0, -2) \\ d_{23}(s^{(5)} \otimes 1 + s^{(3)} \otimes s^{(2)} + s^{(4)} \otimes s) & (\kappa_1, \lambda_1) = (0, -3), \\ 0, & \textit{otherwise} \end{cases}$$

for some $d_{ij} \in \mathbf{k}$ *and* d_{30} *satisfying* $x_{00} d_{30} = 0$.

(4) L has a basis $\{e_0, e_1, e_2\}$ *such that the only nonzero pseudobrackets with this basis are* $[e_0, e_0] = \alpha \otimes_H e_0 + \eta_{00} \otimes_H e_1 + \xi_{00} \otimes_H e_2$, $[e_0, e_1] = (\lambda_1 s \otimes 1 - 1 \otimes s + \kappa_1 \otimes 1) \otimes_H e_1 + \xi_{01} \otimes_H e_2$ *and* $[e_0, e_2] = (\lambda_2 s \otimes 1 - 1 \otimes 2 + \kappa_2 \otimes 1) \otimes_H e_2$ *for some* $\lambda_i, \kappa_i \in \mathbf{k}$, *where either* $\kappa_1 \neq 0$, *or* $\kappa_1 \neq \kappa_2$, *or* $\kappa_1 = \kappa_2 = 0$ *and* $(\lambda_1, \lambda_2) \notin \mathcal{O}_1 \cup \mathcal{O}_2$, η_{00} *is the same as the* η_{00} *in (3), and* ξ_{00} *is the same as in (2),* $\xi_{01} = \beta_t$ *in Lemma 2.4, where* $t = 1 + \lambda_1 - \lambda_2$,

(5) L has a basis $\{e_0, e_1, e_2\}$ *such that the only nonzero pseudobrackets with this basis are* $[e_0, e_0] = \alpha \otimes_H e_0 + \eta_{00} \otimes_H e_1 + \xi_{00} \otimes_H e_2$, $[e_0, e_1] = (\lambda_1 s \otimes 1 - 1 \otimes s) \otimes_H e_1 + \xi_{01} \otimes_H e_2$ *and* $[e_0, e_2] = (\lambda_2 s \otimes 1 - 1 \otimes 2) \otimes_H e_2$ *for some* $(\lambda_1, \lambda_2) \in \mathcal{O}_1$.

(i) In the case when $(\lambda_1, \lambda_2) = (-1, -4)$. *We have* $\eta_{00} = d_{30} s^{(3)} \otimes 1$, $\xi_{01} = x_{31}(s^{(3)} \otimes s - 2s^{(4)} \otimes 1)$ *and* $\xi_{00} = \frac{1}{2} d_{30} x_{31} s^{(3)} \otimes s^{(3)}$ *for some* $d_{30}, x_{31} \in \mathbf{k}$.

(ii) In the case when $(\lambda_1, \lambda_2) = (-1, -5)$. *We have* $\eta_{00} = d_{30} s^{(3)} \otimes 1$, $\xi_{01} = x_{32}(3s^{(5)} \otimes 1 - 3s^{(4)} \otimes s + s^{(3)} \otimes s^{(2)})$ *and* $\xi_{00} = \frac{21}{5} d_{30} x_{32}(s^{(3)} \otimes s^{(4)} - s^{(4)} \otimes s^{(3)})$ *for some* $d_{30}, x_{32} \in \mathbf{k}$.

(iii) In the case when $(\lambda_1, \lambda_2) = (-2, -4)$. *We have* $\eta_{00} = d_{22}(s^{(2)} \otimes s^{(2)} - 3s^{(4)} \otimes 1)$, $\xi_{01} = x_{30} s^{(3)} \otimes 1$ *and* $\xi_{00} = d_{22} x_{30}(s^{(4)} \otimes s^{(2)} - \frac{3}{2} s^{(3)} \otimes s^{(3)})$ *for some* $x_{30}, d_{22} \in \mathbf{k}$.

(iv) In the case when $(\lambda_1, \lambda_2) = (-2, -5)$. *We have* $\eta_{00} = d_{22}(s^{(2)} \otimes s^{(2)} - 3s^{(4)} \otimes 1)$, $\xi_{01} = x_{31}(-4s^{(4)} \otimes 1 + s^{(3)} \otimes s)$ *and* $\xi_{00} = -d_{22} x_{31}(\frac{1}{3} s^{(7)} \otimes 1 + \frac{2}{3} s^{(6)} \otimes s + 2s^{(5)} \otimes s^{(2)} - \frac{16}{15} s^{(4)} \otimes s^{(3)} - \frac{29}{15} s^{(3)} \otimes s^{(4)})$ *for some* $d_{22}, x_{31} \in \mathbf{k}$.

(v) In the case when $(\lambda_1, \lambda_2) = (-3, -5)$. *We have* $\eta_{00} = d_{50}(s^{(5)} \otimes 1 + s^{(4)} \otimes s + s^{(3)} \otimes s^{(2)})$, $\xi_{01} = x_{30} s^{(3)} \otimes 1$ *and* $\xi_{00} = \frac{1}{5} d_{50} x_{30}(s^{(4)} \otimes s^{(3)} - s^{(3)} \otimes s^{(4)})$ *for some* $d_{50}, x_{30} \in \mathbf{k}$.

(6) L has a basis $\{e_0, e_1, e_2\}$ *such that the only nonzero pseudobrackets*

with this basis are $[e_0, e_0] = \alpha \otimes_H e_0 + \eta_{00} \otimes_H e_1 + \xi_{00} \otimes_H e_2$, $[e_0, e_1] = (\lambda_1 s \otimes 1 - 1 \otimes s) \otimes_H e_1 + \xi_{01} \otimes_H e_2$ *and* $[e_0, e_2] = (\lambda_2 s \otimes 1 - 1 \otimes s) \otimes_H e_2$ *for some* $(\lambda_1, \lambda_2) \in \mathcal{O}_2$*, where* ξ_{00} *is the same as the* ξ_{00} *in (4).*

(i) In the case when $(\lambda_1, \lambda_2) = (1, 1)$*. Then* $\xi_{01} = x_{10} s \otimes 1 + x_{00} \otimes 1$ *and* $\eta_{00} = d_{10} s \otimes 1 + d_{00} \otimes 1$*, where* $x_{10}, x_{00}, d_{10}, d_{00} \in \mathbf{k}$ *satisfy* $d_{10} x_{00} = d_{00} x_{10}$*.*

(ii) In the case when $(\lambda_1, \lambda_2) \neq (1, 1)$*. Then either* $\eta_{00} = 0$ *or* $\xi_{01} = 0$*. If* $\xi_{01} \neq 0$*, then* ξ_{01} *is the same as the* ξ_{01} *in (4). If* $\eta_{00} \neq 0$*, then* η_{00} *is same as the* η_{00} *in (4). In addition,* ξ_{00} *is the same as the* ξ_{00} *in (4).*

In the case when the rank of solvable radical I of L is two, $Z_l(L) = 0$ and I is abelian. In the next theorem, $\mathcal{O}_3$ denotes the set $\{(0, -3), (0, -4), (0, -5), (-1, -3), (-1, -4), (-2, -3), (-2, -6), (-5, -6), (-5, -9), (-7, -8), (-7, -9), (-7, -11)\}$, and $\mathcal{O}_4$ denotes the set $\{(0, -1), (0, -2), (-1, -1), (-1, -5), (-2, -2), (-2, -5), (-5, -5), (-7, -7)\}$, $\mathcal{I}$ denotes the set $\{0, -1, -2, -5, -7\}$.

Theorem 4.2. *Suppose L is a Leibniz H-pseudoalgebra of rank three such that* $Z_l(L) = 0$ *and the solvable radical I of L is an abelian Lie H-pseudoalgebra of rank two. Then L is isomorphic to one of the following types.*

(1) A Lie H-pseudoalgebra has a basis $\{e_0, e_1, e_2\}$ *such that the only nonzero pseudobrackets with this basis are* $[e_0, e_0] = \alpha \otimes_H e_0$, $[e_0, e_1] = -((12) \otimes_H 1)[e_1, e_0] = (y_m \otimes s^{(m)}) \otimes_H e_2$ *and* $[e_0, e_2] = -((12) \otimes_H 1)[e_2, e_0] = (2s \otimes 1 - 1 \otimes s + \kappa_2 \otimes 1) \otimes_H e_2$ *for some nonzero* $\kappa_2, y_m \in \mathbf{k}$*.*

(2) A Lie H-pseudoalgebra has a basis $\{e_0, e_1, e_2\}$ *such that the only nonzero pseudobrackets are* $[e_0, e_0] = \alpha \otimes_H e_0 + \eta_{00} \otimes_H e_1 + \xi_{00} \otimes_H e_2$, $[e_0, e_1] = -((12) \otimes_H 1)[e_1, e_0] = (\lambda_1 s \otimes 1 - 1 \otimes s + \kappa_1 \otimes 1) \otimes_H e_1 + \xi_{01} \otimes_H e_2$ *and* $[e_0, e_2] = -((12) \otimes_H 1)[e_2, e_0] = (\lambda_2 s \otimes 1 - 1 \otimes s + \kappa_2 \otimes 1) \otimes_H e_2$*, where* $\lambda_i, \kappa_i \in \mathbf{k}$ *satisfy either* $\kappa_1 \neq 0$*, or* $\kappa_2 \neq 0$*, or* $\kappa_1 = \kappa_2 = 0$ *and* $(\lambda_1, \lambda_2) \notin \mathcal{O}_3 \cup \mathcal{O}_4$*,*

$$\xi_{00} = \begin{cases} d'_{10}\alpha, & (\kappa_2, \lambda_2) = (0, 0) \\ d'_{12}(s \otimes s^{(2)} - s^{(2)} \otimes s) + d'_{20}(s^{(2)} \otimes 1 - 1 \otimes s^{(2)}), & (\kappa_2, \lambda_2) = (0, -1) \\ d'_{13}(s \otimes s^{(3)} - s^{(3)} \otimes s) + d'_{30}(s^{(3)} \otimes 1 - 1 \otimes s^{(3)}), & (\kappa_2, \lambda_2) = (0, -2) \\ d'_{34}(s^{(3)} \otimes s^{(4)} - s^{(4)} \otimes s^{(3)}), & (\kappa_2, \lambda_2) = (0, -5) \\ d'_{36}(s^{(3)} \otimes s^{(6)} - s^{(6)} \otimes s^{(3)} \\ \quad -3(s^{(4)} \otimes s^{(5)} - s^{(5)} \otimes s^{(4)})), & (\kappa_2, \lambda_2) = (0, -7) \\ 0, & \textit{otherwise} \end{cases}$$

for some $d'_{ij} \in \mathbf{k}$.

$$\eta_{00} = \begin{cases} d_{10}\alpha & (\kappa_1, \lambda_1) = (0,0) \\ d_{12}(s \otimes s^{(2)} - s^{(2)} \otimes s) + d_{20}(s^{(2)} \otimes 1 - 1 \otimes s^{(2)}) & (\kappa_1, \lambda_1) = (0,-1) \\ d_{13}(s \otimes s^{(3)} - s^{(3)} \otimes s) + d_{30}(s^{(3)} \otimes 1 - 1 \otimes s^{(3)}) & (\kappa_1, \lambda_1 = (0,-2) \\ d_{34}(s^{(3)} \otimes s^{(4)} - s^{(4)} \otimes s^{(3)}) & (\kappa_1, \lambda_1) = (0,-5) \\ d_{36}(s^{(3)} \otimes s^{(6)} - s^{(6)} \otimes s^{(3)} & \\ \quad -3(s^{(4)} \otimes s^{(5)} - s^{(5)} \otimes s^{(4)})) & (\kappa_1, \lambda_1) = (0,-7) \\ 0 & \textit{otherwise} \end{cases}$$

for some $d_{ij} \in \mathbf{k}$, ξ_{01} *is the same as the* ξ_{01} *in (4) of Theorem 4.1.*

(3) A Lie H-pseudoalgebra has a basis $\{e_0, e_1, e_2\}$ *such that the only nonzero pseudobrackets are* $[e_0, e_0] = \alpha \otimes_H e_0 + \eta_{00} \otimes_H e_1 + \xi_{00} \otimes_H e_2$, $[e_0, e_1] = -((12) \otimes_H 1)[e_1, e_0] = (\lambda_1 s \otimes 1 - 1 \otimes s) \otimes_H e_1 + \xi_{01} \otimes_H e_2$ *and* $[e_0, e_2] = -((12) \otimes_H 1)[e_2, e_0] = (\lambda_2 s \otimes 1 - 1 \otimes s) \otimes_H e_2$, *where* $(\lambda_1, \lambda_2) \in \mathcal{O}_3$. *In addition,* η_{00}, ξ_{00} *and* ξ_{01} *are described as follows.*

(i) In the case when $(\lambda_1, \lambda_2) = (0, -3)$. *Then* $\eta_{00} = d_{10}\alpha$, $\xi_{01} = -(12)\xi_{10} = x'_{31} s^{(3)} \otimes s$ *and* $\xi_{00} = 2d_{10}x'_{31}(s^{(4)} \otimes 1 - 1 \otimes s^{(4)})$ *for some* $d_{10}, x'_{31} \in \mathbf{k}$.

(ii) In the case when $(\lambda_1, \lambda_2) = (0, -4)$. *Then* $\eta_{00} = d_{10}\alpha$, $\xi_{01} = -(12)\xi_{10} = x'_{32}(s^{(4)} \otimes s - s^{(3)} \otimes s^{(2)})$ *and* $\xi_{00} = d_{10}x'_{32}((s^{(5)} \otimes 1 - 1 \otimes s^{(5)}) + 2(s^{(4)} \otimes s - s \otimes s^{(4)}))$ *for some* $d_{10}, x'_{32} \in \mathbf{k}$.

(iii) In the case when $(\lambda_1, \lambda_2) = (0, -5)$. *Then* $\eta_{00} = d_{10}\alpha$, $\xi_{01} = -(12)\xi_{10} = x'_{33}(s^{(5)} \otimes s - 2s^{(4)} \otimes s^{(2)} + s^{(3)} \otimes s^{(3)})$ *and* $\xi_{00} = -2d_{10}x'_{33}(s^{(4)} \otimes s^{(2)} - s^{(2)} \otimes s^{(4)})$ *for some* $d_{10}, x'_{33} \in \mathbf{k}$.

(iv) In the case when $(\lambda_1, \lambda_2) = (-1, -3)$. *Then* $\eta_{00} = d_{12}(s \otimes s^{(2)} - s^{(2)} \otimes s) + d_{20}(s^{(2)} \otimes 1 - 1 \otimes s^{(2)})$, $\xi_{01} = -(12)\xi_{10} = x'_{30} s^{(3)} \otimes 1$ *and* $\xi_{00} = d_{20}x'_{30}(s^{(4)} \otimes 1 - 1 \otimes s^{(4)}) - d_{12}x'_{30}(s^{(4)} \otimes s - s \otimes s^{(4)})$ *for some* $d_{12}, d_{20}, x'_{30} \in \mathbf{k}$.

(v) In the case when $(\lambda_1, \lambda_2) = (-1, -4)$. *Then* $\eta_{00} = d_{12}(s \otimes s^{(2)} - s^{(2)} \otimes s) + d_{20}(s^{(2)} \otimes 1 - 1 \otimes s^{(2)})$, $\xi_{01} = -(12)\xi_{10} = x'_{31}(-2s^{(4)} \otimes 1 + s^{(3)} \otimes s)$ *and* $\xi_{00} = d_{12}x'_{31}(s^{(5)} \otimes s - s \otimes s^{(5)}) - d_{20}x'_{31}((s^{(5)} \otimes 1 - 1 \otimes s^{(5)}) - (s^{(3)} \otimes s^{(2)} - s^{(2)} \otimes s^{(3)}))$ *for some* $d_{12}, d_{20}, x'_{31} \in \mathbf{k}$.

(vi) In the case when $(\lambda_1, \lambda_2) = (-2, -3)$. *Then* $\eta_{00} = d_{13}(s \otimes s^{(3)} - s^{(3)} \otimes s) + d_{30}(s^{(3)} \otimes 1 - 1 \otimes s^{(2)})$, $\xi_{01} = -(12)\xi_{10} = x_{20} s^{(2)} \otimes 1$ *and* $\xi_{00} = d_{13}x_{20}(s^{(4)} \otimes s - s \otimes s^{(4)} + x_{20}d_{30}(s^{(4)} \otimes 1 - 1 \otimes s^{(4)})$ *for some* $d_{13}, d_{30}, x_{20} \in \mathbf{k}$.

(vii) In the case when $(\lambda_1, \lambda_2) = (-2, -6)$. *Then* $\eta_{00} = d_{13}(s \otimes s^{(3)} - s^{(3)} \otimes s) + d_{30}(s^{(3)} \otimes 1 - 1 \otimes s^{(3)})$, $\xi_{01} = -(12)\xi_{10} = x'_{32}(10s^{(5)} \otimes 1 - 5s^{(4)} \otimes s + s^{(3)} \otimes s^{(2)})$ *and* $\xi_{00} = -9d_{30}x'_{32}(s^{(4)} \otimes s^{(3)} - s^{(3)} \otimes s^{(4)}) + d'_{62}(s^{(6)} \otimes$

$s^{(2)} - s^{(2)} \otimes s^{(6)})$ *for some* $d_{13}, d_{30}, x'_{32}, d'_{62} \in \mathbf{k}$.

(viii) In the case when $(\lambda_1, \lambda_2) = (-5, -6)$. *Then* $\eta_{00} = d_{34}(s^{(3)} \otimes s^{(4)} - s^{(4)} \otimes s^{(3)})$, $\xi_{01} = -(12)\xi_{10} = x_{20}s^{(2)} \otimes 1$ *and* $\xi_{00} = d'_{62}((s^{(6)} \otimes s^{(2)} - s^{(2)} \otimes s^{(6)})$ *for some* $d_{34}, x_{20}, d'_{62} \in \mathbf{k}$.

(ix) In the case when $(\lambda_1, \lambda_2) = (-5, -9)$. *Then* $\eta_{00} = d_{34}(s^{(3)} \otimes s^{(4)} - s^{(4)} \otimes s^{(3)})$, $\xi_{01} = -(12)\xi_{10} = x'_{32}(55s^{(5)} \otimes 1 - 11s^{(4)} \otimes s + s^{(3)} \otimes s^{(2)})$ *and* $\xi_{00} = -\frac{40}{7}d_{34}x'_{32}((s^{(8)} \otimes s^{(3)} - s^{(3)} \otimes s^{(8)}) - 5(s^{(7)} \otimes s^{(4)} - s^{(4)} \otimes s^{(7)}) + 10(s^{(6)} \otimes s^{(5)} - s^{(5)} \otimes s^{(6)}))$ *for some* $d_{34}, x'_{32} \in \mathbf{k}$.

(x) In the case when $(\lambda_1, \lambda_2) = (-7, -8)$. *Then* $\eta_{00} = d_{36}((s^{(3)} \otimes s^{(6)} - s^{(6)} \otimes s^{(3)}) - 3(s^{(4)} \otimes s^{(5)} - s^{(5)} \otimes s^{(4)}))$, $\xi_{01} = -(12)\xi_{10} = x_{20}s^{(2)} \otimes 1$ *and* $\xi_{00} = \frac{1}{2}d_{63}x_{20}((s^{(7)} \otimes s^{(3)} - s^{(3)} \otimes s^{(7)}) - 2(s^{(6)} \otimes s^{(4)} - s^{(4)} \otimes s^{(6)}))$ *for some* $d_{36}, x_{20} \in \mathbf{k}$.

(xi) In the case when $(\lambda_1, \lambda_2) = (-7, -9)$. *Then* $\eta_{00} = d_{36}((s^{(3)} \otimes s^{(6)} - s^{(6)} \otimes s^{(3)}) - 3(s^{(4)} \otimes s^{(5)} - s^{(5)} \otimes s^{(4)}))$, $\xi_{01} = -(12)\xi_{10} = x'_{30}s^{(3)} \otimes 1$ *and* $\xi_{00} = \frac{1}{7}d_{36}x'_{30}((s^{(8)} \otimes s^{(3)} - s^{(3)} \otimes s^{(5)}) - 5(s^{(7)} \otimes s^{(4)} - s^{(4)} \otimes s^{(7)}) + 10(s^{(6)} \otimes s^{(5)} - s^{(5)} \otimes s^{(6)}))$ *for some* $d_{36}, x'_{30} \in \mathbf{k}$.

(xii) In the case when $(\lambda_1, \lambda_2) = (-7, -11)$. *Then* $\eta_{00} = d_{36}((s^{(3)} \otimes s^{(6)} - s^{(6)} \otimes s^{(3)}) - 3(s^{(4)} \otimes s^{(5)} - s^{(5)} \otimes s^{(4)}))$, $\xi_{01} = -(12)\xi_{10} = x'_{32}(105s^{(5)} \otimes 1 - 15s^{(4)} \otimes s + s^{(3)} \otimes s^{(2)})$ *and* $\xi_{00} = -\frac{35}{33}d_{36}x'_{32}((s^{(10)} \otimes s^{(3)} - s^{(3)} \otimes s^{(10)}) - 7(s^{(9)} \otimes s^{(4)} - s^{(4)} \otimes s^{(9)}) + 147(s^{(8)} \otimes s^{(5)} - s^{(5)} \otimes s^{(8)}) - 35(s^{(7)} \otimes s^{(6)} - s^{(6)} \otimes s^{(7)}))$ *for some* $d_{36}, x'_{32} \in \mathbf{k}$.

(4) A Lie H-pseudoalgebra has a basis $\{e_0, e_1, e_2\}$ *such that the only nonzero pseudobrackets are* $[e_0, e_0] = \alpha \otimes_H e_0 + \eta_{00} \otimes_H e_1 + \xi_{00} \otimes_H e_2$, $[e_0, e_1] = -((12) \otimes_H 1)[e_1, e_0] = (\lambda_1 s \otimes 1 - 1 \otimes s) \otimes_H e_1 + \xi_{01} \otimes_H e_2$ *and* $[e_0, e_2] = -((12) \otimes_H 1)[e_2, e_0] = (\lambda_2 s \otimes 1 - 1 \otimes s) \otimes_H e_2$, *where* $(\lambda_1, \lambda_2) \in \mathcal{O}_4$,

$$\xi_{00} = \begin{cases} d'_{12}(s \otimes s^{(2)} - s^{(2)} \otimes s) + d'_{20}(s^{(2)} \otimes 1 - 1 \otimes s^{(2)}), & \lambda_2 = -1 \\ d'_{13}(s \otimes s^{(3)} - s^{(3)} \otimes s) + d'_{30}(s^{(3)} \otimes 1 - 1 \otimes s^{(3)}), & \lambda_2 = -2 \\ d'_{34}(s^{(3)} \otimes s^{(4)} - s^{(4)} \otimes s^{(3)}), & \lambda_2 = -5 \\ d'_{36}(s^{(3)} \otimes s^{(6)} - s^{(6)} \otimes s^{(3)} - 3(s^{(4)} \otimes s^{(5)} - s^{(5)} \otimes s^{(4)})), & \lambda_2 = -7, \end{cases}$$

for some $d'_{ij} \in \mathbf{k}$.

(i) In the case when $(\lambda_1, \lambda_2) \in \{(0, -2), (-1, -5), (-2, -5), (-5, -5), (-7, -7)\}$. *Then either* $\eta_{00} = 0$ *or* $\xi_{01} = 0$.

(ii) In the case when $(\lambda_1, \lambda_2) = (0, -1)$. *Then either* $\eta_{00} = 0$, *or* $\xi_{01} = 2x_{11}s^{(2)} \otimes 1 + x_{11}s \otimes s$ *for some* $x_{11} \in \mathbf{k}$.

(iii) In the case when $(\lambda_1, \lambda_2) = (-1, -1)$. *Then* $\eta_{00} = d_{12}(s \otimes s^{(2)} - s^{(2)} \otimes s) + d_{20}(s^{(2)} \otimes 1 - 1 \otimes s^{(2)})$ *and* $\xi_{01} = x_{10}s \otimes 1 + x_{00} \otimes 1$, *where* $x_{10}, x_{00}, d_{12}, d_{20} \in \mathbf{k}$ *satisfy* $d_{20}x_{10} + d_{12}x_{00} = 0$.

(iv) In the case when $(\lambda_1, \lambda_2) = (-2, -2)$. *Then* $\eta_{00} = d_{13}(s \otimes s^{(3)} - s^{(3)} \otimes s) + d_{30}(s^{(3)} \otimes 1 - 1 \otimes s^{(3)})$ *and* $\xi_{01} = x_{10}s \otimes 1 + x_{00} \otimes 1$, *where* $x_{10}, x_{00}, d_{13}, d_{30} \in \mathbf{k}$ *satisfy* $d_{30}x_{10} + d_{13}x_{00} = 0$.

In the case when the rank of the solvable radical I is two, the rank of $Z_l(L)$ is one and I is abelian. Let

$$\xi_{10} = \sum_{j\geq 1} \frac{1}{j+1}(ix_{j+1\ i-1} - \kappa_1 x_{j+1\ i} - (i+j)x_{1\ i+j-1} - \kappa_1 x_{1\ i+j})s^{(i)} \otimes s^{(j)}$$

$$+ \sum_{p=1}^{t}\sum_{j=1}^{p} \frac{(1-\lambda_2+(\lambda_1+1)j)z_{p0}}{j+1} s^{(p-j)} \otimes s^{(j)} + \sum_{i=0}^{t} z_{i0}s^{(i)} \otimes 1. \qquad (7)$$

Then we have the following

Theorem 4.3. *Let L be a nonsolvable Leibniz H-pseudoalgebra with a rank two abelian ideal I and a rank one left annihilator ideal $Z_l(L)$. Then L has a basis $\{e_0, e_1, e_2\}$ such that $I = He_1 + He_2$ and $Z_l(L) = He_2$. Moreover, L is isomorphic to one of the following types.*

(1) The only nonzero pseudobrackets with the basis $\{e_1, e_2, e_3\}$ are $[e_0, e_0] = \alpha \otimes_H e_0 + \eta_{00} \otimes_H e_1 + \xi_{00} \otimes_H e_2$, $[e_0, e_1] = -(12)[e_1, e_0] = (\lambda_1 s \otimes 1 - 1 \otimes s + \kappa_1 \otimes 1) \otimes_H e_1 + x_{00} \otimes 1 \otimes_H e_2$ *for some* $x_{00}, \lambda, \kappa_2 \in \mathbf{k}$ *satisfying* $(\lambda_1 + 1)x_{00} = 0$, *where* η_{00} *is the same as (2) of Theorem 4.2 and* ξ_{00} *is the same as (2) of Theorem 4.1.*

(2) The only nonzero pseudobrackets with the basis $\{e_1, e_2, e_3\}$ are $[e_0, e_0] = \alpha \otimes_H e_0 + \eta_{00} \otimes_H e_1 + \xi_{00} \otimes_H e_2$, $[e_0, e_1] = -((12) \otimes_H 1)[e_1, e_0] = (\lambda_1 s \otimes 1 - 1 \otimes s + \kappa_1 \otimes 1) \otimes_H e_1$, $[e_0, e_2] = (\lambda_2 s \otimes 1 - 1 \otimes s + \kappa_2) \otimes_H e_2$ *for some* $\lambda_i, \kappa_i \in \mathbf{k}$, *where* $\kappa_1 \neq \kappa_2$, η_{00} *and* ξ_{00} *are the same as (1).*

(3) The only nonzero pseudobrackets with the basis $\{e_1, e_2, e_3\}$ are $[e_0, e_0] = \alpha \otimes_H e_0 + \xi_{00} \otimes_H e_2$, $[e_0, e_1] = (\lambda_1 s \otimes 1 - 1 \otimes s + \kappa_1 \otimes 1) \otimes_H e_1 + \xi_{01} \otimes_H e_2$, $[e_1, e_0] = (s \otimes 1 - \lambda_1 \otimes s - \kappa_1 \otimes 1) \otimes_H e_1 + \xi_{10} \otimes_H e_2$, $[e_0, e_2] = (\lambda_2 s \otimes 1 - 1 \otimes s + \kappa_1 \otimes 1) \otimes_H e_2$ *for some* $\lambda_1, \lambda_2, \kappa_1 \in \mathbf{k}$, *where* $\xi_{00} = 0$ *is the same as (2),* $\xi_{01} = \sum_{ij} x_{ij}s^{(i)} \otimes s^{(j)}$ *is the same as (3) of Theorem 4.2, and* ξ_{10} *is given by (7), where* $z_{i0} = \frac{(-\kappa_1)^{t-i}}{(t-i)!} z_{t0}$ *for* $0 \leq i \leq t-1$.

(4) The only nonzero pseudobrackets with the basis $\{e_1, e_2, e_3\}$ are $[e_0, e_0] = \alpha \otimes_H e_0 + \eta_{00} \otimes_H e_1 + \xi_{00} \otimes_H e_2$, $[e_0, e_1] = -((12) \otimes_H 1)[e_1, e_0] = (\lambda_1 s \otimes 1 - 1 \otimes s) \otimes_H e_1$, $[e_0, e_2] = (\lambda_2 s \otimes 1 - 1 \otimes s) \otimes_H e_2$ *for some* $(\lambda_1, \lambda_2) \notin \{(0,0), (0,-1), (-1,-1), (-1,-2), (-2,-2), (-5,-5)\}$, *where* η_{00} *and* ξ_{00} *are the same as (1).*

(5) The only nonzero pseudobrackets with the basis $\{e_1, e_2, e_3\}$ are

$[e_0, e_0] = \alpha \otimes_H e_0 + \eta_{00} \otimes_H e_1 + \xi_{00} \otimes_H e_2$, $[e_0, e_1] = (\lambda_1 s \otimes 1 - 1 \otimes s) \otimes_H e_1 + \xi_{01} \otimes_H e_2$, $[e_1, e_0] = (s \otimes 1 - \lambda_1 \otimes s) \otimes_H e_1 + \xi_{10} \otimes_H e_2$, $[e_0, e_2] = (\lambda_2 s \otimes 1 - 1 \otimes s) \otimes_H e_2$ *for some* $(\lambda_1, \lambda_2) \in \{(0,0), (0,-1), (-1,-1), (-1,-2), (-2,-2), (-5,-5)\}$.

(i) If $(\lambda_1, \lambda_2) = (0,0)$, *then* $\eta_{00} = d_{10}\alpha$, $\xi_{01} = x_{10}s \otimes 1 + x_{00} \otimes 1$, $\xi_{10} = z_{10}s \otimes 1 + (z_{10} - \frac{1}{2}x_{10}) \otimes s$ *and* $\xi_{00} = \frac{1}{2}x_{10}d_{10}s \otimes 1 - x_{00}d_{10} \otimes 1$ *for some* $d_{10}, x_{10}, x_{00}, z_{10} \in \mathbf{k}$.

(ii) If $(\lambda_1, \lambda_2) = (0,-1)$, *then* $\eta_{00} = d_{10}\alpha$, $\xi_{01} = x_{11}(\frac{3}{4}s^{(2)} \otimes 1 + s \otimes s)$, $\xi_{10} = -x_{11}(\frac{1}{4}s^{(2)} \otimes 1 + s \otimes s + 1 \otimes s^{(2)})$ *and* $\xi_{00} = -\frac{1}{2}x_{11}d_{10}s \otimes s + \frac{1}{4}x_{11}d_{10} \otimes s^{(2)}$ *for some* $d_{10}, x_{11} \in \mathbf{k}$.

(iii) If $(\lambda_1, \lambda_2) = (-1,-1)$, *then* $\eta_{00} = d_{12}(s \otimes s^{(2)} - s^{(2)} \otimes s) + d_{20}(s^{(2)} \otimes 1 - 1 \otimes s^{(2)})$, $\xi_{01} = x_{10}s \otimes 1 + x_{00} \otimes 1$, $\xi_{10} = z_{10}s \otimes 1 + (z_{10} - \frac{1}{2}x_{10}) \otimes s$ *and* $\xi_{00} = z_{10}d_{20}(s^{(2)} \otimes 1 - 1 \otimes s^{(2)}) + d'_{30}s^{(3)} \otimes 1 + z_{10}d_{12}s \otimes s^{(2)}$ *for some* $d_{12}, d_{20}, d'_{30}, x_{10}, x_{00}, z_{10} \in \mathbf{k}$.

(iv) If $(\lambda_1, \lambda_2) = (-1,-2)$, *then* $\eta_{00} = d_{12}(s \otimes s^{(2)} - s^{(2)} \otimes s) + d_{20}(s^{(2)} \otimes 1 - 1 \otimes s^{(2)})$, $\xi_{01} = x_{20}s^{(2)} \otimes 1$, $\xi_{10} = -x_{20}\Delta(s^{(2)})$ *and* $\xi_{00} = (2d'_{22} - 3d_{12}x_{20})s^{(4)} \otimes 1 + d'_{22}s^{(2)} \otimes s^{(2)} - x_{20}d_{20}(3s^{(3)} \otimes 1 + s^{(2)} \otimes s - \frac{3}{2} \otimes s^{(3)})$ *for some* $d_{12}, d_{20}, d'_{22}, x_{20} \in \mathbf{k}$.

(v) If $(\lambda_1, \lambda_2) = (-2,-2)$, *then* $\eta_{00} = d_{13}(s \otimes s^{(3)} - s^{(3)} \otimes s) + d_{30}(s^{(3)} \otimes 1 - 1 \otimes s^{(3)})$, $\xi_{01} = x_{10}s \otimes 1 + x_{00} \otimes 1$, $\xi_{10} = z_{10}s \otimes 1 + (z_{10} - \frac{1}{2}x_{10}) \otimes s$ *and* $\xi_{00} = (\frac{1}{2}x_{10}d_{30} - d_{13}x_{00})s^{(3)} \otimes 1 + (d_{13}x_{10} - 3d'_{22})s^{(4)} \otimes 1 + d'_{22}s^{(2)} \otimes s^{(2)} + \frac{1}{2}d_{13}x_{10}s \otimes s^{(3)}$ *for some* $d_{13}, d_{30}, d'_{22}, x_{10}, x_{00} \in \mathbf{k}$.

(vi) If $(\lambda_1, \lambda_2) = (-5,-5)$, *then* $\eta_{00} = d_{34}(s^{(3)} \otimes s^{(4)} - s^{(4)} \otimes s^{(3)})$, $\xi_{01} = 0$, $\xi_{10} = z_{10}s \otimes 1 + z_{10} \otimes s$ *and* $\xi_{00} = 0$ *for some* $d_{34}, z_{10} \in \mathbf{k}$.

In the case when the rank of the solvable nonabelin radical I of L is equal to two. Then I is a solvable Leibniz H-pseudoalgebra with maximal derived series. Thus I is described by Lemma 2.1. We can assume that

$$\begin{cases} [e_0, e_0] = \alpha \otimes_H e_0 + \eta_{00} \otimes_H e_1 + \xi_{00} \otimes_H e_2, \\ [e_0, e_1] = \eta_{01} \otimes_H e_1 + \xi_{01} \otimes_H e_2, \quad [e_1, e_0] = \eta_{10} \otimes_H e_1 + \xi_{10} \otimes_H e_2, \\ [e_0, e_2] = \eta_{02} \otimes_H e_1 + \xi_{02} \otimes_H e_2, \quad [e_2, e_0] = \eta_{20} \otimes_H e_1 + \xi_{20} \otimes_H e_2 \\ [e_1, e_1] = \xi_{11} \otimes_H e_2 \quad [e_1, e_2] = \xi_{12} \otimes_H e_2, \quad [e_2, e_1] = \xi_{21} \otimes_H e_2 \\ [e_2, e_2] = 0, \end{cases} \tag{8}$$

where $\alpha = s \otimes 1 - 1 \otimes s$, either $\xi_{12} = -(12)\xi_{21} = \sum_{i=0}^{n} b_i s^{(i)} \otimes 1 \neq 0$ and $\xi_{11} = 0$ for some $b_i \in \mathbf{k}$, or $\xi_{12} = \sum_{i=0}^{n} b_i s^{(i)} \otimes s$ and $\xi_{21} = \xi_{00} = 0$ for some $b_i \in \mathbf{k}$, or $\xi_{12} = \xi_{21} = 0$ and $0 \neq \xi_{11} \in H^{\otimes 2}$.

Theorem 4.4. *Suppose a Leibniz H-pseudoalgebra L has a basis $\{e_0, e_1, e_2\}$ such that the pseudobracket is given by (8), where $\xi_{12} = -(12)\xi_{21} = \sum_{i=0}^{n} b_i s^{(i)} \otimes 1 \neq 0$ and $\xi_{11} = 0$ for some $b_i \in \mathbf{k}$. Then L is a Lie H-pseudoalgebra with $\eta_{02} = \eta_{20} = 0$. Moreover, L is isomorphic to one of the following Leibniz H-pseudoalgebras.*

(1) The Leibniz H-pseudoalgebra L has a basis $\{e_1, e_2, e_3\}$ such that possibly nonzero pseudobrackets with this basis are $[e_0, e_0] = \alpha \otimes_H e_1$, $[e_0, e_2] = A(s) \otimes 1 \otimes_H e_2$, $[e_1, e_2] = b_0 \otimes 1 \otimes_H e_2$ for some nonzero $b_0 \in \mathbf{k}$ and some $A(s) \in H$.

(2) The Leibniz H-pseudoalgebra L has a basis $\{e_1, e_2, e_3\}$ such that possibly nonzero pseudobrackets with this basis are $[e_0, e_0] = \alpha \otimes_H e_1$, $[e_0, e_1] = -((12) \otimes_H 1)[e_1, e_0] = -1 \otimes s \otimes_H e_1$, $[e_0, e_2] = -1 \otimes s \otimes_H e_2$ and $[e_1, e_2] = b_0 \otimes 1 \otimes_H e_2$ for some nonzero $b_0 \in \mathbf{k}$.

Theorem 4.5. *Suppose a Leibniz H-pseudoalgebra L has a basis $\{e_0, e_1, e_2\}$ such that the pseudobracket is given by (8), where $\xi_{12} = \sum_{i=0}^{n} b_i s^{(i)} \otimes 1 \neq 0$, $\xi_{21} = \xi_{11} = 0$. Then $\eta_{02} = \eta_{20} = \xi_{20} = 0$ and $\eta_{10} = -(12)\eta_{01}$. Moreover, L is isomorphic to one of the following Leibniz H-pseudoalgebras.*

(1) L has a basis $\{e_0, e_1, e_2\}$ such that the nonzero pseudobrackets with basis are $[e_0, e_0] = \alpha \otimes_H e_0$, $[e_1, e_1] = \xi_{11} \otimes_H e_2$, $[e_1, e_2] = [e_2, e_1] = 0$ for some nonzero $\xi_{11} \in H^{\otimes 2}$.

(2) L has a basis $\{e_0, e_1, e_2\}$ such that the nonzero pseudobrackets with basis are $[e_0, e_0] = \alpha \otimes_H e_0 + (d'_{12}(s \otimes s^{(2)} - s^{(2)} \otimes s) + d'_{20}(s^{(2)} \otimes 1 - 1 \otimes s^{(2)})) \otimes_H e_2$, $[e_0, e_1] = -((12) \otimes_H 1)[e_1, e_0] = -1 \otimes s \otimes_H e_1 + (x_{10} s \otimes 1 + x_{00} s \otimes 1) \otimes_H e_2$, $[e_0, e_2] = -((12) \otimes_H 1)[e_2, e_0] = -(s \otimes 1 + 1 \otimes s) \otimes_H e_2$, $[e_1, e_1] = y_{10} \alpha \otimes_H e_2 \neq 0$ for some $d'_{12}, d'_{20}, x_{00}, x_{10}, y_{10} \in \mathbf{k}$.

(3) L has a basis $\{e_0, e_1, e_2\}$ such that $[e_0, e_0] = \alpha \otimes_H e_0 + (d'_{12}(s \otimes s^{(2)} - s^{(2)} \otimes s) + d'_{20}(s^{(2)} \otimes 1 - 1 \otimes s^{(2)})) \otimes_H e_2$, $[e_0, e_1] = -((12) \otimes_H 1)[e_1, e_0] = (\frac{1}{2} s \otimes 1 - 1 \otimes s) \otimes_H e_1$, $[e_0, e_2] = -((12) \otimes_H 1)[e_2, e_0] = -(s \otimes 1 + 1 \otimes s) \otimes_H e_2$, $[e_1, e_1] = y_{20}(s^{(2)} \otimes 1 - 1 \otimes s^{(2)}) \otimes_H e_2 \neq 0$, $[e_1, e_2] = [e_2, e_1] = 0$ for some $d'_{12}, d'_{20}, y_{20} \in \mathbf{k}$.

(4) L has a basis $\{e_0, e_1, e_2\}$ such that $[e_0, e_0] = \alpha \otimes_H e_0 + d'_{23}(s^{(2)} \otimes s^{(3)} - s^{(3)} \otimes s^{(2)}) \otimes_H e_2$, $[e_0, e_1] = -((12) \otimes_H 1)[e_1, e_0] = -1 \otimes s \otimes_H e_1 + (x_{31} s^{(3)} \otimes s + x_{22} s^{(2)} \otimes s^{(2)}) \otimes_H e_2$, $[e_0, e_2] = -((12) \otimes_H 1)[e_2, e_0] = -(3s \otimes 1 + 1 \otimes s) \otimes_H e_2$, $[e_1, e_1] = y_{12}(s \otimes s^{(2)} - s^{(2)} \otimes s) \otimes_H e_2 \neq 0$, $[e_1, e_2] = [e_2, e_1] = 0$ for some $c_{23}, x_{01}, x_{40}, x_{31}, x_{22}, y_{12} \in \mathbf{k}$.

(5) L has a basis $\{e_0, e_1, e_2\}$ such that $[e_0, e_0] = \alpha \otimes_H e_0$, $[e_0, e_1] =$

$-((12)\otimes_H 1)[e_1,e_0] = (\frac{2}{3}s\otimes 1 - 1\otimes s)\otimes_H e_1$, $[e_0,e_2] = -((12)\otimes_H 1)[e_2,e_0] = -(\frac{5}{3}s\otimes 1 + 1\otimes s)\otimes_H e_2$, $[e_1,e_1] = y_{30}((s^{(3)}\otimes 1 - 1\otimes s^{(3)}) + \frac{1}{2}(s\otimes s^{(2)} - s^{(2)}\otimes s))\otimes_H e_2 \neq 0$, $[e_1,e_2] = [e_2,e_1] = 0$ *for some* $y_{30}\in \mathbf{k}$.

(6) L has a basis $\{e_0,e_1,e_2\}$ *such that* $[e_0,e_0] = \alpha\otimes_H e_0$, $[e_0,e_1] = -((12)\otimes_H 1)[e_1,e_0] = (\lambda_1 s\otimes 1 - 1\otimes s + \kappa\otimes 1)\otimes_H e_1$, $[e_0,e_2] = -((12)\otimes_H 1)[e_2,e_0] = (\lambda_2 s\otimes 1 - 1\otimes s + 2\kappa\otimes 1)\otimes_H e_2$ *for some nonzero* κ, $[e_1,e_1] = \xi_{11}\otimes_H e_2 \neq 0$, $[e_1,e_2] = [e_2,e_1] = 0$, *where* $\lambda_i\in\mathbf{k}$, ξ_{11} *is one of the following cases:*

(i) $\xi_{11} = y_{10}(s\otimes 1 - 1\otimes s)$, *where* $2\lambda_1 - \lambda_2 = 1$;

(ii) $\xi_{11} = y_{20}((s^{(2)}\otimes 1 - 1\otimes s^{(2)}) - \kappa\alpha)$, *where* $\lambda_1 = \frac{1}{2}$, $\lambda_2 = -1$;

(iii) $\xi_{11} = y_{12}(\kappa(s^{(2)}\otimes 1 - 1\otimes s^{(2)}) + (s\otimes s^{(2)} - s^{(2)}\otimes s) - \frac{\kappa^2}{2}\alpha)$, *where* $\lambda_1 = 0$, $\lambda_2 = -3$;

(iv) $\xi_{11} = y_{30}((s^{(3)}\otimes 1 - 1\otimes s^{(3)}) + \frac{1}{2}(s\otimes s^{(2)} - s^{(2)}\otimes s) - \frac{3}{2}\kappa(s^{(2)}\otimes 1 - 1\otimes s^{(2)}) - \frac{3\kappa^2}{4}\alpha)$, *where* $\lambda_1 = \frac{2}{3}$.

Theorem 4.6. *Suppose a Leibniz H-pseudoalgebra L has a basis* $\{e_0,e_1,e_2\}$ *such that the pseudobracket is given by (8), where* $\xi_{12} = \xi_{21} = 0$ *and* $\xi_{11} = \sum_{i=0}^{n}\sum_{j=0}^{p_i} y_{ij}s^{(i)}\otimes s^{(j)} \neq 0$. *Then L is isomorphic to one of the following Leibniz H-pseudoalgebras.*

(i) L has a basis $\{e_0,e_1,e_2\}$ *such that* $[e_0,e_0] = \alpha\otimes_H e_0 + \xi_{00}\otimes_H e_2$, $[e_1,e_1] = \alpha\otimes_H e_1$, $[e_0,e_2] = (\lambda_1 s\otimes 1 - 1\otimes s + \kappa_1\otimes 1)\otimes_H e_2$, $[e_0,e_1] = [e_1,e_0] = [e_1,e_2] = [e_2,e_i] = 0$ *for some* $\lambda_1,\kappa_1\in\mathbf{k}$, *where*

$$\xi_{00} = \begin{cases} x_{00}\otimes 1, & (\kappa_1,\lambda_1) = (0,1)\\ x_{30}s^{(3)}\otimes 1, & (\kappa_1,\lambda_1) = (0,-1)\\ x_{22}(s^{(2)}\otimes s^{(2)} + \frac{3}{2}s^{(3)}\otimes s), & (\kappa_1,\lambda_1) = (0,-2)\\ x_{23}(s^{(2)}\otimes s^{(3)} + 4s^{(3)}\otimes s^{(2)} + 2s^{(4)}\otimes s), & (\kappa_1,\lambda_1) = (0,-3)\\ 0, & \text{otherwise.} \end{cases}$$

Thus L is a direct sum of Virasoro the Lie conformal algebra He_1 *and a non-simsimple, non-solvable Leibniz H-pseudoalgebra* $He_0\oplus He_2$ *of rank two.*

(ii) $L = He_0\oplus He_1\oplus He_3$ *is a direct sum of Lie H-pseudoalgebras, where* He_1 *and* He_0 *are Virasoro Lie conformal algebras and* He_2 *is an abelian Lie conformal algebra.*

Now, suppose the rank of I is one. Then either $Z_l(L) = 0$ or $Z_l(L) = I$. If $Z_l(L) = I$, then L has a basis $\{e_0,e_1,e_2\}$ such that $[e_0,e_0] = \alpha\otimes_H e_0 + \xi_{00}\otimes_H e_2$, $[e_1,e_1] = \alpha\otimes_H e_1 + \xi_{11}\otimes_H e_2$, $[e_0,e_1] = \xi_{01}\otimes_H e_2$, $[e_1,e_0] =$

$\xi_{10} \otimes_H e_2$, $[e_0, e_2] = \xi_{02} \otimes_H e_2$, $[e_1, e_2] = \xi_{12} \otimes e_2$ and $[e_2, e_i] = 0$ for $0 \leq i \leq 2$.

Theorem 4.7. *Suppose L is a Leibniz H-pseudoalgebra of rank three, $Z_l(L) = I$ and the rank of I is one. Then L is one of the following types:*

(i) L has a basis $\{e_0, e_1, e_2\}$ such that $[e_0, e_0] = \alpha \otimes_H e_0 + \xi_{00} \otimes_H e_2$, $[e_1, e_1] = \alpha \otimes_H e_1$, $[e_0, e_2] = (\lambda_1 s \otimes 1 - 1 \otimes s + \kappa_1 \otimes 1) \otimes_H e_2$, $[e_0, e_1] = [e_1, e_0] = [e_1, e_2] = [e_2, e_i] = 0$ for some $\lambda_1, \kappa_1 \in \mathbf{k}$, where

$$\xi_{00} = \begin{cases} x_{00} \otimes 1, & (\kappa_1, \lambda_1) = (0, 1) \\ x_{30} s^{(3)} \otimes 1, & (\kappa_1, \lambda_1) = (0, -1) \\ x_{22}(s^{(2)} \otimes s^{(2)} + \frac{3}{2} s^{(3)} \otimes s), & (\kappa_1, \lambda_1) = (0, -2) \\ x_{23}(s^{(2)} \otimes s^{(3)} + 4s^{(3)} \otimes s^{(2)} + 2s^{(4)} \otimes s), & (\kappa_1, \lambda_1) = (0, -3) \\ 0, & \textit{otherwise}. \end{cases}$$

Thus L is a direct sum of Virasoro Lie conformal algebra He_1 and a non-simsimple, non-solvable Leibniz H-pseudoalgebra $He_0 \oplus He_2$ of rank two.

(ii) $L = He_0 \oplus He_1 \oplus He_3$ is a direct sum of Lie H-pseudoalgebras, where He_1 and He_0 are Virasoro Lie conformal algebras and He_2 is an abelian Lie conformal algebra.

Finally, we determine Lie H-pseudoalgebras of rank three, whose solvable radicals are of rank one and $Z_l(L) = 0$. Suppose L is a Lie H-pseudoalgebra of rank three and the rank of its solvable radical is one.

Theorem 4.8. *Suppose L is a Leibniz H-pseudoalgebra of rank three, $Z_l(L) = 0$ and the rank of its solvable radical is one. Then L has a basis $\{e_0, e_1, e_2\}$ such that $[e_0, e_0] = \alpha \otimes_H e_0$, $[e_1, e_1] = \alpha \otimes_H e_1$, $[e_0, e_2] = -((12) \otimes_H 1)[e_2, e_0] = (\lambda s \otimes 1 - 1 \otimes s + \kappa \otimes 1) \otimes_H e_2$, $[e_0, e_1] = [e_1, e_2] = [e_2, e_2] = 0$ for some $\lambda, \kappa \in \mathbf{k}$.*

Acknowledgements

We thank the referees sincerely for his/her careful reading, helpful comments and suggestions about exposition and mathematical issues in this paper.

References

[1] B. Bakalov, A. D'Andrea and V. G. Kac, *Theory of finite pseudoalgebras*, Adv. Math., 162(2001), 1–140.

[2] B. Bakalov, V. G. Kac and A. A. Voronov, *Cohomology of conformal algebras*, Comm. Math. Phys., 200(1999), 561–598.

[3] S. Cheng and V. G. Kac, *Conformal modules*, Asian J. Math. 1 (1997), 181–193.
[4] V. Kac, *Vertex algebras for beginners*, volume 10 of University Lecture Series. American Mathematical Society, Providence, RI, second edition, 1998.
[5] A. Retakh, *Unital associative pseudoalgebras and their representations*, J. Algebra, 277(2004), 769–805.
[6] C. Roger and J. Unterberger, *The Schrödinger-Virasoro Lie group and algebra: Representation theory and cohomological study*, Annales Henri Poincar (7–8) (2006), 1477–1529.
[7] M. E. Sweedler, *Hopf algebras*, Math. Lecture Note Series, Benjamin, New York, 1969.
[8] Z. Wu, *Leibniz H-Pseudoalgebras*, J. Algebra, 437(2015), 1–33.
[9] Z. Wu, *Schrödinger-Virasoro Lie H-Pseudoalgebras*, arxiv.1809.03295. [math.QA]